WiMAX Technology
and Network Evolution

D1445751

WiMAX Technology and Network Evolution

Edited by
KAMRAN ETEMAD
MING-YEE LAI

IEEE
COMMUNICATIONS
SOCIETY

The ComSoc Guides to Communications Technologies
Nim K. Cheung, *Series Editor*
Thomas Banwell, *Associate Series Editor*
Richard Lau, *Associate Series Editor*

IEEE PRESS

A JOHN WILEY & SONS, INC., PUBLICATION

Library of Congress Cataloging-in-Publication Data:

Etemad, Kamran.
    WIMAX technology and network evolution / Kamran Etemad, Ming-Yee Lai.
       p. cm.
    Includes bibliographical references.
    ISBN 978-0-470-34387-6 (cloth)
    1. Wireless metropolitan area networks. 2. IEEE 802.15 (Standard) I. Lai, Ming-Yee, 1952– II. Title.
    TK5105.87.E86 2010
    004.67—dc22   2009054416

Printed in Singapore.

10 9 8 7 6 5 4 3 2 1

Contents

■ Preface

As the evolution of wireless technologies continues toward realization of mobile broadband access to content-rich multimedia services, the communication industry is going through a significant paradigm shift. There is an ever increasing emphasis on architecture simplicity and cost lowering, as well as leveraging the ongoing success and innovations in Internet-based protocols and applications. Mobile WiMAX, which has emerged from computer and Internet ecosystems, is one of the first technologies leading this change of paradigm.

Despite widespread interest in and debate on mobile WiMAX, there is no comprehensive end-to-end description of the technology available to the industry and academia. This book is planned and organized to provide an accurate, complete, and objective description of mobile WiMAX technology. The breadth and depth of the material is carefully balanced to cover a wide range of questions on WiMAX as a new wireless technology while emphasizing key technical concepts and design principles. Each chapter was developed by selected subject-matter experts who have been directly involved as leading contributors to this technology in the IEEE 802.16 Working Group and/or the WiMAX Forum.

The book is organized into 20 chapters as shown in the figure on page xviii.

Chapter 1 provides an overview of the WiMAX standardization and certification process and development in IEEE 802.16 and in the WiMAX Forum. It also presents the high-level evolution road map for the WiMAX technology. Chapters 2 through 5 focus on the mobile WiMAX air interface based on IEEE 802.16 standards, whereas Chapters 6 through 17 articulate the key concepts of WiMAX end-to-end architectures, protocols, and services, including both over-the-air and network aspects. Based on this functional organization, some of the important WiMAX features, such as security, mobility, quality of service (QoS), and multicast and broadcast service (MBS), which are briefly described in earlier overview chapters, are described more comprehensively in the dedicated chapters.

More specifically, Chapter 2 describes the physical (PHY) and media access control (MAC) layers of WiMAX Release 1.0 in detail, whereas Chapter 3 summarizes key improvements and enhancements introduced in Release 1.5. Chapter 4

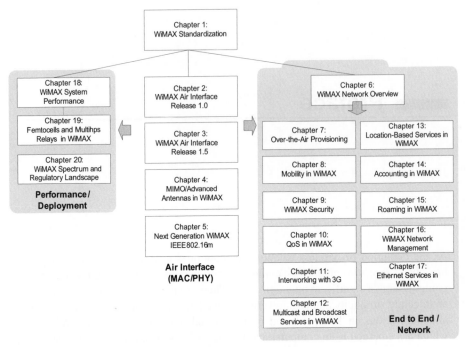

Organization of the book.

provides an overview of advanced antenna systems, which play a key role in improving radio-link-layer efficiency and provide high data throughput and system capacity. Chapter 5 presents a technical overview of IEEE 802.16m standard, which is the basis for the next-generation mobile WiMAX radio technology in Release 2.0 and also is a candidate for ITU International Mobile Telecommunications—Advanced (IMT-ADV) technologies. Chapter 7 describes over-the-air provisioning and activation in mobile WiMAX, which is a key feature to enable the open retail device distribution model. Mobility, security, and QoS, which constitute the essential end-to-end working of mobile WiMAX technology and require coherent air interface and network specifications, are covered in Chapters 8, 9, and 10, respectively. Chapter 11 addresses WiMAX interworking with 3G cellular networks and technologies, such as UMTS/HSPA and CDMA2000/1X-EVDO. The interworking enables seamless overlay and integration of WiMAX systems with existing mobile operators' networks.

Multicast and broadcast services and location-based services, which are advanced network features introduced in network Release 1.5 and are expected to be key enablers for new services and usage models, are covered in Chapters 12 and 13, respectively. Chapters 14 through 16 cover the important WiMAX network aspects in accounting, roaming, and network management. Although mobile WiMAX is primarily IP-based, the architecture can also support Ethernet-based services, which are described in Chapter 17.

The last three chapters describe different perspectives on performance and deployment issues. Chapter 18 presents an overall structure of an end-to-end application performance simulator followed by radio performance evaluation focusing on link-level and system-level performance. Chapter 19 describes two advanced radio network solutions—femtocells and multihop relays—to improve coverage and system capacity. Finally, Chapter 20 highlights the spectrum allocation and regulatory issues that have direct influence on WiMAX deployments and global adoption.

The editors and contributing authors dedicate the result of this teamwork to our colleagues, families, and friends, and hope that this book provides a helpful educational reference for those who seek to learn about WiMAX technology and prepare for future related innovations.

KAMRAN ETEMAD
MING-YEE LAI

Potomac, Maryland
Short Hills, New Jersey
January 2010

Contributors

SASSAN AHMADI, *Intel*
MEHDI ALASTI, *consultant*
REZA AREFI, *Intel*
WAYNE BALLANTYNE, *Motorola*
YONG CHANG, *Samsung Electronics*
JOEY CHOU, *Intel*
JOHN DUBOIS, *WiMAX Forum*
KAMRAN ETEMAD, *Intel*
PERETZ FEDER, *Alcatel-Lucent*
RAMANA ISUKAPALLI, *Alcatel-Lucent*
YERANG HUR, *Posdata*
BONG-HO KIM, *Posdata*
MING-YEE LAI, *Telecordia Technologies*
JICHEOL LEE, *Samsung Electronics*
QUIGHUA LI, *Intel*
XINTIAN EDDIE LIN, *Intel*
AVI LIOR, *Bridgewater Systems*
SEMYON B. MIZIKOVSKY, *Alcatel-Lucent*
BEHNAM NEEKZAD, *Clearwire*
MASOUD OLFAT, *Clearwire*
CHIRAG PATEL, *WiMAX Forum*
MAXIMILIAN RIEGEL, *Nokia Siemens Networks*
WONIL ROH, *Samsung Electronics*
SHAHAB SAYEEDI, *Motorola*
JOSEPH R. SCHUMACHER, *Motorola*

AVISHAY SHRAGA, *Intel*
JAYNE STANCAVAGE, *Intel*
JERRY SYDIR, *Intel*
POUYA TAAGHOL, *Intel*
SHILPA TALWAR, *Intel*
RAKESH TAORI, *Samsung*
MUTHAIAH VENKATACHALAM, *Intel*
LIMEI WANG, *Huawei Technologies*
HASSAN YAGHOOBI, *Intel*
SHU-PING YEH, *Stanford University*
JUNGNAM YUN, *Posdata*
JIANZHONG (CHARLIE) ZHANG, *Intel*
PEIYING ZHU, *Huawei Technologies*

Acronyms

1XRTT	Single-Carrier Radio Transmission Technology
3DES	Triple DES
3GPP	3rd Generation Partnership Project
3GPP2	3rd Generation Partnership Project 2
3WHS	Three-Way Handshake
A-A-AMAP	Assignment A-MAP
A-MAP	Advanced MAP
A-GPS	Assisted Global Positioning System
A-PCEF	Access Policy Control Enforcing Fucntion
A-PREAMBLE	Advanced Preamble
AAA	Authentication, Authorization, Accounting
AAS	Advanced Antenna Systems
AAV	Alternative Access Vendor
AES	Advanced Encryption Standard
AF	Application Function, Assured Forwarding
AFC	Automatic Frequency Control
AK	Authentication Key
AKA	Anonymity Key
AKA	Authentication and Key Agreement
AMC	Adaptive Modulation and Coding
AMF	Authenticated Message Field
AN	Access Network
ANDSF	Access Network Discovery and Selection Function
AO	Authentication Option
AP	Access Point
APT	Asia-Pacific Telecommunity
AQM	Active Queue Management
ARQ	Automatic Repeat reQuest
ASMG	Arab Spectrum Management Group

ASN	Access Service Network
ASN-GW	Access Service Network Gateway
ASP	Application Service Provider
ATU	Access Terminal, African Telecommunications Union
AV	Authentication Vector
AWG	Application Working Group
AWGN	Additive White Gaussian Noise
BA	Binding Acknowledgement
BBERF	Bearer Binding and Event Reporting Function
BCG	Band Class Group
BE	Best Effort
BEK	Bootstrap Encryption Key
BEM	Block Edge Mask
BER	Bit Error Rate
BGCF	Breakout Gateway Control Function
BLER	Block Error
BML	Business Management Layer
BNG	Broadband Network Gateway
bps	bit per second
BPSK	Binary Phase Shift Keying
BS	Base Station
BSC	Base Station Controller
BSHO	Base Station Initiated Handover
BS ID	Base Station Identifier
BSS	Business Support System
BTC	Block Turbo Code
BU	Binding Update
BWA	Broadband Wireless Access
C/No	Carrier-to-Noise Density
C-PCEF	Core Policy Control Enforcing Function
C-SAP	Control Service Access Point
C-SM	Collaborative Spatial Multiplexing
CA	Certificate Authority
CAC	Call Admission Control
CALEA	Communications Assistance for Law Enforcement Act
Cap-Ex	Capital Expenditure
CAPL	Contractual Agreement Preference List
CBC	Cipher Block Chaining
CBR	Constant Bit Rate
CC	Convolutional Coding, Chase Combining
CCM	Counter-Mode Encryption (CTR) with Cipher Block Chaining Message Authentication Code (CBC-MAC)
CDD	Cyclic Delay Diversity
CDF	Charging Distribution Function, Cumulative Distribution Function

CDMA	Code Division Multiple Access
CEPT	European Conference of Postal and Telecommunications Administrations
CID	Connection Identifier
CINR	Carrier-to-Interference and Noise Ratio
CITEL	Inter-American Telecommunication Commission
CMAC	Cipher-based Message Authentication Code
CMIP	Client Mobile IP
CoA	Care of Address
CoRe	Constellation Rearrangement
CP	Cyclic Prefix
CPE	Customer Premise Equipment
CPS	Common Part Sublayer
CQI	Channel Quality Information
CQICH	Channel Quality Information Channel
CRC	Cyclic Redundancy Check
CRU	Contiguous Resource Unit
CS	Convergence Sublayer
CSCF	Call Session Control Function
CSG	Closed Subscriber Group
CSI	Channel State Information
CSM	Collaborated Spatial Multiplexing
CSN	Connectivity Service Network
CTC	Convolutional Turbo Coding
CUI	Chargeable User Identity
CWG	Certification Working Group
D-TDOA	Downlink Time Difference of Arrival
DB	Database
dB	decibel
dBm	decibels of power, w.r.t. 1 milliwatt
DCD	Downlink Channel Descriptor
DES	Data Encryption Standard
DF	Don't Fragment
DHCP	Dynamic Host Configuration Protocol
DiffServ	Diffrentiated Services
DL	Downlink
DLFP	Downlink Frame Prefix
DM	Device Management
DNS	Domain Name System
DOCSIS	Data-Over-Cable-Service Interface Specification
DPF	Data Path Function
DPI	Deep Packet Inspection
DPID	Data Path IDentification
DRMD	Device-Reported Metrics and Diagnostics
DRU	Distributed Resource Unit

DSA	Dynamic Service Addition
DSA-ACK	Dynamic Service Addition Acknowledgement
DSA-REQ	Dynamic Service Addition Request
DSA-RSP	Dynamic Service Addition Response
DSC	Dynamic Service Change
DSCP	DiffServ Code Point
DSD	Dynamic Service Deletion
DSL	Digital Subscriber Line
DSx	Dynamic Service Addition, Change, or Deletion
DIUC	Downlink Interval Usage Code
E-CSCF	Emergency Call Session Control Function
E-MBS	Enhanced Multicast Broadcast Service
E-UTRAN	Evolved UTRAN (3GPP)
E2E	End-to-End
EAP	Extensible Authentication Protocol
EAP-AKA	EAP Authentication and Key Agreement
EAP-TLS	EAP Transport-Layer Security
EAP-TTLS	EAP Tunneled Transport-Layer Security
ECINR	Effective Carrier-to-Interference and Noise Ratio
EESM	Exponential-Effective SINR Mapping
EDE	Encrypt-Decrypt-Encrypt
EDGE	Enhanced Data rates for GSM Evolution
EF	Expedited Forwarding
EH	Extended Header
EIK	EAP Integrity Key
EML	Element Management Layer
EMS	Element Management System
EMSK	Extended Master Session Key
EPC	Evolved Packet Core
EPDG	Evolved Packet Data Gateway
EPS	Evolved Packet System
ERTPS	Enhanced Real-Time Polling Service
ES	Emergency Service
ESM	Effective SINR Mapping
eTOM	Enhanced Operations Map
EVC	Ethernet Virtual Circuits
EVDO	EVolution Data Optimized
EVM	Vector Magnitude
FA	Foreign Agent
FBBS	Fast BS Switching
FCH	Frame Control Header
FCC	Federal Communications Commission
FDD	Frequency Division Duplexing
FD-FDD	Full-Duplex Frequency Division Duplex
FDMA	Frequency Division Multiple Access

FEC	Forward Error Correction
Femto-AP	Femto Access Point (or FAP)
FFR	Fractional Frequency Reuse
FFT	Fast Fourier Transform
FID	Flow Identifier
FIFO	First In, First Out
FTP	File Transfer Protocol
FUSC	Full Usage of Subchannels
GARP	Generic Attribute Registration Protocol
GDOP	Geometric Dilution of Precision
GERAN	GSM EDGE Radio Access Network
GGSN	Gateway GPRS Support Node
GMH	Generic MAC Header
GPRS	General Packet Radio Service
GPS	Global Positioning System
GRE	Generic Route Encapsulation
GRWG	Global Roaming Working Group
GSM	Global System for Mobile communications
GT	Guard Time
GTP	GPRS Tunneling Protocol
GW	Gateway
H-NSP	Home Network Service Provider
H-NSP-ID	Home Network Service Provider Identifier
HA	Home Agent
HAAA	Home AAA
HARQ	Hybrid Automatic Repeat reQuest
HARQ-CC	Hybrid Automatic Repeat reQuest for Convolutional Code
HCSN	Home Connectivity Service Network
HD-FDD	Half-Duplex Frequency Division Duplex
HHO	Hard Handover
HLR	Home Location Register
HMAC	keyed-Hash Message Authentication Code
HO	Handover
HoA	Home Address
HODI	Handover Data Integrity
HSPA	High Speed Packet Access (3GPP)
HSS	Home Subscriber Server
HTTP	Hypertext Transfer Protocol
I-CSCF	Interrogating Call Session Control Function
IAS	Internet Access and Services
IASP	Internet Application Service Provider
ICMP	Internet Control Message Protocol
ICT	Information and Communication Technologies
ID	Identifier
IE	Information Element

IEEE	Institute of Electrical and Electronics Engineers
IETF	Internet Engineering Task Force
IGMP	Internet Group Management Protocol
IIOT	Infrastructure Interoperability Testing
IM	Instant Messaging
IMS	IP Multimedia Subsystem
IMSI	International Mobile Subscriber Identity
IMT	International Mobile Telecommunications
IntServ	Integrated Services
IOT	Interoperability Testing
IP	Internet Protocol
IP-CAN	IP Connectivity Access Network
Ipsec	Internet Protocol Security
IPTV	Internet Protocol TV
IPv4	Internet Protocol Version 4
IPv6	Internet Protocol Version 6
IR	Incremental Redundancy
IRP	Integration Reference Point
IS	Information Services
ISF	Initial Service Flow
ISM	Industrial, Scientific, and Medical
ISP	Internet Service Provider
ITU	International Telecommunications Union
ITU-D	International Telecommunications Union-Development Sector
ITU-R	International Telecommunications Union-Radiocommunications Sector
ITU-T	International Telecommuncations Union-Telecom Standardization Sector
IWK	Interworking
JTG 5-6	Joint Task Group 5-6
KEK	Key-Encryption-Key
KPI	Key Performance Indicators
L2	Layer 2
L3	Layer 3
LAES	Lawfully Authorized Electronic Surveillance
LBO	Local Breakout
LBS	Location-Based Service
LCID	Logical Channel IDentifier
LDAP	Lightweight Directory Access Protocol
LDPC	Low Density Parity Check
LEC	Local Exchange Carrier
LI	Lawful Interception
LRU	Logical Resource Unit
LTE	Long-Term Evolution (3GPP)

M-SAP	Management Service Access Point
M-WiMAX	Mobile WiMAX
MAC	Media Access Control, Message Authentication Code
MAC CPS	MAC Common Part Sublayer
MAN	Metropolitan Area Network
MBS	Multicast Broadcast Service
MCBCS	Multicast Broadcast Service
MCID	Multicast Connection IDentifier
MCS	Modulation and Coding Scheme
MDHO	Macro Diversity Handover
MEH	Multiplexing Extended Header
MGW	Media Gateway
MIB	Management Information Base
MIESM	Mutual Information Effective SINR Mapping
MIMO	Multiple Input, Multiple Output
MIOT	Mobile Interoperability Testing
MIP	Mobile IP
MISO	Multiple Input, Single Output
MLD	Multicast Listener Discovery
MLRU	Minimum A-MAP Logical Resource Unit
MMSE	Minimum Mean Square Error
MN	Mobile Node
MO	Management Object
MOB_NBR-ADV	Mobile Neighbor Advertisement
MOB_PAG-ADV	Mobile Paging Advertisement
MOB_SCN-REP	Mobile Scanning Result Report
MOB_SCN-REQ	Mobile Scanning Interval Allocation Request
MOB_SCN-RSP	Mobile Scanning Interval Allocation Response
MOB_SLP-RSP	Mobile Sleep Response
MOB_TRF-IND	Mobile Traffic Indication
MP3	MPEG-1 Audio Layer 3
MPEG	Moving Picture Experts Group
MR	Multicast Router
MRC	Maximal Ratio Combining
MRS	Mobile Relay Station
MRTR	Maximum Reserved Traffic Rate
MS	Mobile Station (also referred to as "device" in this book
MS ID	Mobile Station Identifier
MSCHAP	Microsoft's Challenge Handshake Authentication Protocol
MSHO	Mobile Station Initiated Handover
MSK	Master Session Key
MSTR	Maximum Sustained Traffic Rate
MTU	Maximum Transmission Unit
MU-MIMO	Multiple Use-MIMO
MVNO	Mobile Virtual Network Operator

MWG	Marketing Working Group
NAI	Network Access Identifier
NAP	Network Access Provider
NAP ID	Network Access Provider Identifier
NAS	Network Access Server
NCS	Network Control System
NCT	Network Conformance Testing
ND&S	Network Discovery and Selection
NE	Network Element
NGMN	Next-Generation Mobile Networks
NMS	Network Management System
NRM	Network Reference Model, Network Resources Models
NRTPS	Non-Real-Time Polling Service
NSP	Network Service Provider
NSP ID	Network Service Provider Identifier
NWG	Network Working Group
NWIOT	Network Interoperability Testing
OAM	Operations, Administration, Maintenance
OCS	Online Charging Systems
OFCS	Offline Charging Systems
OFDM	Orthogonal Frequency Division Multiplexing
OFDMA	Orthogonal Frequency Division Multiple Access
OFCOM	Office of Communications
OFUSC	Optional Full Usage of Subchannels
OMA	Open Mobile Alliance
OMA DM	Open Mobile Alliance Device Management
OPEX	Operating Expense
OSI	Open System Interconnection
OSS	Operations Support System
OTA	Over-The-Air
P-CSCF	Proxy Call Session Control Function
P-GW	Packet Data Network Gateway
P2P	Peer to Peer
PA	Power Amplifier
PA-PREAMBLE	Primary Advanced Preamble
PAPR	Peak to Average Power Ratio
PC	Paging Controller
PCC	Policy and Charging Control
PEDB	Pedestrian B type (ITU channel model)
PCEF	Policy-Control Enforcing Function
PCINR	Physical-Carrier-to-Interference-and-Noise Ratio
PCRF	Policy and Charging Rules Function
PCT	Protocol Conformance Testing
PDA	Personal Digital Assitant
PDF	Policy Distribution Function

PDFID	Packet Data Flow ID
PDP	Policy Decision Point, Packet Data Protocol
PDU	Protocol Data Unit (802.16), Packet Data Unit (WiMAX Forum)
PEP	Policy Enforcement Point
PF	Policy Function
PHB	Per-Hop Behavior
PHS	Packet Header Suppression
PHY	Physical Layer
PICS	Implementation Conformance Statement
PKI	Public Key Infrastructure
PKM	Privacy Key Management
PMC	Power control Mode Change
PMI	Preferred Matrix Index
PMIP	Proxy Mobile IP
PMK	Pairwise Master Key
PMP	Point to MultiPoint
PN	Pseudonoise, Packet Number
POA	Point of Activation
POM	Point of Manufacturing
POS	Point of Sale
PPP	Point to Point Protocol
pps	packet per second
PRF	Pseudorandom Function
PRU	Physical Resource Unit
PS	Policy Server
PSC	Power Saving Class
PSD	Power Spectrum Density
PTP	Precision Timing Protocol
PTT	Push to Talk
PUSC	Partial Usage of Subchannels
QAM	Quadrature Amplitude Modulation
QoS	Quality of Service
QPSK	Quadrature Phase-Shift Keying
RADIUS	Remote Authentication Dial-In User Service
RAPL	Roaming Agreement Preference List
RNG	Ranging
RAN	Radio Access Network
RCC	Regional Commonwealth in the Field of Communications
RCT	Radio Conformance Testing
RCP	Ranging Cyclic Prefix
RED	Random Early Detection
REG	Registration
RF	Radio Frequency
RG	Residential Gateway

RK	Root Key
RLAN	Radio Local Area Networks
RNC	Radio Network Controller
RNG	Ranging
ROHC	Robust Header Compression
RP	Ranging Preamble
RPT	Radiated Performance Testing
RR	Radio Regulations
RR-REQ	Resource Request
RRCM	Radio Resource Controller and Management
RRM	Radio Resource Management
RRP	Registration Reply
RRQ	Registration Request
RS	Relay Station
RSA	Rivest-Shamer-Adleman
RSSI	Received Signal Strength Indicator
RTD	Round-Trip Delay
RTG	Receive/Transmit Transition Gap
RTP	Real-time Transport Protocol
RTPS	Real-Time Polling Service
RTT	Round-Trip Time
RU	Resource Unit
RWG	Regulatory Working Group
S-GW	Serving Gateway
S-SFH	Secondary Superframe Header
S-CSCF	Serving Call Session Control Function
S-OFDMA	Scalable Orthogonal Frequency Division Multiple Access
SA	Security Association
SA-PREAMBLE	Secondary Advanced Preamble
SADC	Southern African Development Community
SAE	System Architecture Evolution (3GPP)
SAID	Security Association Identifier
SAP	Service Access Point
SBC	Subscriber Station Basic Capabilitity
SDFID	Service Data Flow ID
SDMA	Spatial Division Multiple Access
SDU	Service Data Unit
SE	Sprectral Efficiency
SF	Service Flow
SFA	Service Flow Authorization
SFF	Signaling Forwarding Function
SFH	Superframe Header
SFID	Service Flow IDentifier
SFM	Service Flow Management
SGSN	Serving GPRS Support Node

SHA	Secure Hash Algorithm
SII	Service Information Identity
SIM	Subscriber Identity Module
SINR	Signal to Interference and Noise Ratio
SIP	Session Initiation Protocol
SISO	Single Input, Single Output
SKU	Stock Keeping Unit
SLA	Service Level Agreement
SLRU	Subband LRU
SLS	System Level Simulation
SM	Spatial Multiplexing
SML	Service Management Layer
SNR	Signal-to-Noise Ratio
SNMP	Simple Netework Management Protocol
SOA	Service-Oriented Architecture
SOHO	Small Office/Home Office
SON	Self-Organizing Networks
SPI	Security Parameter Index
SPID	Subpacket Identifier
SPR	Subscriber Profile Repository
SPWG	Service Provider Working Group
SQN	Sequence Counter
SS	Subscriber Station, Security Sublayer, Solution Set
SSL	Secure Sockets Layer
SSRTG	Subscriber Station Receive/Transmit Transition Gap
SSTTG	Subscriber Station Transmit/Receive Transition Gap
STC	Space–Time Coding
STTD	Space–Time Transmit Diversity
STFC	Space–Time/Frequency Coding
SU-MIMO	Single User-MIMO
SUB	Subscriber
SVD	Singular Value Decomposition
TCP	Transmission Control Protocol
TCXO	Temperature-compensated Crystal Oscillator
TDD	Time Division Duplexing
TDL	Tapped Delay Line
TDM	Time Division Multiplex
TDMA	Time Division Multiple Access
TDOA	Time Difference of Arrival
TEK	Traffic Encryption Key
TEV	Testing Equipment Vendor
TLS	Transparent LAN Service
TLV	Type Length Value
TMN	Telecommunications Management Network
ToS	Type of Service

TP	Traffic Priority
TSC	Technical Steering Committee
TTG	Transmit/Receive Transition Gap
TUSC1	Tile Usage Subchannels 1
TUSC2	Tile Usage Subchannels 2
TWG	Technical Working Group
UATI	Unicast Access Terminal Identifier
UCD	Uplink Channel Descriptor
UDP	User Datagram Protocol
UE	User Equipment
UGS	Unsolicited Grant Service
UHF	Ultra High Frequency
UICC	User Identity Credentials Card
UIUC	Uplink Interval Usage Code
UL	Uplink
UMTS	Universal Mobile Telecommunications System
URI	Uniform Resource Identifier
URL	Uniform Resource Locator
USI	Universal Services Interface
USIM	Universal Subscriber Identity Module
UTRA	Universal Terrestrial Radio Access
UTRAN	UMTS Terrestrial Radio Access Network
V-NSP	Visited Network Service Provider
V-NSP-ID	Visited Network Service Provider Identifier
VAAA	Visited Authentication Authorization and Accounting
VC	Virtual Circuit
VEHA	Vehicle A Type (ITU channel model)
VLAN	Virtual Local Area Network
VNSP	Visited Network Service Provider
VoIP	Voice-over-Internet Protocol
VPN	Virtual Private Network
VR	Variable Rate
VSP	VoIP Service Provider
WAPECS	Wireless Access Policy for Electronic Communications Services
WCDMA	Wideband CDMA
WFAP	WiMAX Femto Access Point
WIB	WiMAX Initial Bootstrap
WiMAX	Worldwide Interoperability for Microwave Access
WMAN	Wireless Metropolitan Area Network
WRI	WiMAX Roaming Exchange
WRX	WiMAX Roaming Interface
WSIM	WiMAX Subscriber Identity Module
XML	Extensible Markup Language

CHAPTER 1

WiMAX STANDARDIZATION OVERVIEW

KAMRAN ETEMAD AND MING-YEE LAI

1.1 INTRODUCTION

The Worldwide Interoperability for Microwave Access, abbreviated and most often cited as WiMAX™, is a wireless technology that provides standard-based solutions for broadband data access to fixed, portable, and mobile devices. WiMAX end-to-end specifications and products are built based on the IEEE 802.16 standards, which define the air-interface protocols.

WiMAX has gained growing acceptance worldwide since the inception of WiMAX-certified products in 2005. While the initial WiMAX specifications and products addressed the fixed-broadband access market, the industry quickly turned its focus to enable mobility to significantly expand the market opportunities and deployment models. Mobile WiMAX is now being positioned and accepted as an IMT2000 technology and a promising contender for IMT-advanced or the so-called "4G systems." This technology is being deployed on a large scale in many countries and regions including the United States, Japan, Russia, and India, and, according to WiMAX Forum tracking data [1], as of April 2009 there have been 599 deployments in 147 countries.

The mobile WiMAX air interface is designed based on orthogonal frequency division multiple access (OFDMA) and advanced multiantenna technologies, which are accepted by the industry at large to be the key building blocks for next-generation wireless access technologies. Mobile WiMAX also relies on IP network architecture with a focus on simplicity and Internet-based open protocols which are also the new trends and requirements among operators for next generation networks. WiMAX is well positioned to realize the convergence of fixed and mobile broad-

WiMAX Technology and Network Evolution. Edited by Kamran Etemad and Ming-Yee Lai
Copyright © 2010 the Institute of Electrical and Electronics Engineers, Inc.

band access in one simple network. Although the WiMAX technology is primarily focused on IP protocols and solutions, it also supports Ethernet as an option, which has limited usage but importance to some fixed-access deployments.

In addition to its focus on OFDMA, multiantenna and IP-based technologies and despite many technical and business challenges, the WiMAX Forum emphasizes retail device distribution model enablement and end-to-end infrastructure "plug and play" interoperability to lower costs of products and networks. In the traditional subscription model, user devices go through operator-specific testing procedures and distribution chains. In the open retail model, embedded WiMAX devices, including consumer electronics, are sold through conventional consumer electronics stores, and no longer require operator-specific validation and sales distribution. Following and extending the success of the WiFi model into wide-area networks, the retail device distribution model also helps in reducing the cost of customer acquisition by removing the subsidy of devices that most operators pay in today's wireless business models.

The term WiMAX was created by the WiMAX Forum™, which was formed in June of 2001 as a worldwide consortium focusing on the global adoption of WiMAX technology. The WiMAX Forum is chartered to promote and enable end-to-end standards-based WiMAX solutions and their interoperability. To achieve these goals, the WiMAX Forum has created working groups that address end-to-end technical specifications, testing and certification, marketing, global roaming, application, and regulatory issues. On end-to-end technical specifications, the WiMAX Forum specifies the WiMAX network standards and develops air-interface system profiles.

The WiMAX air interface is developed in IEEE 802.16. Mobile WiMAX based on IEEE 802.16e-2005 was formally approved by the International Telecommunication Union (ITU) in 2008 as an International Mobile Telecommunications-2000 (IMT-2000) technology for third-generation (3G) wireless communications. Mobile WiMAX based on the developing IEEE 802.16m is being considered for inclusion in the IMT-Advanced for the fourth-generation (4G) broadband wireless communications.

WiMAX technology development and evolution rely on cooperative and complementary efforts in the WiMAX Forum and in IEEE 802.16. This chapter provides an overview of WiMAX standardization as it relates to IEEE 802.16 and WiMAX Forum activities, and also presents an overview of the WiMAX road map in the evolution of wireless technologies.

1.2 IEEE 802.16 WORKING GROUP STRUCTURE AND STANDARDS

The IEEE 802.16 Working Group, established by the IEEE Standards Board in 1999, has developed and published several versions of air-interface standards for wireless metropolitan area networks (WirelessMAN) with focus on the media access control (MAC) and physical (PHY) layers specifications. The initial versions of the standard 802.16/a/d focused on fixed access, and the later versions (802.16e-2005 [2] and 802.16-2009 [3]) include many new features and functionalities needed to support enhanced QoS, security, and mobility. The 802.16 Working Group is currently focusing on the specification for next-generation systems under the 802.16m Task Group. A recent 802.16 work group organization is shown in Figure 1.1.

The following summarizes the history of IEEE 802.16 specifications as they relate to WiMAX technology:

- 802.16 is the first version of the line-of-sight (LOS) air-interface standard with frequency range 10–66 GHz for fixed wireless access, completed in December 2001.

- 802.16a, the first version of the non-line-of-sight (NLOS) air-interface standards with frequency less than 11 GHz for fixed wireless access, was completed in January 2003.

- 802.16-2004 or 802.16d, completed in June 2004, was the first version of this standard that was considered by the WiMAX Forum to enable fixed broadband access or the so-called "fixed WiMAX." The IEEE 802.16d-based systems do not support mobility but they have been deployed as the last-mile wireless broadband access alternative to cable and DSL.

- 802.16e-2005, also referred to as 802.16e, is an amendment to 802.16-2004 to add mobility and other MAC and PHY enhancements; it was completed in December 2005. This version of the standards, which is the basis for Mobile

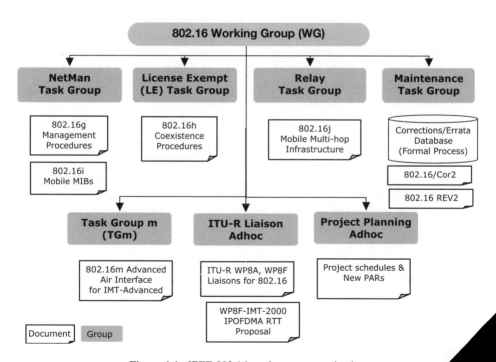

Figure 1.1. IEEE 802.16 work group organization.

WiMAX Release 1.0 products, also adds scalable OFDMA and enhanced multiple input multiple output (MIMO) schemes plus some new MAC features such as hybrid automatic repeat request (ARQ) and multicast and broadcast services.

- 802.16-2009, previously called 802.16-REV2 before its finalization, is a revision of IEEE 802.16 that combines 802.16e-2005 along with corrigendum 2 and other pertinent amendments produced by IEEE 802.16 task groups, including 802.16i (management information bases—MIBs—for fixed WiMAX), 802.16f (management procedures), and 802.16g (MIBs for mobile WiMAX), into one specification. This revision of this standard, which is the reference for Mobile WiMAX System Profiles Release 1.0/1.5 and includes clarifications and enhancements in frequency division multiplexing (FDD) mode of operation and other MACs, was completed in May 2009.

- 802.16m, the latest version of the IEEE 802.16 standards, is being developed as a candidate for IMT-advanced technologies and includes major enhancements in PHY and MAC to improve high mobility support, user experience, and system efficiency. 802.16m is planned for completion in the fourth quarter of 2010.

Besides the 802.16e-2005 corrigendum 2 (802.16 Cor2 in Figure 1.1), 802.16m, 802.16g, and 802.16i mentioned above, 802.16h works on the coexistence procedures between WiMAX and other license-exempt devices, such as 2.4 GHz Wi-Fi.

The IEEE 802.16 standards define the structure of the PHY layer (layer 1) and MAC layer (layer 2) operations that occur between mobile subscriber stations (SS/MS) and base stations (BS) over the air, where base stations are at the edge of network interfacing the rest of radio access and connectivity/core network elements. The upper layer signaling over the air as well as the network architecture and protocols behind the base stations, which are required for an end-to-end specification, are considered to be outside the scope of IEEE 802.16 standards and need to be defined elsewhere.

In addition, IEEE 802.16 MAC and PHY specifications are developed with emphasis on flexibility and, thus, have allowed many optional features and modes of operation that may enable various implementation alternatives to BSs and SS/MSs. Without specifying interoperability system profiles, which limits optional configurations and features, such implementation alternatives may result in incompatible products.

The WiMAX Forum addresses the above issues as it undertakes the task of end-to-end interoperability specification by defining air-interface system profiles [4], network architectures [5], and certification frameworks. The next section describes the role and organization of the WiMAX Forum.

UM OVERVIEW

 was established in 2003 to promote and enable deployment of oadband access technology based on the IEEE 802.16 standards. l, the WiMAX Forum initiated several long-term projects and programs to complement IEEE 802.16 standardization by defining system profiles as well as protocol and radio conformance testing specifications to be used as basis for industry-wide certification of BSs and SS/MSs. In addition to developing testing specifications, the WiMAX Forum has also established certification laboratories and is managing conformance and interoperability testing to ensure that all WiMAX-certified products across different implementations work seamlessly with one another.

To extend the scope of interoperability beyond the air interface and based on service provider's requirements, the WiMAX Forum also defines end-to-end network architecture and protocol specifications consistent with IEEE 802.16 and WiMAX system profile air-interface definitions. The WiMAX Forum fosters a strong WiMAX ecosystem involving several hundred member companies that contribute to chipsets, radio and network subsystems, user device and computing systems, management systems, software applications, and systems integration.

1.3.1 WiMAX Forum Organization

The WiMAX Forum's organization, working groups, and their charters are shown in Figure 1.2 [6]. The key functions for individual working groups are described in the following:

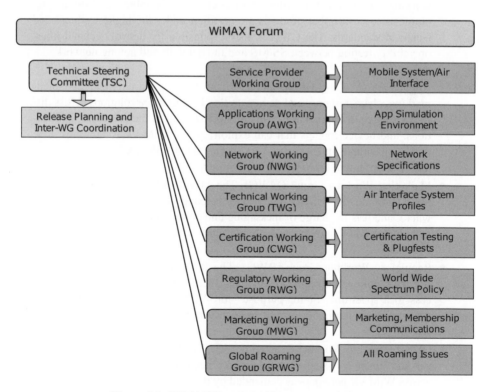

Figure 1.2. WiMAX Forum working groups and charters.

- Service Provider Working Group (SPWG). The SPWG is responsible for the development of Stage 1 requirements on air-interface and network systems.

- Network Working Group (NWG). The NWG's charter is to develop Stage 2 network architecture specifications and Stage 3 network procedures and protocols based on Stage 1 requirements. NWG also includes the Network Interoperability Testing (NWIOT) Task Group, which is in charge of defining network-level interoperability testing specifications. The network specification efforts in WiMAX Forum involve interactions with other standards organizations, such as the Internet Engineering Task Force (IETF), Third Generation Partnership Project (3GPP), Third Generation Partnership Project 2 (3GPP2), Broadband Forum (previously DSL Forum), and Open Mobile Alliance (OMA).

- Technical Working Group (TWG). The TWG develops technical conformance specifications, system profiles, and certification test suites for the air interface, complementary to the IEEE 802.16 standards, primarily for the purpose of interoperability and certification of IEEE 802.16 standards-compliant SS/MSs and BSs.

- Certification Working Group (CWG). The CWG manages the WiMAX Forum Certification Program through the selection and oversight of certification test labs, the evaluation of testing options, the integration of key technical working group system profiles, testing-related deliverables, and organizing plugfests to ensure that products are certified under the highest ethical and technical standards. The CWG is also responsible for network-related interoperability testing between SS/MS and the network and among network elements.

- Application Working Group (AWG). The AWG's charter includes benchmarking, characterizing, and demonstrating best-practice solutions across different classes of applications. This group also studies the use cases, applications, and device models that can exploit unique WiMAX capabilities. One notable accomplishment in the AWG is creating a WiMAX PHY/MAC simulation environment for WiMAX applications.

- Marketing Working Group (MWG). The MWG's charter is to drive worldwide adoption of WiMAX-enabled broadband connectivity anytime, anywhere, and it is in charge of messaging and communications of WiMAX developments and deployments both internally within the WiMAX Forum and externally.

- Regulatory Working Group (RWG). The RWG takes charge of worldwide spectrum policy and regulatory matters, including promoting worldwide access to a spectrum "fit for purpose" for WiMAX Forum certified systems with sufficient harmonization of frequency ranges to facilitate economies of scale.

- Global Roaming Working Group (GRWG). The GRWG addresses the technical and business issues and framworks needed to enable global roaming across WiMAX networks as demanded by the marketplace. Based on this objective, the group develops WiMAX roaming agreement templates, roaming guidelines, roaming interfaces, and roaming tests to expedite and ease the implementation of WiMAX global roaming.

- The Technical Steering Committee or Technical Plenary has been formed to help with better cross-working-group coordination and decision making on strategic issues, such as release planning and working group process guidelines, which impact multiple working groups.

1.3.2 WiMAX Forum Specification Development Process

The specifications produced by the WiMAX Forum follow a coordinated process among the WiMAX Forum working groups, as shown in Figure 1.3 [7].

In the requirement phase, the SPWG develops Stage 1 service providers' requirements on network and roaming as well as air-interface coordination with IEEE 802.16 WG.

The Stage 2 and Stage 3 air-interface specifications are developed by the IEEE 802.16 WG. The Stage 3 air-interface specification from IEEE 802.16 WG is used as a basis for the WiMAX Forum TWG's air-interface system profile, Protocol Implementation Conformance Statement (PICS) document, and air-interface test specification in the specification phase.

The NWG develops Stage 2 and Stage 3 Network specifications, which are used for the Network Conformance Statement document and network test specification in the specification phase. Similarly, the GRWG develops Stage 2 and Stage 3 roaming specifications, which are used for the Roaming Conformance Statement document and roaming test specification in the specification phase.

The air-interface test specification, together with the network test specification and the roaming test specification, is used in implementation in products and test equipment by the CWG, equipment vendors, and testing equipment vendors (TEVs), followed by validation under CWG's auspices in the implementation phase. After the implementation phase, the CWG conducts certification of WiMAX products and systems. The next section provides more details on the certification process.

1.3.3 WiMAX Forum Certification

WiMAX is backed with an established certification program managed by the CWG. This certification program will enable network and device interoperability, drive infrastructure costs down, open the technology to roaming, and lead to more device types and distribution models.

The WiMAX Forum certification framework, which was initially focused on MAC and PHY layer testing, includes test suites to cover the following [8]:

- Radio conformance testing (RCT) to test the conformance on the over-the-air physical (PHY) layer protocol.
- Protocol conformance testing (PCT) to test the conformance on the over-the-air media access control (MAC) layer.

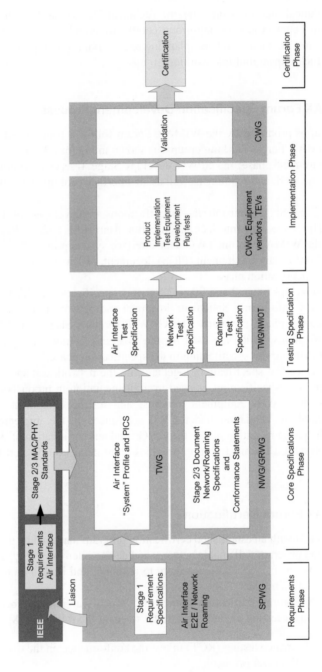

Figure 1.3. WiMAX specification process flow.

- Interoperability testing (IOT) or mobile interoperability testing (MIOT) to test interoperability of mobile station MS/SS (abbreviated as device), and base station (BS).

The WiMAX Forum certification process for the initial test modules on WiMAX BS and MS/SS products is shown in the Figure 1.4.

Conformance testing is undertaken to ensure that the equipment has correctly implemented the specifications and is compliant with the standards—IEEE 802.16 standards at MAC/PHY layers and WiMAX NWG specification at the upper layer. Conformance testing typically requires emulators against which devices and BSs are tested. Interoperability testing, on the other hand, is undertaken to verify that equipment from different vendors work within the multivendor network setup using real equipment. To pass certification, each BS needs to interoperate with a minimum of three devices from other vendors. Similarly, each MS needs to work with a minimum of two base stations from other vendors within the same certification profile (defined by spectrum band, channelization, and duplexing).

Although conformance testing (Step 2 in Figure 1.4) is relatively straightforward, interoperability testing (Step 3 in Figure 1.4) can cause delays during the initial phase of testing for a profile, since different vendors are often required to make changes to their products, but may not be able to do so within the anticipated time frame.

The detailed tasks involved in developing conformance testing or interoperability testing in the certification process are:

- Development of test scripts for validation. The first task is the development of test scripts, which describe individual tests to be performed on the vendor's

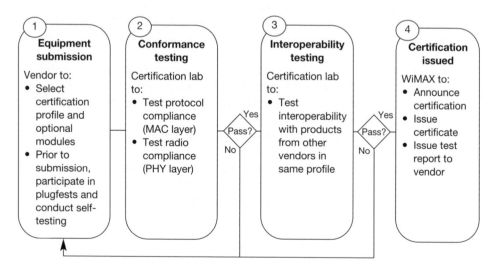

Figure 1.4. WiMAX Forum's certification process flow.

product to ensure that it meets the mandatory functionality requirements for certification.

- Lab preparation. The lab preparation task involves validating test scripts, sourcing relevant test tools, opening the lab, preparing test hardware, integrating test scripts into the test hardware, and, finally, accepting vendor equipment.
- Validation testing of test scripts and test bed. Before a test script is released for use in the lab (test bed), it must be tested. If all three vendors successfully pass validation testing, then the script is considered valid.
- Certification testing. The final task in the test step is certification testing

In the traditional subscription model, operators maintain tight control of the devices allowed on their network and the distribution of these devices. The primary reason for this is due to concerns that allowing unapproved devices on their network could have a negative impact on system performance and functionality. Many operators spend a lot of time and money validating devices in their labs and certifying them for approved use in their network. These devices are then sold by the operators with a considerable subsidy. On the contrary, the open retail model allows the devices that have been tested in the WiMAX certification labs to be sold through electronic retail stores with no subsidy cost by the operator. The WiMAX certification program significantly removes the operators' burden on device and infrastructure testing. The WiMAX technology and ecosystem also promotes open network interfaces and infrastructure interoperability to enable low-cost plug-and-play deployments and expansions of multivendor networks. To pave the way to achieve this long sought-after goal in telecommunications, an industry-wide WiMAX network interoperability testing (NWIOT) program has been established and is progressing.

With the objective of expanding the certification to address interoperability of generic retail distributed devices roaming across any WiMAX network, the following new testing modules were developed and added in late 2009:

- Radiated performance testing (RPT) to determine over-the-air radio performance of SS/MS (but not BS) from the PHY layer metrics, during normal operation, in the presence of near-field impairments due to objects (head, hands, or desktop) typically found near the device.
- Network conformance testing (NCT) to test the network layer (layer 3) signaling and messaging to and from the device for security, mobility, and other network functions needed to ensure the device will behave properly when connected to a WiMAX network. NCT also includes testing of over-the-air (OTA) provisioning and activation, which is key for subsequent operation of a new device in a network.

In addition to the above, the CWG is also working on infrastructure interoperability testing (IIOT) to test interoperability at the defined reference points within the network for interaccess service network (ASN), intra-ASN, and ASN-to-connectivity service network (CSN) messaging. The IIOT specifications are developed by the NWIOT-TG in the NWG and the first set of tests were completed in early 2010.

1.4   WiMAX TECHNOLOGY HIGH-LEVEL ROAD MAP

The growing demand for broadband Internet applications is driving the need for higher data rates with more cost-effective transport mechanisms. The wireless industry is in transition toward offering cost-effective broadband mobility access that provides service transparency with wired broadband (e.g., xDSL or cable). Unlike the migration from 2G to 3G, the transition to 4G is more than a simple technology upgrade and it requires significant changes to backhauls, radio sites, core networks, network management, services paradigm, and mobile device distribution model. In this transition, new hardware and software solutions are needed at the MS, BS, and across the network, shifting the design focus from low-rate circuit-switched and voice-based systems to optimize high-rate multimedia traffic. All candidate future technologies rely on OFDMA-based air-interface designs supported by IP-based transport and core networks. Figure 1.5 shows the evolution of air interfaces at high levels.

In this evolution, Mobile WiMAX is positioned to be a solution for migration of exiting 2G/2.5G networks in countries where 3G deployments are delayed or skipped and also for matured CDMA-based 3G networks, which are already offering high-speed data services and are looking for a long-term OFDMA-based and

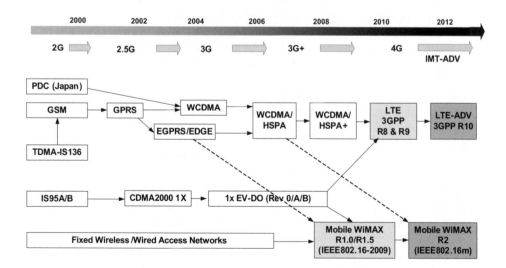

Figure 1.5. Evolution of mobile wireless technologies.

scalable solution. Mobile WiMAX has also been considered for Greenfield deployment by many new operators or those offering fixed access to add mobility to their services and improve efficiency of their networks.

In this book, we skip the early versions of WiMAX technology, which addressed the fixed network applications, and focus on the Mobile WiMAX systems and their evolution in the near and long term. Figure 1.6 shows the high-level view of this evolution and captures different key elements of Mobile WiMAX standardization and certification [9].

The Mobile WiMAX System Profile Release 1.0, based on 802.16e-2005, was completed in late 2006. The radio-level certification of products started in 2007. The WiMAX system profile development and certification takes a phased approach to address deployment priorities and vendor readiness.

System Profile Release 1.0 includes all 802.16e-2005 mandatory features and also requires some of the optional features needed for enhanced mobility and QoS support. This system profile is based on the OFDMA radio technology and enables downlink and uplink multiple input multiple output (MIMO) as well as beamforming features. System Profile Release 1.0 is defined only for time division duplex (TDD) mode of operation with focus on 5 MHz and 10 MHz bandwidths in several band classes in 2.3 GHz, 2.5 GHz, and 3.5 GHz bands, but it also includes the 8.75 MHz bandwidth specifically for Korea.

The WiMAX certification for System Profile Release 1.0 started with a Wave 1 subset, excluding MIMO and a few optimization features, to enable early market deployment. This was followed by full Wave 1 and Wave 2, which progressively add more and more feature tests over time based on vendor and testing tool availability. The early phases of certification were also limited to MAC and PHY layer conformance and interoperability testing between BS and SS/MS.

Meanwhile, the specification of WiMAX Forum Network Release 1.0 was completed in December 2007. Based on that, network-level device conformance testing as well as infrastructure interoperability testing projects were initiated. The goal is to ensure end-to-end interoperability of WiMAX devices with the networks and also ensure multivendor plug-and-play network infrastructure deployment. Network Release 1.0 defines the basic architecture for IP-based connectivity and services while supporting all levels of mobility.

For meeting operators' needs to deploy advanced services, take on new market opportunities, and be more competitive with other evolved 3G systems, the WiMAX Forum developed interim releases for both system profile and network specifications without major modifications to the IEEE 802.16 standards.

The work on Network Release 1.5 network specifications was started in parallel with that for System Profile Release 1.5. Network Release 1.5 includes dynamic QoS, over-the-air provisioning for open retail distribution model, support for advanced network services (e.g., multicast/broadcast service, location-based service, emergency service, and lawful interception), and commercial grade voice over IP (VoIP). More recently, new work items were initiated, as part of Network Release 1.6, and are planned for completion in the early 2010 to cover network enhancements, such as femtocell, IP4/IP6 evolution, emergency telecommunications service (ETS), measurements, and device diagnostics.

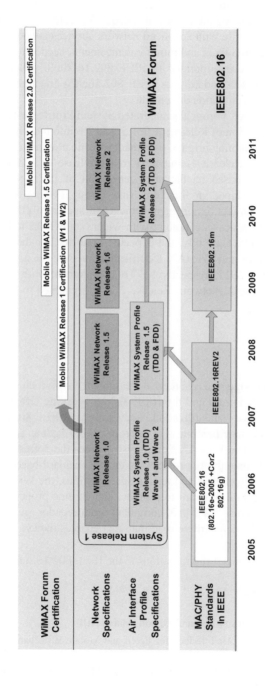

Figure 1.6. Mobile WiMAX technology and network evolution road map.

13

The System Profile Release 1.5 work item was initiated to enable Mobile WiMAX in new spectra, including frequency division duplex (FDD) bands, to address a few MAC efficiency improvements needed for technology competitiveness, and to align system profile with advanced network services supported by Network Release 1.5. All required fixes and minor enhancements needed to support System Profile Release 1.5 are incorporated in IEEE 802.16-2009, which combines the IEEE 802.16-2004 base standard plus IEEE 802.16e/f/g amendments and related corrigenda into one specification document.

Recently, the WiMAX Forum has integrated System Profile Releases 1.0 and 1.5 and Network specification of Releases 1.0, 1.5, and 1.6 into one end-to-end System Release 1, while maintaining original document specification numbers.

Following System Release 1, the next major release of Mobile WiMAX is Release 2. It will be based on the next generation of IEEE 802.16 standards, which are being developed in the 16m Task Group (TGm) of IEEE 802.16. WiMAX Release 2 targets major enhancements in spectrum efficiency, latency, and scalability of the access technology with wider bandwidths in the challenging spectrum environments. Currently, the expected time line for the formal completion of IEEE 802.16m and WiMAX Certification of Release 2 products are late mid 2010 and early 2011, respectively. To complement with development of System Profile Release 2, Network Release 2 is also planned to be completed by early 2011.

1.5 SUMMARY

The WiMAX technology and standard development involve close collaborations between the WiMAX Forum and the IEEE 802.16 Working Group. Although the MAC/PHY air-interface standardization is developed by IEEE 802.16, the product conformance, profile specification, and interoperability testing are carried out by the WiMAX Forum. The WiMAX Forum also defines the corresponding network architecture and interoperability protocols to enable standard-based end-to-end solutions. In addition to technical and testing specifications, the WiMAX Forum has established a certification program to promote and assume proper operation of all WiMAX devices across WiMAX-compliant multivendor networks. Mobile WiMAX has defined a competitive road map that enables the evolution of existing mobile and fixed networks.

1.6 REFERENCES

[1] WiMAX Forum Industry Research Report, March 2010.
[2] IEEE 802.16e-2005: Air Interface for Fixed and Mobile Broadband Wireless Access Systems, Dec. 2005.
[3] IEEE 802.16-2009: Air Interface for Fixed and Mobile Broadband Wireless Access Systems, May 2009.

[4] WiMAX Forum™ Mobile System Profile Release 1.5, Version 1, Approved Specifications, August 2009.

[5] WiMAX Forum Network Architecture Stage 2 and 3—Release 1.5, Version 1, Approved Spefications, Nov. 2009.

[6] WiMAX Forum Web site: http://www.wimaxforum.org/about/working-groups.

[7] WiMAX Forum Documentation Structure and Identification, A11-001, 2009.

[8] The WiMAX Forum Certified Program, September 2008.

[9] Kamran Etemad, Overview of Mobile WiMAX Technology and Evolution, *IEEE Communications Magazine,* Oct. 2008.

CHAPTER 2

OVERVIEW OF MOBILE WiMAX AIR INTERFACE IN RELEASE 1.0

KAMRAN ETEMAD, HASSAN YAGHOOBI, AND MASOUD OLFAT

2.1 INTRODUCTION

This chapter provides an overview of MAC and the PHY layer in IEEE 802.16 standards [1] as it pertains to Mobile WiMAX System Profile Release 1.0 [3]. The system profile and associated conformance and interoperability specifications in Mobile WiMAX Release 1.0 are primarily based on IEEE 802.16e-2005 [1] with selective enhancements from IEEE 802.16-2009 [2]. It should be mentioned that IEEE 802.16 defines multiple PHY modes and many optional PHY and MAC features that are not included and productized in Mobile WiMAX and, therefore, will not be covered or elaborated in this chapter. The objective here is to highlight the key features, design concept, and protocols used as building blocks for this air interface, while referring readers to respective specifications for further details. WiMAX Forum has also developed some enhancements in System Profile Release 1.0 as part of an interim Release 1.5 [4,5,6] which will be covered in the next chapter.

The high-level MAC/PHY protocol structure for Mobile WiMAX as specified in IEEE 802.16-2005/2009 is shown in Figure 2.1. This protocol structure consists of a scalable OFDMA PHY and a flexible MAC layer design composed of three sublayers; namely: convergence sublayer (CS), MAC common part sublayer (MAC CPS), and security sublayer (SS).

Sections 2.1 and 2.3 present an overview of PHY and MAC layer features in Release 1.0, respectively, which is the basis for all initial mobile WiMAX deployments. Section 12.1 provides a summary list of Release 1.0 system profile features as well as the relevant target certification profile band classes.

WiMAX Technology and Network Evolution. Edited by Kamran Etemad and Ming-Yee Lai
Copyright © 2010 the Institute of Electrical and Electronics Engineers, Inc.

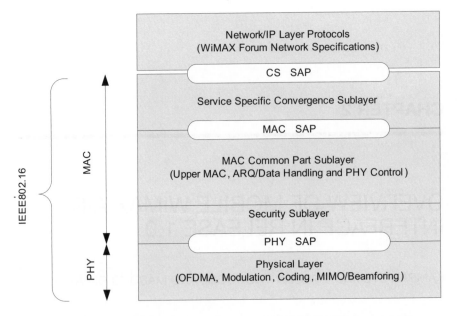

Figure 2.1. Air interface protocol structure in Mobile WiMAX Release 1.0/1.5.

2.2 OVERVIEW OF MOBILE WiMAX PHY LAYER (RELEASE 1.0)

This section describes the PHY layer of Release 1.0, starting with channel bandwidth and symbol structure, followed by frame structure and channelization, as well as PHY layer procedures.

2.2.1 Channel Bandwidths and OFDMA Symbol Structure

IEEE 802.16 standards support five different PHY modes [1] including the scalable OFDMA PHY that is adopted as the only physical layer mode for Mobile WiMAX systems in Releases 1.0 and 1.5.

WirelessMAN-OFDMA was originally used as one of the major PHY layer modes in IEEE 802.16-2004. In this mode, the number of tones was initially 2048 (2K FFT size) regardless of channel bandwidth (starting from 1.25 MHz up to 20 MHz). On the other hand, in WirelessMAN-OFDM mode, where the number of tones are fixed to 256 (256 FFT size), smaller channel bandwidth results in smaller tone spacing and, therefore, more sensitivity to the impairments caused by mobility. When there is relative motion between the transmitter and receiver, a Doppler shift of the RF carrier results in frequency error. Also, any error in the oscillators either at the transmitter or at the receiver can result in residual frequency error.

Consequently, to have consistent performance across all channel bandwidths, the concept of scalable OFDMA (S-OFDMA) was introduced in IEEE 802.16-2005 to keep the tone spacing constant by changing the FFT size with bandwidth. For example, for channel bandwidths of 1.25, 5, 10, or 20 MHz, FFT sizes of 128, 512, 1024, and 2048 are used, respectively, to maintain common tone spacing. The FFT size of 256 is not selected intentionally to avoid overlap with WirelessMAN-OFDM, where a single FFT size of 256 is used. For other channel bandwidths such as 3.5, 7, and 8.75 MHz, the FFT sizes are selected such that deviation from default tone spacing is minimized.

The mobile devices can discover the FFT size and bandwidth of operation through scanning, spectral detection during initial network entry, or by inspecting the broadcasting message, including the neighbor advertisement by the serving base station facilitating the handoff. Note that, due to fixed tone spacing, many base band parameters, such as OFDM symbol duration, cyclic shift, OFDM symbol duration, and frame duration, remain unchanged. Consequently, reimplementation of many unnecessary bandwidth-dependent parameters is eliminated. This is the principal of designing scalable OFDMA used in Mobile WiMAX [1].

In OFDM(A) systems, the cyclic prefix is used to provide for immunity against multipath fading, maintaining orthogonality among tones (by converting the linear convolution to cyclic convolution in the time domain), and increasing the tolerance of the receiver to symbol time synchronization error. The values allowed for cyclic prefix (CP) in IEEE 802.16e-2005 are $\frac{1}{16}$, $\frac{1}{8}$, $\frac{1}{4}$, and $\frac{1}{2}$ of the OFDM symbol duration. In this case, the value of the CP is decided by the base station (BS), and the mobile station (MS) recognizes the value of the CP at the time of network entry. However, Mobile WiMAX release 1.0 and Release 1.5 have chosen the value of $\frac{1}{8}$ for the cyclic prefix.

A band class defines a particular frequency range (which may consist of multiple subbands within that band) as supported by the BS. In Mobile WiMAX, each band class is associated with supported channel bandwidth and duplexing mode in addition to a channel roster (center frequency list). For example, the U.S. 2.6 GHz WiMAX network is operating in the 2.496–2.690 GHz spectrum with 10 MHz channel bandwidth in TDD mode and channel raster of 250 KHz. This configuration is based on Mobile WiMAX Release 1.0 Certification Profile Band Class Group (BCG) 3.A. For more information on Mobile WiMAX Release 1.0 band classes, please refer to Section 12.1.

In order to increase the efficiency of the OFDM system and to allow overlapping between adjacent channels without imposing interference as a result of out-of-band emission of the adjacent channel, OFDM systems typically utilize oversampling. Oversampling is also used for other reasons, such as decreasing the peak-to-average power ratio (PAPR) or meeting target spectral emission masks. In general, if the oversampling ratio is n, the OFDM sampling frequency F_s and the tone spacing are calculated using equations (2.1) and (2.2):

$$\text{Sampling frequency } F_s = \text{floor} (n \times \text{BW}/8000) \times 8000 \qquad (2.1)$$

$$\text{Tone Spacing} = F = \frac{\text{floor} (n \times \text{BW} / 8000) \times 8000}{FFT} \qquad (2.2)$$

The objective here is to make the sampling frequency a multiple of 8 KHz. For the channel bandwidths that are a multiple of 1.25 MHz, the oversampling ratio is 28/25 or 1.12.

In this section, we use BCG 3.A as an example to describe the Mobile WiMAX frame structure; however, the technology supports other band classes, as described in Section 2.4. For BCG 3.A, where both channel bandwidths of 5 and 10 MHz are required, the oversampling ratio is 1.12, whereas the FFT size is 512 for 5 MHz and 1024 for 10 MHz, and, therefore, the tone spacing is 10.94 KHz for both channel bandwidths. As mentioned, the cyclic prefix guard interval is selected to be ⅛ of the useful symbol duration. Table 2.1 summarizes the PHY layer parameters for the channel bandwidth that are multiples of 1.25 MHz.

2.2.2 TDD Frame Structure

The IEEE 802.16e-2005 PHY supports both time division duplexing (TDD) and frequency division duplexing (FDD) operations. The FDD mode also defines the half duplex FDD mode to support lower complexity terminals in which one radio front unit is time shared for receive and transmit. Mobile WiMAX Release 1.0 is based on TDD mode only, but Release 1.5 supports both TDD and FDD systems. IEEE 802.16-2005 supports different values for size of TDD S-OFDMA frame duration. The choice of frame duration is a trade-off between the delay and latency performances (such as handoff delay, etc.), and overhead performances. Mobile WiMAX has chosen the value of 5 msec for the frame size.

Figure 2.2 illustrates the OFDMA frame structure for the TDD mode, where each 5 msec radio frame is flexibly divided into downlink (DL) and uplink (UL) subframes. The DL and UL subframes are separated by small transmit/receive and receive/transmit transition gaps (TTG and RTG, respectively) to prevent DL and UL transmission collisions.

As seen in Figure 2.2, the S-OFDMA frame is mainly composed of four parts: a DL subframe, a UL subframe, and two gaps between the DL to UL subframes (TTG) and UL to DL subframes (RTG), transmit/receive transition gap and receive/transmit transition gap, respectively. These gaps are accommodated for BS to

Table 2.1. WiMAX OFDMA parameters

Parameters	Values			
Bandwidth (MHz)	1.25	5	10	20
Sampling frequency, F_s (MHz)	1.4	5.6	11.2	22.2
Sample time at the receiver ($1/F_s$) (nsec)	714	178.5	89.25	44.625
FFT size	128	512	1024	2048
Tone spacing (F_s/FFT size) (KHz)	10.94			
Useful symbol time (1/tone Spacing) (μs)	91.43			
Guard time (symbol time/8) (μs)	11.43			
OFDMA symbol time (symbol time + guard time) (μs)	102.94			

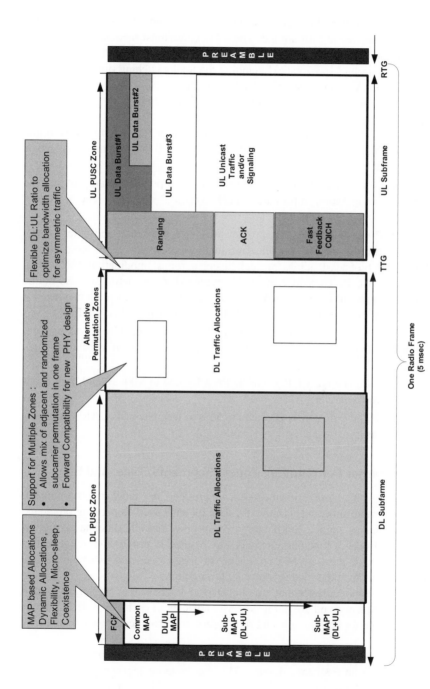

Figure 2.2. TDD frame structure.

21

switch transmission to reception (TTG) or vice versa (RTG), and MS switches otherwise. Each of these values must be greater than 50 nsec subscriber station TTG and RTG (SSTTG and SSRTG), as defined by the standard, the maximum time needed for the device to turn around. The TTG value must include the round-trip delay in addition to the SSTTG. The Mobile WiMAX has chosen the sum of RTG and TTG to be around 163.7 nsec. Excluding the total switching time from 5 msec frame, 47 OFDMA symbols can be fit into each frame.

This frame structure defines the following physical channels:

- *Preamble.* The broadcast channel using the first OFDM symbol of the frame in downlink; used by the MS initial and handover-related scanning as well as PHY synchronization to the DL.
- *Frame Control Header (FCH).* This follows the preamble and provides the frame configuration information such as MAP message length, coding scheme, and usable subchannels.
- *DL-MAP and UL-MAP.* This provides resource allocation and other control information for the DL and UL subframes ,respectively. MAP messages are typically broadcast across the cell using a robust modulation and coding scheme (MCS). To reduce the MAP overhead, the system may also define one or more multicast sub-MAPs that can carry traffic allocation messages at higher MCS levels for users closer to the BS and with higher CINR.
- *UL Ranging.* The UL ranging subchannel is allocated for MS to perform closed-loop time, frequency, and power adjustment, as well as bandwidth requests.
- *UL CQICH.* The UL CQICH channel is allocated to MSs to feed back channel state information.
- *UL ACK.* The UL ACK is allocated to MSs to feed back DL HARQ acknowledgements.

2.2.3 Subcarrier Permutation, Zones, Segments, Tiles, and Bins

In an OFDMA system, tones or subcarriers are grouped into subchannels. In the DL or UL, one or more subchannels carry the data for each user. Depending on the usage model, channel model, morphology, or specific multiple antenna configurations, various forms of tone permutations may be used to provide optimum performance. In principal, there are two approaches to divide tones into subchannels. One is distributed and the other one is adjacent permutation.

In distributed mode, tones that are allocated to a subchannel are spread around the channel frequency and are not necessarily adjacent to each other. This mode is more appropriate for both mobile and fixed users, and provides frequency diversity gain.

In adjacent permutation mode, the tones allocated to each subchannel are adjacent to each other in frequency domain. This case is more appropriate for adaptive coding and modulation, and provides better channel estimation capabilities. This permutation mode allows instantaneous feedback from the receiver to the transmitter and, therefore, it is a better choice for fixed, portable and low-speed mobility. Normally, this mode is used for beamforming and closed-loop MIMO techniques. Figure 2.3 shows the difference between the two permutation modes.

IEEE 802.16e-2005 has defined several distributed subcarrier permutation modes that differ from each other in number of pilots per subchannel or the number of symbols per slot. These modes are partial usage of subchannels (PUSC) (both segmented, and with all subchannels), full usage of subchannels (FUSC), optional PUSC (OPUSC), optional FUSC (OFUSC), tile usage of subchannels 1 (TUSC1), and Tile Usage of Subchannels 2 (TUSC2). The only adjacent permutation mode is called Adaptive modulation coding (AMC). Some of these modes are only used in DL and some are used both in DL and UL. PUSC is the only mode that is mandatory for the standard and all other modes are optional. PUSC and AMC modes are described here briefly. Interested readers are referred to the standard [7] and [1] for more detailed information about these modes as well as those permutation modes that are not covered here.

PUSC was originally designed to allow virtual frequency reuse. In the networks in which frequency reuse one ($n = 1$) is used, every sector might cause interference on adjacent sectors both in UL and DL. However, if frequency reuse of, say, three is used, the amount of interference is significantly reduced at the expense of larger spectrum usage. This luxury may not be available to all operators due to the scarcity of spectrum and tight government regulations. By using PUSC, Mobile WiMAX allows frequency reuse of one to be virtually mapped to a frequency reuse of three through disjoint segregation of subcarriers into three segments for three sectors. This eliminates the need for carrier management, while spectral mask is defined for only one carrier, and, at the same time, frequency diversity is achieved.

When PUSC is used in the DL, each symbol is divided into several clusters (depending on the FFT size) having 14 adjacent tones each; the clusters are renumbered and divided into six major groups, each group having either 24 or 16 clusters.

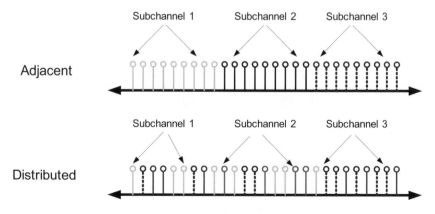

Figure 2.3. Adjacent and distributed OFDMA permutation.

In each cluster, tones are allocated to pilots first, and then the remaining tones are allocated to data subchannels. These major groups are allocated to segments or sectors (by default, group 0 to sector 0, group 2 to sector 1, and group 4 to sector 2). If there are only three sectors, the rest of groups are allocated similarly to the same three sectors as well. In this case, each sector will get two major groups, which is one-third of the useful data tones, and the adjacent sectors use nonoverlapping frequency segments, thus virtually creating a frequency reuse of three within one RF carrier.

A DL PUSC slot includes two adjacent OFDM symbols and 24 data tones per symbol. As mentioned before, pilots had been already preallocated within each cluster.

In the UL, the concept of tiles are used, where a tile is a 4×3 structure constructed from 8 data and 4 pilot tones as depicted in Figure 2.4. Each slot in UL PUSC is composed of six tiles, not necessary adjacent in frequency domain. This results in 48 data tones and 24 pilot tones per slot.

Tables 2.3 and 2.4 provide some details of configuration parameters associated with PUSC in the downlink and uplink.

The PUSC in WiMAX Systems is particularly distinguished by the method by which the tones are allocated to subchannels and by the structure and density of pilot tones. Therefore, PUSC mode can be used as segmented PUSC (virtual frequency reuse or so-called "PUSC 1/3") or without segmentation—the so-called "PUSC with all subchannels," in which all major groups are allocated to a sector or all sectors. Figure 2.5 shows these two configurations. If an operator is able to use three separate channels for a real frequency reuse of 3, then, depending on frequency planning, it might make sense to use PUSC with all subchannels, as segmented PUSC obviously reduces the system throughput since it reduces the effective bandwidth.

The only adjacent permutation used in Mobile WiMAX is AMC, which is mainly used for advanced antenna systems such as beamforming or when link adaptation is essential. Normally, when the mobile speed is low, the instantaneous feedback

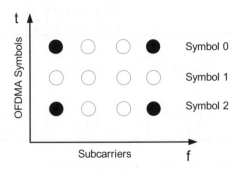

Figure 2.4. Uplink PUSC tile.

Table 2.2. PUSC DL parameters

Parameter	2K FFT	1K FFT	512 FFT	128 FFT
Number of DC carriers	1	1	1	1
Number of left guard tones	184	92	46	22
Number of right guard tones	183	91	45	21
Number used	1681	841	421	85
Number of tones per cluster	14	14	14	14
Number of clusters	120	60	30	6
Number of data tones per SC per symbol	24	24	24	24
Number of SC	60	30	15	3

can be exploited and, therefore, AMC could be considered as an appropriate permutation. AMC permutation is the same for UL and DL. In AMC, in each OFDM symbol, pilots have fixed locations in the frequency domain at each symbol.

AMC is defined based on the concept of bands and, hence, the name AMC band is applied to the AMC allocations. Each band consists of four consecutive rows of bins. Each bin consists of nine consecutive tones in the frequency domain, with a pilot located in the middle of a bin (Figure 2.6).

AMC subchannels consist of six contiguous bins, either in time or frequency (with N bins by M symbols, and $N \times M = 6$) in the same band. This means that there are three configurations for AMC permutations, 2×3, 3×2, and 1×6 (Figure 2.7). In AMC, each symbol is composed of 96 subchannels as opposed to 70 subchannels in PUSC. This is due to the smaller number of pilots in AMC relative to PUSC. In general, in AMC, channel estimation reliability (due to less pilot density), and frequency diversity is traded off with throughputs and, therefore, is more appropriate for low-speed environments.

Following preamble, the rest of the frame is divided into zones. A zone is a group of OFDMA symbols with the same permutation scheme (such as PUSC or AMC). The first zone following the preamble, where FCH and DL/UL MAPs are transmitted, is always a PUSC zone. Within each zone, there are segments consisting of one or more subchannels, each representing single instance of a MAC data unit.

Table 2.3. PUSC UL parameters

Parameter	2K FFT	1K FFT	512 FFT	128 FFT
Number of DC carriers	1	1	1	1
Number of used tones	1681	841	409	97
Number of guard tones (left, right)	184,183	92,91	52,51	16,15
Number of subchannels	70	35	17	4
Number of tones per subchannel	48	48	48	48
Number of tiles	420	210	102	24
Number of subcarriers per tile per symbol	4	4	4	4
Number of tiles per subchannel	6	6	6	6

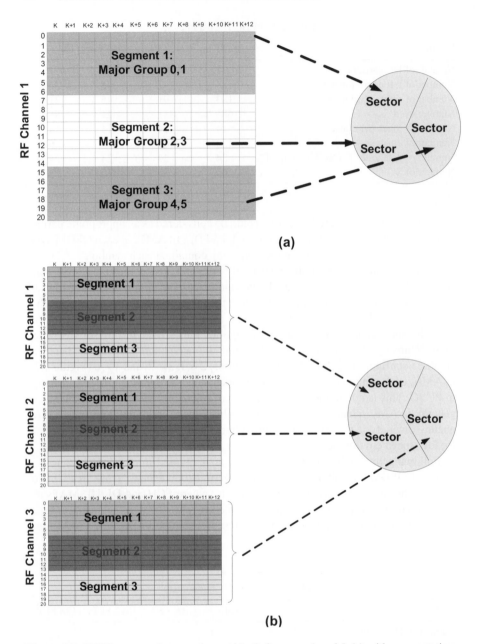

Figure 2.5. PUSC permutation may be used both for reuse 1 and 3 (a) with segmentation and (b) all subchannels.

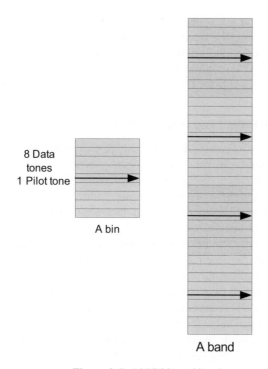

Figure 2.6. AMC bin and band.

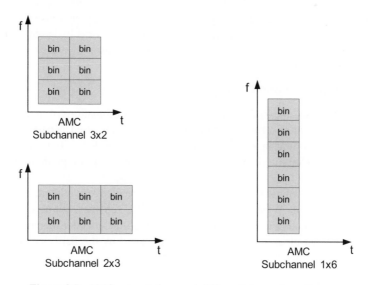

Figure 2.7. AMC permutations and different bin configurations.

For PUSC in a DL or UL zone, slots span over two and three OFDMA symbols, consecutively; therefore, the number of symbols in the DL zone must be odd, whereas the number of symbols in the UL zone must be a multiple of 3. As a result, if there are a total of 47 symbols in 11 PUSC configurations, only a limited number of combinations are possible for DL:UL separation, such as 41:6, 35:12, or 29:18.

2.2.4   Fractional Frequency Reuse (FFR)

The concept of virtual frequency reuse 3 is useful when there is a shortage of spectrum in some area and/or the band planning is not trivial, and the operator requires an effective reuse-3 configuration to avoid interference. However, Mobile WiMAX allows a combination of these two modes (reuse 1 and virtual reuse 3) to maximize the spectrum efficiency. This mode is called "fractional frequency reuse," wherein the mobile stations are categorized into two groups, depending on the level of interference they experience. The PUSC segmentation can be configured in a way that mobiles with low interference (close to the BS) operate in the zone with all subchannels (reuse 1), whereas for the mobiles that experience high interference (e.g. cell-edge users) each cell or sector operates in the segmented PUSC mode. This way, the actual frequency reuse is reuse 1 ($N = 1$) and, therefore, the operator does not have to be concerned about the availability of three channels and the band planning in the same area, while the cell-edge devices experience the interference similar to a reuse-3 configuration. At the same time, since the main contribution for overall sector throughput comes from high-SINR devices (those close to the BS), using reuse 1 mode prevents a throughput reduction for them and, therefore, the overall sector throughput is not significantly impacted. Figure 2.8 depicts the fractional frequency reuse mode. In this figure, F_{S1}, F_{S2}, and F_{S3} represent different sets of PUSC segments in the same frequency channel. The segment reuse planning can be dynamically optimized and configured across sectors or cells based on network load and interference conditions on a frame-by-frame basis.

2.2.5   Modulation and Coding

In Mobile WiMAX, the PHY layer defines various combinations of modulation and coding rates, providing a fine resolution of data rates to be used as part of link adaptation. Channel coding in Mobile WiMAX includes randomization, forward error correction (FEC), bit interleaving, and modulation, as well as repetition of orders 2, 4, and 6.

The FEC adds redundancy to the signal for error recovery without requiring retransmission. Mobile WiMAX includes several FEC schemes, out of which tail biting convolutional coding (see Figure 2.9), in which the encoder memory is initialized with the last block being encoded, is mandated by the IEEE standard. Other FEC schemes defined by IEEE 802.16e-2005 are:

- Zero tailing convolutional coding (CC), which appends a zero tail byte at the end of each burst to initialize the Viterbi decoder
- Convolutional turbo coding (CTC), used in conjunction with HARQ to miti-

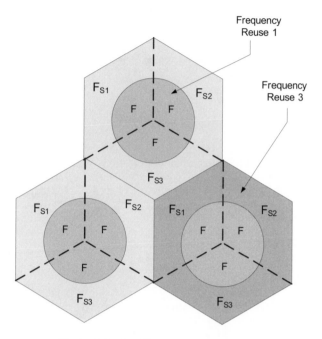

Figure 2.8. Fractional frequency reuse.

gate the effect of channel impairment and interference

- Block turbo code (BTC), which is the product of two simple component codes that are binary extended Hamming codes or parity check codes
- Low-density parity check (LDPC) codes, which provide competitive performance, very close to Shannon capacity, with reduced complexity. They are block codes, which are well suited for bursty packet-switched data services.

Although IEEE 802.16e-2005 identifies tail biting convolutional coding (CC) as the only mandatory FEC mode, the Mobile WiMAX system profile requires all compliant equipment to also support CTC, especially when HARQ is used. CC is mainly used for FCH and DL and UL MAP, whereas CTC is used in the data region [3].

The modulation schemes used in WiMAX Release 1.0/1.5 are QPSK, 16QAM, and 64QAM, with different data rates. Due to the limitation of the power and the error vector magnitudes (EVM) of the mobile devices, in the UL, 64QAM is only mandatory in the DL. Table 2.4 shows the data rates corresponding to different modulation schemes for different FEC modes.

Table 2.5 shows the data rates for the 5 MHz channel with 512 FFT and the 10 MHz channel with 1024 FFT, assuming partially utilized subchannels (PUSC). The frame duration is 5 milliseconds. Each frame has 48 OFDM symbols, with 44 OFDM symbols available for data transmission. The highlighted values indicate

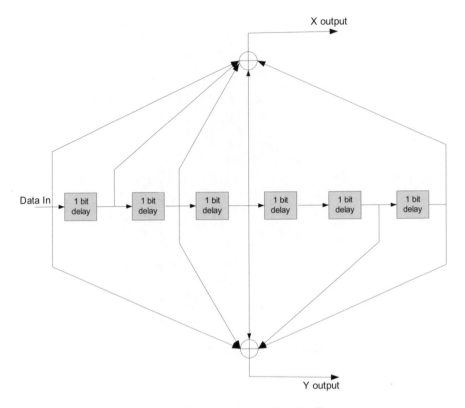

Figure 2.9. Tail biting convolutional coding.

data rates for optional 64QAM in the UL.

2.2.6   Advanced Antenna and MIMO

Mobile WiMAX is the first cellular technology that has commercialized multiple antenna solutions, especially MIMO in a large-scale cellular deployment. In this section, we provide a high-level overview of advanced antenna systems in Mobile WiMAX; a more detailed description of these techniques is presented in Chapter 4.

In general, multiple antennas in wireless networks may provide one or more of

Table 2.4. Coding rates for different modulation and FEC schemes in WiMAX

	CC-Tail-Biting	BTC	CTC	LPDC
QPSK	1/2, 3/4	1/2, 3/4	1/2, 3/4	1/2, 2/3, 3/4, 5/6
15 QAM	1/2, 3/4	1/2, 3/4	1/2, 3/4	1/2, 2/3, 3/4, 5/6
64 QAM	1/2, 2/3, 3/4	1/2, 3/4	1/2,2/3,3/4, 5/6	1/2, 2/3, 3/4, 5/6

Table 2.5. Mobile WiMAX PHY rates [PUSC, DL:UL (22:15), 11 symbol overhead, MIMO 2 × 2]

Modulation	Code rate	5 MHz Channel		10 MHz Channel	
		Downlink rate, Mbit/s	Uplink rate, Mbit/s	Downlink rate, Mbit/s	Uplink rate, Mbit/s
QPSK	1/2 CTC, 6x	0.53	0.27	1.06	0.56
	1/2 CTC, 4x	0.79	0.41	1.59	0.84
	1/2 CTC, 2x	1.59	0.82	3.17	1.68
	1/2 CTC, 1x	3.17	1.63	6.34	3.36
	3/4 CTC	4.75	2.45	9.50	5.04
16 QAM	1/2 CTC	6.34	3.26	12.67	6.72
	3/4 CTC	9.50	4.90	19.01	10.08
64 QAM	1/2 CTC	9.50	4.90	19.01	10.08
	2/3 CTC	12.67	6.53	25.34	13.44
	3/4 CTC	14.26	7.34	28.51	15.12
	5/6 CTC	15.84	8.16	31.68	16.80

the following three advantages: (1) SNR improvement through beamforming array gain, (2) link reliability and coding enhancement through diversity and coding gain, and (3) capacity improvement using multiplexing gain. IEEE 802.16 provides all three gains by allowing the following multiple antenna modes:

- Simple receiver diversity achieved by using multiple antennas at the receiver with sufficient spacing is used for combating fading. The receiver uses either maximal ratio combining (MRC) or minimum mean square error (MMSE) combining. Normally, MRC has a significant advantage over MMSE, but it is more difficult to implement. Receive diversity can be done both in UL and DL.
- Transmit diversity modes by space time/frequency coding (STFC) are used for combating fading. In this case, transmit diversity of orders 2, 3, and 4 are supported by the standard.
- Beamforming by advanced antenna systems (AAS) is used to increase the coverage and reduce the outage probability
- Capacity enhancement by MIMO schemes exploiting spatial multiplexing (SM) techniques.
- Combination of beamforming and STFC or beamforming and MIMO by closed-loop MIMO mode or using precoding, mainly for low-speed mobiles providing very high throughput gains and coverage enhancements.

Space time coding (STC) yields diversity gain without channel knowledge at the transmitter by coding across antennas (space) and across time and/or frequency. This mode could provide capacity improvement as well as better spectral efficiency. The simplest STC method is the Alamouti scheme (2 Tx, 1 Rx antenna, or 2 × 1).

In spatial multiplexing (SM), the transmit data stream is split into several parallel streams, which are transmitted from different antennas simultaneously, along with signal processing at the receiver to detect the streams. It provides capacity gain at no additional power or bandwidth consumption cost. Each stream has a lower data rate. The main computational complexity lies in the separation step.

To optimally utilize channel characteristics, Mobile WiMAX is employing adaptive MIMO switching between the two MIMO modes, STC and SM. The adaptive switching means that when a user is in a region where its SINR is above a threshold and the correlation factor of its channel matrix is low, the SM mode is automatically enabled and, as a result, capacity gain is achieved. However, when the SINR for a MS is low, or the channel condition number is high, STC is used to maximally realize the coverage gain.

The switching recommendation can be made either at the mobile station and then fed back to the BS using some feedback mechanisms, and/or it can be made by the BS directly. In either case, it is the BS that makes the final decision about the switching. The decision process is not specified by the standard and is left for implementation design.

Beamforming is another mode of multiple antenna transmission in IEEE 802.16, where signals from multiple antenna elements are transmitted with different gains and phases to collectively form a composite beam. By changing the gain and phase of the signal at each antenna, one can steer the beam. This can maximize the transmitted energy toward the desired mobile station, increase the coverage, provide the diversity gain, mitigate noise and interference, and improve signal quality (SINR). In addition to simple switched beams, Mobile WiMAX supports different types of transmit beamforming such as single-stream beamforming (using spatial channel correlation), in which no restriction on the antenna separation is applied. Beamforming can also be used in an urban environment enabling interference nulling, where interference toward a particular mobile station is effectively nulled or eliminated. The later mode requires small antenna separation of half a wavelength and is mainly appropriate to morphologies without many paths. Another beamforming mode is multistream beamforming or SDMA, in which interference toward multiple mobile stations is eliminated at the same time.

In order to realize the beamforming gain and regardless of the beamforming type, support of coherent signal processing at the receiver is required. At the same time, the transmitter requires instantaneous channel knowledge to determine the beamforming antenna gains applied to the transmitted signal. Therefore, beamforming is mainly appropriate for low-speed mobility usage models.

Mobile WiMAX has different mechanisms to provide BS with instantaneous channel knowledge. Such knowledge may be obtained at the BS by estimating the channel in the UL and applying that to the DL, relying on TDD channel reciprocity. Alternatively, the channel information can be provided to the BS through instantaneous feedback from the mobile station through either dedicated pilots (digital feedback) or channel sounding (analog feedback) [1].

Closed-loop MIMO or MIMO plus beamforming are two modes of operation for providing a high level of capacity and coverage gain in low-mobility environments. These modes are enabled through application of precoding, wherein a precoding matrix W is multiplied by the transmission matrix (A, B, or C). BS uses channel quality indication fed back from the mobile station to calculate the precoding matrix W. Closed-loop MIMO can be implemented in different ways: (a) selection of a number of antennas among several antennas, (b) antenna grouping, (c) codebook method (CQICH feedback), and (d) using UL channel sounding (analog precoding).

In UL, due to the restriction on the power and size of mobile devices such as hand-held terminals, transmission from multiple antennas is problematic. First, the MS typically cannot afford to have two power amplifiers, which requires higher power consumption and makes the device more complex and, hence, more expensive. Moreover, an efficient MIMO transmission requires considerable antenna separation, which is not always possible given the size of mobile devices. For these reasons, the IEEE 802.16 standard has devised a collaborative MIMO in the UL, where two devices each transmitting on one antenna are selected by the BS to collaborate and share a common time-frequency resource for their UL transmission. This is similar to sending two streams from two antennas with a difference that the two antennas are on two different mobile stations, of course with significant spatial separations. The only requirement is that pilots used by the two MSs must be orthogonal. The processing at the BS is very similar to the processing of a two-branch MIMO UL transmission, with minimal or zero impact on the MS complexity. Note that unlike DL MIMO, UL collaborative MIMO does not provide any enhancement on user throughput, but improves overall UL sector throughput. See Chapter 4 for more details about MIMO features in IEEE 802.16 standards.

2.2.7 DL Physical Channels

In IEEE 802.16, physical channelization is very simple. Only few types are defined as most channelization is logical and achieved through MAC messages transmitted using the same type of physical layer resources and traffic. This section describes the DL control channel structure and the next section focuses on the UL.

Preamble and Frame Control Header

In the DL subframe, the first symbol is the preamble. The preamble is one OFDM symbol containing a known pseudonoise (PN) sequence carrying a particular ID called IDCell (a number from 0 to 31), and a segment ID (from 0 to 2) with different PAPR characteristics. The total number of sequences is 114. It is used for initial timing and carrier synchronization, network scanning, and cell identification. It can be used for a coarse channel estimation, as well as signal-to-noise and interference (SINR) calculation.

The second symbol after the preamble starts with a mandatory PUSC zone. This zone starts with the frame control header (FCH), which contains a fixed number of bits (24 bits) and is located in the first and second OFDMA symbol after the preamble. It consists of one PUSC slot, using QPSK1/2 modulation with repetition 4. It includes the downlink frame prefix (DLFP). The DLFP contains the "used subchannel bitmap," which indicates which subchannel group (segment in PUSC mode) is used in this frame, the burst profile for the DL MAP following the FCH (e.g., coding and repetition), and the length of the DL MAP.

Downlink and Uplink MAP and Sub-MAPs

A broadcast MAC message called "DL MAP" is transmitted in the downlink, right after the FCH, to transmit the "map" of DL allocations to all mobile stations (hence the name DL MAP). Moreover, it defines all control information, contains the synchronization field, and could include allocation for other BSs. The basic element of the DL MAP is an information element (IE) called DL MAP IE that specifies allocations to mobile stations within a particular frame (the concept of connections per mobile station and connection ID (CID) will be explained later). The IE is per connection per mobile station and includes CID (unicast, multicast, or broadcast), downlink interval usage code (DIUC) to define the type of downlink burst type, OFDMA symbol offset (in units of OFDMA symbols) at which the burst starts, subchannel index (the lowest index subchannel used for carrying the burst), power boost applied to the allocated data subcarriers, number of OFDMA symbols used to carry the uplink burst, and repetition code used for the allocation.

DL MAP always follows the FCH in every frame, and its length is identified in the FCH message. The UL MAP, however does not need to be included in all frames. If present, it should be always transmitted as the first PDU on the burst described by the first DL MAP IE of the DL MAP.

If a frame or its next frame includes a UL subframe (which is typically the case), the next message is called UL MAP and includes the UL MAP defining UL allocations for all UL bursts transmitted by mobile stations. Like DL MAP, the main information element is called UL MAP IE, which contains the CID, uplink interval usage code (UIUC) to define type of UL access and the burst type associated with the access, OFDMA symbol offset at which the burst starts, subchannel offset that is the lowest index subchannel used to carry the burst, number of OFDMA symbols used to carry the uplink burst, the number of subchannels with subsequent indices, the duration in units of OFDMA slots for the burst, and the repetition code used inside the allocated burst.

Note that there are other types of DL MAPs and UL MAPs, such as compressed MAP, normal MAP, HARQ MAP, and so on, to optimize the size of DL and UL MAP so as to reduce the corresponding overhead. Some of these MAPs are appropriate for specific situations, such as when HARQ is used. For details on structure and usage case corresponding to these MAPs, interested readers are referred to the standard text [1].

In this section, different standard MAPs are briefly summarized along with their usage cases, advantages, and disadvantages. The DL/UL MAP message structure is presented first, and then other MAP formats such as compact MAP, normal MAP, and HARQ MAPs are described briefly.

DL MAP and UL MAP are mainly designed to introduce the location of DL or UL data allocation to subscribers in the DL or UL subframes. They are MAC messages that define burst start times and frequency (TDMA and FDMA) for each allocation.

However, they contain more information elements to define control message.

Table 2.6 shows the general format and the contents of the DL MAP. Note that this table only includes the main fields in DL MAP. Similarly, the main structure of UL MAP (not including all details) and UL MAP IE are shown in Table 2.7.

HARQ MAP

IEEE 802.16e-2005 defines the HARQ MAP message for HARQ-enabled mobile stations. This MAP is only used by BSs supporting HARQ, for mobile stations supporting HARQ.

The following modes of HARQ are supported by the HARQ DL MAP IE:

Table 2.6. DL MAP Structure

Field	Size in bits	Comments
PHY synchronization field	32	
Frame number	8	The frame number, incremented by 1 Mode 2^{24}
Frame duration code	24	A code that indicates the frame duration
DCD count	8	Shows value of the configuration change count of the DCD
Base station ID	48	Identifies the BS. The least significant 24 bits is programmable. The most significant 24 bits is operator ID.
Number of OFDMA symbols	8	For TDD, the number of OFDMA symbols in the DL subframe including the preamble (e.g., 48 symbols per frame)
DL MAP IE		Defines all rectangular allocations
DIUC (or extended or extended-2 DIUC)	Variable	DIUC or extended or extended-2 DIUC used for the burst
N_CID	8	Number of connection IDs contained in this IE for each connection ID
CID or RCID of each connection ID N_CID	16 or less	RCID is the reduced CID used when SUB DL UL MAP is used
OFDMA symbol offset	8	The offset of the OFDMA symbol in which the burst starts
Subchannel offset	6 or 8	The lowest index OFDMA subchannel used for carrying the burst
Boosting	3	Power boost applied to the allocation's data subcarriers
Number of OFDMA symbols (or triple symbol)	5 or 7	Number of OFDMA symbols (or multiples of 3 symbols if AMC 2×3 permutation)
Number of subchannels	6	The number of subchannels used to carry the burst
Repetition coding Indication	2	Repetition code used inside the allocated burst

Table 2.7. UL MAP Structure

Field	Size in bits	Comments
UCD count	8	Shows value of the configuration change count of the UCD
Allocation start time	48	Effective start time of the UL allocation defined by the UL MAP
UL MAP IE		Defines all rectangular allocations in UL
CID	16	The basic or broadcast/multicast CID of the mobile depending of the value of UIUC
UIUC (or extended or extended-2 UIUC) or other UIUCs	Variable	UIUC or extended or extended-2 UIUC used for the burst. Others UIUCs are used for ranging, CDMA allocation, PAPR reduction, etc.
Duration	10	Indicates the duration, in units of OFDMA slots, of the allocation
Repetition coding Indication	2	Repetition code used inside the allocated burst
Slot offset in case of AMC zone	12	Offset from start of the AAS or AMC zone for this allocation, specified in slots

1. Chase combining HARQ for all FEC types (HARQ chase). In this mode, the burst profile is indicated by a DIUC.
2. Incremental Redundancy HARQ with CTC (HARQ IR). In this mode, the burst profile is indicated by the parameters NEP and NSCH.
3. Incremental Redundancy HARQ for Convolutional Code (HARQ CC-IR)

The IE may also be used to indicate a non-HARQ transmission. The HARQ DL MAP IE defines one or more two-dimensional data regions (a number of symbols by a number of subchannels). These allocations are further partitioned into bursts, termed subbursts, by allocating a specified number of slots to each burst. Subbursts of a data region only support the same HARQ mode. The number of slots is indicated by duration or NSCH fields. The slots are allocated in a frequency-first order, starting from the slot with the smallest symbol number and smallest subchannel, and continuing to slots with increasing subchannel number. When the edge of the allocation is reached, the symbol number is increased by a slot duration, as depicted in Figure 2.10. Each subburst is separately encoded.

The HARQ MAP message includes compact DL (or UL) MAP IE and defines the access information for the DL and UL burst of HARQ-enabled mobile stations. These IEs are designed to reduce the overhead caused by the DL/UL MAP. These messages are sent without a generic MAC header. BS may broadcast multiple HARQ MAP messages using multiple bursts after the MAP message. Each HARQ MAP message has a different modulation and coding rate, which is described by the DL-MAP IE along with the location of the bursts. The general format of HARQ MAP is described in Table 2.8.

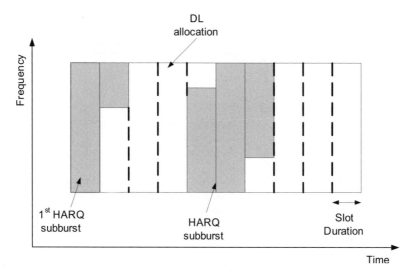

Figure 2.10. HARQ DL allocation.

Compressed MAP

In addition to the normal DL MAP and UL MAP formats, IEEE 802.16 also introduced the concept of "compressed MAP" to reduce the size of DL/UL MAP overhead. The presence of compressed MAP is indicated by the generic MAC header of the MAP burst with a combination, which is invalid for a standard header in the downlink. A compressed UL MAP may appear only after compressed DL MAP, indicated by a bit in the compressed DL MAP data structure. Like the normal MAP, the compressed MAP must occur directly after the DL frame prefix. A 32 bit CRC is appended to the end of the message. The general format of a compressed DL MAP is shown in Table 2.9.

As shown in Table 2.9, the compressed MAP reduces the length of the normal MAP by removing the General MAC header, considering only one CRC, and re-

Table 2.8. HARQ MAP

Field	Size in bits	Comments
Compact UL MAP appended	1	Shows if compact UL MAP is included
MAP message length	9	Length of HARQ MAP in bytes
DL IE count	8	Number of DL IEs in the burst
For each DL IE burst		
Compact DL IE	variable	Includes the DL burst allocation
If compact UL MAP appended		Identifies the BS. The least significant 24 bits is programmable. The most significant 24 bits is operator ID.
Compact UL MAP	variable	Includes the UL burst allocation

Table 2.9. Compressed DL MAP

Field	Size	Comments
Compressed Map indicator	3	
UL MAP appended	1	
Map message length	11	
PHY synchronization field	32	
DCD count	8	Shows the value of the configuration change count of the DCD
Operator ID	8	
Sector ID	8	
Number of OFDMA symbols	8	For TDD, the number of OFDMA symbols in the DL subframe including the preamble (e.g., 48 symbols per frame)
DL MAP counts		
For each DL IE		
DL MAP IE	variable	

moving the burst profiles (DIUC and UIUC). The SUB DL UL MAP message shall appear in a compressed form, in which the generic MAC header is omitted. Figure 2.11 illustrate an example of DL and UL compressed MAPs in a TDD frame.

Sub-MAPs

In Mobile WiMAX, the BS may reduce the overhead by using sub-MAP messages within the MAP message. Although the main MAP is a broadcast message, which is sent with the most robust modulation and coding and received across the BS's coverage area, the sub-MAPs can be multicast to groups of users with similar geome-

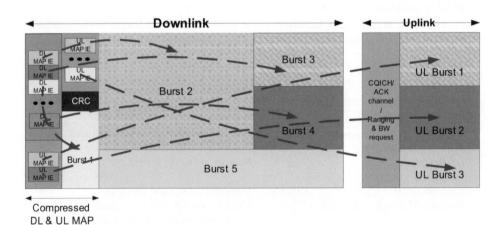

Figure 2.11. Compressed DL and UL MAPs.

tries and long-term channel conditions using optimized modulation and coding scheme (MCS). Using sub-MAPs reduces the overhead associated with common control and traffic assignment messaging, which is the major part of the MAP, while maintaining the reliability of broadcast and common control messaging that is highly desired. Figure 2.12 provides an example of sub-MAP messages.

In Figure 2.12, the BS defines sub-MAP 1 and sub-MAP 2 within the MAP message. The MAP message is modulated using rate ½ QPSK with six repetitions, sub-MAP 1 is modulated with rate ½ QPSK with four repetitions, and sub-MAP 2 is modulated with rate ½ QPSK with two repetitions. Based on channel quality indication (CQI) feedback from the mobile stations, BS is able to transmit assignment messages to the mobile stations by matching the CQI feedback to the sub-MAP MCS. For dynamic scheduling, the use of sub-MAPs is an effective tool for MAP overhead reduction.

Downlink and Uplink Channel Descriptor (DCD and UCD) Message

DCD and UCD are MAC broadcast messages transmitted by the BS at a periodic interval to describe the PHY characteristics of DL and UL channels. The mobile station needs to receive the DCD/UCD before any data session with the BS. Therefore, after the registration and synchronization with the BS, the mobile station must wait to receive the DCD/UCD to learn the physical characteristic (coding, modulation, etc.) of the DL and UL channels.

The DL burst profiles in DCD consists of CINR for each MCS level, BS EIRP, maximum receive sensitivity for initial ranging, TTG and RTG, H-ARQ ACK delay for DL burst, HO type and parameters, and so on.

Similar burst profile parameters included in UCD are CINR for each MCS level, uplink center frequency (primarily used for FDD operation), UL allocated subchannel bitmap, ranging parameters, Band AMC parameters, H-ARQ ACK delay for UL burst, UL open loop power control parameters, and so on.

2.2.8   UL Physical Channels

In the UL, the first three symbols (or part of the first three symbols) are used for UL control signals. There are three major control signals sent in the UL. They are ranging/BW request channel, ACK channel, and channel quality indicator channel (CQICH).

Ranging/BW Request Channel

Ranging is a process used by the MS to synchronize to the downlink frequency and time, and by the BS to adjust all MS transmissions for aligned UL subframe reception at the BS. The ranging process is achieved by the MS transmitting a signal and the BS responding with required adjustments (closed loop).

Bandwidth request is also used by the MS to request UL allocation for data transmissions in the UL.

There are two types of ranging. Ranging at initialization (during network entry), called initial ranging, and ranging in a periodic basis, called periodic ranging,

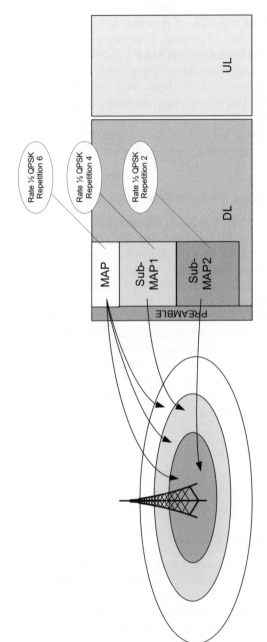

Figure 2.12. Use of sub-MAPs.

to ensure continuous connection of mobile to the network. The channels allocated for ranging and bandwidth request are shared between all mobile stations. Considering the contention based nature of the allocations, for initial ranging, periodic ranging, and bandwidth request, the mobile station selects a CDMA code from a set of codes (subsets of codes dedicated to each functionality) and encodes its UL ranging signal or bandwidth request transmission with the CDMA codes. This might require the mobile station to invoke a random back-off process to avoid further collision.

Ranging is also needed in handoff to adjust the three aforementioned parameters with the target BS. However, in this case, using a possible back-off process will increase the handoff delay significantly. As a result, the target BS allocates a dedicated ranging channel within the first three UL OFDMA symbols to the mobile station to expedite the handoff process. This process is called the fast-ranging process.

UL ACK Channel

The UL ACK Channel is a dedicated PHY channel within the first three UL OFDMA symbols used for sending the acknowledgement or NACK of the DL HARQ packets (subpackets) by the mobile station. This channel is only supported by mobile stations supporting HARQ. One ACK channel occupies one half subchannel, which consists of three pieces of 4×3 UL tile in the case of PUSC. The content of the ACK channel is one bit with value of 0 or 1 depending on whether ACK or NACK is to be transmitted to the BS. This one bit is encoded into a length 3 codeword using an 8-ary alphabet for the error protection, and is orthogonally modulated with QPSK symbols.

CQICH Channel

Fast-feedback slots are assigned to individual MSs when the BS requires information from the MSs in a very timely manner. Information concerning CINR, MIMO, antenna, and spatial multiplexing are all examples of time-critical information that can be considered for the fast-feedback channel. A portion of the beginning of the UL subframe is allocated for this purpose. Orthogonal encoding is used to allow for multiple MSs to transmit simultaneously within the same slot.

Fast-feedback channels are individually allocated in a unicast manner to mobile stations for transmission of PHY-related information that requires fast response. This physical channel is dedicated to a mobile station for sending fast feedback periodic CINR reports (physical CINR or effective CINR). Allocating these subchannels to the mobile stations is performed using the CQICH IE (CQICH allocation IE or CQICH control IE). The CQICH allocation IE may indicate whether the reported metric applies to the preamble or to a specific permutation zone or to the pilots. For the report on the preamble, the BS can request the mobile station to report the CINR based on the measurement from the preamble for different frequency reuse factors. For the report on specific permutation zones, the CQICH allocation IE indicates the report-type configuration, which includes the zone for which the CINR is estimated. The zone is identified by its permutation type (PUSC, segmented PUSC, AMC AAS, FUSC, etc.), and pseudorandom binary sequence (PRBS) ID. If the CQICH is allocated through the CQICH control IE, the mobile station uses the measurement configuration defined in the latest CQICH allocation IE. The feedback period is defined and configured by the BS.

Two types of feedback could be provided by the MS: effective CINR (or ECINR) and physical CINR (or PCINR). PCINR is the actual CINR measured over the preamble, pilots, or permutation zones, averaged and transmitted to the BS through the CQICH channel. ECINR is the CQI that is interpreted as the mobile station recommendation that best meets the specified target error rate for the remaining period until the next scheduled CQI report.

Each fast-feedback slot consists of one OFDMA slot using QPSK modulation on the 48 data subcarriers it contains and can carry up to 4 bits of data.

The feedback can be sent using the enhanced fast-feedback channel as well. The enhanced fast-feedback slots are individually allocated to an MS for transmission of PHY-related information that requires fast response from the MS. Each enhanced fast-feedback slot consists of one OFDMA slot using QPSK modulation on the 48 data subcarriers it contains and can carry a data payload of 6 bits. The enhanced fast-feedback channel can be allocated using enhanced CQICH allocation IE, can be used for band AMC, and can allocate up to 6 bits CQICH channels.

The rest of the DL and UL frames are divided into subchannels and then allocated to users for data burst transmission.

2.2.9 PHY Control Sublayer

Ranging

Ranging provides several functions including bandwidth request, timing and power control, periodic maintenance, and handover. The initial ranging, which utilizes special CDMA ranging codes, is used by MSs that wish to synchronize with the system for the first time. Default time and power parameters are used to initiate communications and then adjusted until they meet acceptance criteria. These parameters are updated periodically during a maintenance interval. A ranging channel is defined in the UL and composed of one to eight adjacent subchannels. A PN code sequence is chosen by the MS and this sequence is modulated onto the subcarriers. The PN sequence provides a spreading gain that allows multiple ranging transmissions to be simultaneously received by the BS, thus defining this technique as CDMA ranging.

Link Adaptation and HARQ

Link Adaptation. Efficient and accurate link adaptation is required to ensure WiMAX multirate performance given the time-varying nature of mobile radio channels. The IEEE 802.16 standard includes multiple report mechanisms and metrics to support DL link adaptation. Link adaptation mechanisms should address the dynamic nature of interference, knowing that loading is not constant and interference is not reciprocal. The WiMAX link adaptation mechanism is insensitive to changes in parameters and channel model, and takes mobility into account.

The link adaptation process is performed by sending SINR measured at the mobile receiver through the CQICH. These SINR values are averaged over the subchannels, over the PUSC slots, or the AMC band. They are reported through the fast-feedback allocation subheader (MAC layer dedicated packets) using either CQICH dedicated channels, or CQICH enhanced allocation dedicated channels. Both support 4 bits and 6 bits, where the 6 bits case supports multiple CQICH and band AMC report.

How these reports are used at the BS is mainly implementation based, and not specified in the standard.

The IEEE 802.16 allows the mobile station to measure the CINR in three different regions:

1. CINR measurement on preamble, which has the advantage that is sent in each and every frame but does not reflect the noise and interference in the data zone because of the preamble structure. It is sited specifically for reuse 1 and reuse 3 systems and not other reuse patterns (such as 6 or more).

2. CINR measurement on pilots, which reflects the data zone interference between cells better than CINR measurement on preamble. However, it does not reflect the boosted data zone, AAS, or changing load of data zones.

3. CINR measurement on data, which is the best reflection of the data zone interference level and does not need the compensation of pilot/preamble boosting. However, it is prone to variations of interference level (due to change of loading, boosting, and AAS), and it is more complicated to calculate and report.

As was mentioned earlier, Mobile WiMAX supports two types of feedback. The first is physical CINR (or PCINR), in which the actual measured and averaged CINR at the MS is fed back to the BS. At the BS the scheduler uses these values to determine the actual power level as well as the modulation and coding scheme (MCS) for the corresponding mobile station. The second type is called effective CINR (or ECINR), in which the MS, according to its own channel estimation and the waterfall curve (BER vs. CINR curve) identifies the best MCS and proposes the MCS value to the BS, rather than sending the actual CINR value. Even in this case, it is up to BS's scheduler to accept or reject this MCS value proposed by the MS.

HARQ. In Mobile WiMAX, an automatic repeat request (ARQ) can be done both in the physical layer and in the MAC layer. ARQ in the physical layer is called hybrid ARQ or hybrid automatic repeat request (HARQ). It is faster and more reliable than MAC-layer ARQ. HARQ mitigates the effect of channel impairment and interference, increases the data throughput by the multiple parallel data streams, enhances the robustness of transmitted bursts, efficiently resolves the problem of channel quality difference, and overcomes the adaptation error of the AMC in fading channels.

A MAC-level ARQ is used by the receiver to provide feedback on successfully received and missing blocks of data. An ARQ mechanism greatly reduces the error rate at the expense of adding a time delay. The major differences between ARQ and HARQ are that ARQ discards previously transmitted data, whereas HARQ combines the previous and retransmitted data to gain time diversity. HARQ also makes use of faster responding physical layer ACK and NACK packets, transmitted from the receiving device. HARQ is typically set to operate with approximately 10% packet error rate, minimizing delays in repeat packet transmission, which has a dramatic effect on data throughput. HARQ is one crucial element in the Mobile WiMAX that can deliver high performance with mobility speeds in excess of 120 km/h.

HARQ is supported both in DL and UL. It builds on the fast-uplink feedback channels supported in IEEE 802.16 OFDMA mode. HARQ utilizes the time diversity achieved through fast retransmission and combining of previously erroneously decoded packets and retransmitted packets (see Figure 2.13). In case of a NACK, HARQ retransmits more redundancy and the receiver combines them with the previously sent data to achieve better SNR and coding gain versus the time-varying channel.

Two types of HARQ implementations are envisioned in IEEE 802.16. The first type is chase combining (CC), in which, in case of NACK, the transmitter resends the same copy of the packets. The MCS has to be the same for repeated packets and, therefore, the implementation of this mode is relatively simple.

The second HARQ type is called incremental redundancy (IR) HARQ, in which, in case of NACK, the transmitter sends part of code word that may be different from the previous first transmission (subpackets). Unlike HRAQ chase, the MCS level can be adapted for each subpacket. HARQ IR provides additional coding gain and better SNR relative to HARQ chase. It also provides more flexibility to adapt to the time variation of the channel. However, it requires more DL or UL MAP overhead, is more complex in implementation, and requires more memory at the receiver.

Mobile WiMAX requires HARQ chase combining in Release 1.0, and is investigating the inclusion of HARQ incremental redundancy for future releases.

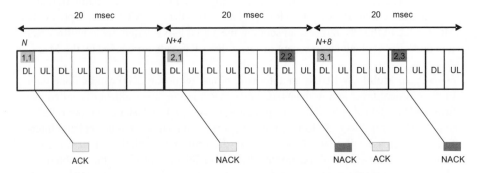

Figure 2.13. Example of hybrid ARQ process in IEEE 802.16.

Power Control

Power control is employed to adjust MS power so that UL signal strength received at the BS falls within an acceptable range for the BS receiver without introducing additional unwanted interference. The ideal level depends on assigned modulation/coding rates.

Power control is required at the MS in order to combat propagation changes, the near–far problem, fast fading (changes over frames), lognormal fading (changes over hundreds of frames), power law changes (changes over thousands of frames) as distance to BS changes, and interference changes.

There are two general types of power control supported in IEEE 802.16 and Mobile WiMAX, namely, open loop and closed loop power control.

Closed loop power control is mainly used at the network entry process and is shifted to open loop control when the data session starts. Closed loop power control could be used during periodic ranging and bandwidth request as well. In closed loop power control, the MS adjusts power based on instructions from the BS processing ranging messages, PMC-RSP messages, power control IE, UL MAP fast tracking IE, and fast power control. All of these schemes have high overheads and require power adjustment every nth frame in order to track fast fading, in which significant signal level fluctuations occur every n frames.

In open loop power control, the MS adjusts power based on an estimate of propagation loss, noise plus interference supplied by the BS in the DL MAP, required CINR for the assigned modulation/coding/repetition rate, and offset from the BS. For Mobile WiMAX Release 1.0, accurate path loss can be calculated using DL RSSI measurement of the preamble at the MS as a representative of UL propagation loss.

If any packets were received from MS in the previous frame, fast power control can be achieved. Normally, open loop power control is performed in the data region and is faster than closed loop power control.

Open loop power control can be done both in passive and active mode. In passive UL open loop power control, the mobile station does not add the correction term for MS-specific power offset, and the offset is not controlled by the MS. In active UL open loop power control, the mobile station does add the correction term for MS-specific power offset. The offset is controlled by the MS.

Mobile WiMAX system profile requires both closed loop power control and passive open loop power control.

2.3 OVERVIEW OF MOBILE WiMAX MAC LAYER (RELEASE 1.0)

The objective of this section is to highlight the key functional MAC protocols and mobile station states and procedures to support mobility, power saving, and quality of service. Details on some procedures and features that impact end-to-end functionalities are further discussed in relevant sections. For example QoS, security, mobility, and multicast/broadcast services are discussed in respective chapters that present these functions both from air link and end-to-end perspectives.

The WiMAX air-interface MAC layer consists of three sublayers: convergence sublayer (CS), MAC common part sublayer (MAC CPS), and security sublayer (SS); see Figure 2.14.

The functional blocks in the CPS may be logically classified into upper MAC functions responsible for mobility control and resource management, and the lower MAC functions that focus on control and support for the physical channels defined by the PHY. Although not formally separated in the standard, one may also classify functions into the control plane and data plane functions.

The upper MAC functional group includes protocols procedures related to radio resource control and mobility related functions such as

- Network discovery, selection, and entry
- Paging and idle mode management
- Radio resource management
- L2 Mobility management and handover protocols
- QoS, scheduling, and connection management
- Multicast and broadcast services (MBS)

On the control plane, the lower MAC functional group includes features related to layer 2 security, sleep mode management, as well as link control and resource allocation and multiplexing functions.

The PHY control block in the MAC layer handles PHY signaling such as ranging, measurement/feedback (CQI), and HARQ ACK/NACK in support of link adaptation.

The following subsections describe the functions within each sublayer of MAC and also provide a high-level overview of MAC states and procedures.

2.3.1   Convergence Sublayer

The convergence sublayer (CS) is the interface between the IEEE 802.16e MAC layer and network layer functions and it is responsible for:

- Classification and mapping of the upper layer SDUs into appropriate MAC connections over the air interface since MAC and PHY layers have no visibility of higher layer addresses
- Header compression (PHS or ROHC) for some applications such as VoIP (optional)
- Protocol data unit (PDU) formatting

IEEE 802.16 specifies two types of CSs: ATM CS and Packet CS (IPv4, IPv6, and Ethernet). The type of CS determines the fields that may appear in QoS classification rules. IP CS carries IP packets over the IEEE 802.16 MAC frames and permits a combination of the five tuple in TCP/UDP and IP headers [i.e., source and

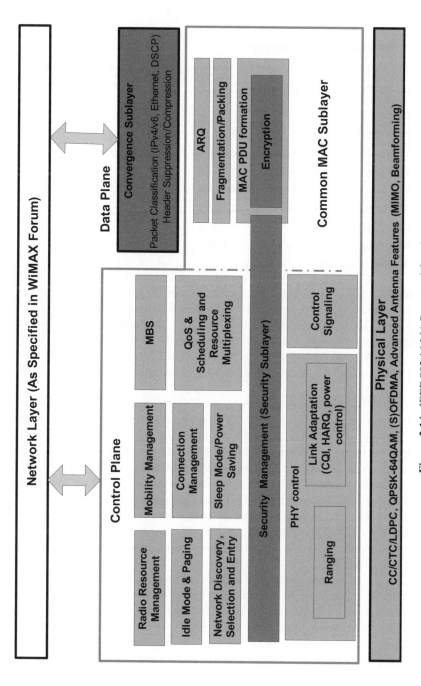

Figure 2.14. IEEE 802.16 MAC protocol functions.

destination IP address, source and destination port number, and differntiated service code point (DSCP)] to determine the classification rules.

2.3.2   Common MAC Sublayer

The MAC common part sublayer (MAC CPS) has an essential role in QoS support over the air interface and it performs

- Access control
- MAC PDU operations
- QoS provisioning, MAC connection management, and packet scheduling
- Call admission control (CAC)
- Bandwidth management, bandwidth request, and bandwidth allocations
- Automatic repeat request (ARQ) support
- Mobility support
- Multicast and broadcast services (MBS)
- Modulation and coding selection

The WiMAX air interface MAC is connection oriented with full QoS and security support features capable of concatenation, fragmentation and reassembly, and packing of the higher layer "service data units" (SDUs—data units exchanged between adjacent layers). The higher layer SDUs coming to the MAC CPS are assembled to create the MAC PDU. Multiple SDUs may be carried on a single MAC PDU, or a single SDU may be fragmented to be carried over multiple MAC PDUs; see Figure 2.15.

Each MAC PDU begins with a 6 byte generic MAC header (GMH) followed by variable-length payload information and an optional 4 byte CRC, with maximum total size of up to 2048 bytes.

The MAC layer establishes a number of MAC layer unidirectional logical links between the base station and the mobile station over the air interface known as MAC connections, each of which is identified in a sector by a 16 bit "connection ID" (CID). A MAC connection with a set of QoS parameters is represented by a service flow (SF), and a MAC connection with a set of security features is represented by a security association (SA). In IEEE 802.16, each MS is identified with a 48-bit IEEE MAC address and each BS has a 48-bit base station ID (24 bits reserved as operator ID.)

All the QoS attributes of a service over the air interface (i.e., packet latency and jitter, throughput, packet loss, etc.) is controlled by MAC management, scheduler, and ARQ modules in the MAC CPS layer. This layer provisions and maintains the QoS of each MAC connection. Admission control may honor or deny the request of a certain service by checking the subscriber's profile and the required resources against the available network resources to guarantee the committed QoS. The scheduler allocates bandwidth among MAC connections in terms of scheduled transmissions.

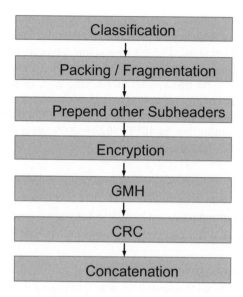

Figure 2.15. MAC PDU operation.

Each scheduled transmission is defined by uplink or downlink MAPs describing which MAC connection (on the DL) or which MS (on the UL) can transmit, when to transmit, how much time and how many subchannels, and which modulation and coding to be used for the transmission. A bandwidth request mechanism based on the request/grant mechanism allows data services to request uplink transmission when they have data to send. There are many opportunities for the device and the network vendors to enhance the system QoS by implementing proprietary algorithms for call admission control, scheduling, bandwidth request, and so on.

The automatic repeat request (ARQ) module retransmits the lost MAC PDUs to control the packet loss and to mitigate TCP window size changes due to over-the-air errors. ARQ is disabled for real-time applications, which are loss tolerant and delay sensitive, and is enabled for data applications.

2.3.3 Security Sublayer

The security sublayer (SS) secures over-the-air transmissions, protects data from theft of service, and performs user and device authentication. The security schemes involved in the WiMAX network security architecture are authentication (including device authentication and user authentication), data encryption (including data confidentiality and integrity), access control and key management.

There are two main protocols within the security sublayer: an encapsulation protocol for encrypting data packets and a privacy and key management protocol (PKM). The data encapsulation protocol is used for securing packet data across the network. This protocol defines a set of supported cryptographic suites—pairings of data encryption and authentication algorithms and the rules for applying those algorithms to a MAC PDU payload. The key management protocol (PKM) provides the secure distribution of keying data from the BS to the SS. Through this key management protocol, the MSs and BSs synchronize keying data. In addition, the BS uses the protocol to enforce conditional access to network services.

IEEE 802.16-2004 supported one-way MS authentication to BS based on device certificates/RSA operation, but in IEEE 802.16-2009 several enhancements are made as part of PKMv2, including:

- Three-party protocol—supplicant, authenticator, and authentication derver
- Authentication, data integrity, and OTA data encryption
- Framework for group multicast and MBS key management
- Model similar to WiFi; necessary to support NAP/NSP separation, mobility, and roaming
- Mutual authentication—device and/or user
- RSA- and EAP-based framework; mostly EAP-based will be deployed
- Support for multiple credential types, thanks to EAP support
- Strong crypto support—AES, CMAC-AES

2.3.4   MS States and MAC Procedures

IEEE 802.16 MAC defines multiple states and substates for the MS to optimize radio resource utilization and power saving.

After power is turned on, the MS goes to initialization states and prepares for initial ranging and registration, which are captured as access state and, depending on data availability, may go to connected state or idle state; see Figure 2.16. The following is a summary of the MS state transitions and major procedures within each state.

Initialization

Following the power on, the MS enters the initialization state in which it scans for available BSs in the area and attempts to synchronize with a target BS and acquire system and access related parameters. The initial scanning is performed according to a preferred RF channel list and initial cell selection is based on power level received on the preamble of the OFMA frame. Once the candidate BS is selected, the MS proceeds with synchronizing to the selected BS's preamble and acquiring the broadcast system configuration information before entering the access state for network entry. In this process, the MS determines the DL and UL parameters using the frame control header (FCH) and management data bursts, the DL channel descriptor (DCD), and the UL channel descriptor (UCD.) Following this system and access parameter acquisition, the MS has the necessary information to transmit on a UL ranging slot.

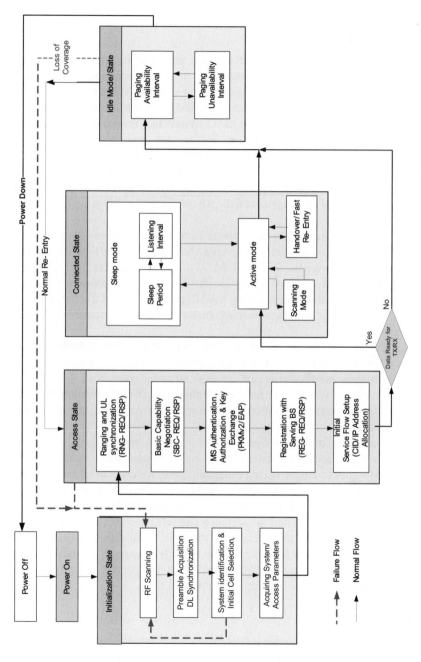

Figure 2.16. Mobile station state transition in IEEE 802.16.

The MS may return back to the scanning step in case it fails to perform the actions required to complete the initialization state or realizes that the selected BS cannot be accessed.

Access and Network Entry

After successful completion of the initialization state, that is, after cell selection, synchronization, and acquisition of all essential DL and UL parameters, the MS enters the access state. Upon first entering the access state, the MS performs the ranging process using the initial ranging code, as specified by the access parameters of the BS. Each MS has a 48-bit universal MAC address that is used during the initial ranging and subsequent authentication process. The MS randomly selects an available slot from the list contained in the UL-MAP and sends a ranging request (RNG-REQ) message. The BS replies with a message containing the power, timing, and frequency adjustments required by the requesting MS. In the RNG-RSP, the MS will also be given a basic/management CID, which is used for following MAC message exchanges.

Once the MS's timing and transmit power have been properly adjusted, that is, when initial ranging is successfully completed, the MS and BS perform basic capability negotiation using a SBC-REQ/RSP MAC message. Here, the MS and BS indicate the optional features they support so that they can be enabled and used as needed. Following capability negotiation, the MS proceeds to perform the authentication and authorization process, which involves a mutual authentication of the MS and BS networks.

Next, the MS performs the registration process using a REG-REQ/RSP MAC message, where the MS is also given a transport CID. Using this transport connection, the IP address assignment is performed as an upper layer exchange with the DHCP server, typically in the core network. At this point, the MS is managed by the BS and additional timing and power adjustments are preformed during a maintenance cycle called periodic ranging.

Upon successful completion of access state operation, the MS goes to the idle state/mode or to the connected state. If there is data immediately available to exchange, the MS goes to the connected state; otherwise, it moves to idle state monitoring and waiting for paging messages or an MS-originated bandwidth request. If the access state operation fails (network entry is not successful), the MS goes back to the initialization state to restart the scanning and cell selection procedure.

Signals from the MS are distinguished at the PHY using different codes called the permutation base (similar to the separation of the BS signals using the preamble ID.) Unlike WLAN, which uses a MAC address, each data burst is logically marked for a specific receiver using a connection identifier (CID).

Idle Mode Operation

The idle state is defined to provide power saving for the mobile stations and also to release network resources when they not needed for users' communication. During the idle mode, the MS does not have an interactive connection, does not perform periodic ranging, and may not be registered at specific BS but it is periodically available for DL broadcast signaling or traffic. In this mode, paging is needed to reach MSs.

Following a slotted paging principal, the MS in idle mode performs power saving by switching between paging-available mode and paging-unavailable mode.

In paging-available mode the MS monitors broadcast messages as well as MOB_PAG-ADV for any incoming paging messages. If the MS is paged, it will transition to the access state for its network reentry and to respond to paging messages.

In paging-unavailable mode, the MS does not need to monitor the downlink channels in order to reduce its standby power consumption. While in this mode, the MS may transition to the access state if triggered by a bandwidth request or location update.

A mobile station in idle state, moving from one BS's coverage area to another, does not perform any ranging or handover processes with the target BS. However, when wide area mobility is supported, the MS is required to perform a location update procedure whenever it crosses a paging group boundary or if triggered by other events such as expiration of the idle mode timer or MS power-down. The paging group in IEEE 802.16 is similar in concept to location area in conventional 3G systems and it refers to a group of BSs within which the MS can move without updating the network of its location, which would cause the network to concurrently broadcast the MS's paging messages across the paging group when the MS is initially paged.

When the MS is in idle mode, those already registered can perform an optimized or fast network reentry to ensure fast connection setup.

Connected Mode Operations

The connected state is the state in which the MS is registered and interacts with a serving BS on the control or data plane. In the connected state, the MS maintains at least one connection with the BS as established during the access state; additional transport connections may be established as required by services offered to the MS. In addition, to reduce power consumption of the MS during user data exchange, the MS or BS can request a transition to sleep mode. Also, the MS can scan a neighbor cell's signal to reselect a cell that provides robust and reliable services. Therefore, the connected state consists of three modes: sleep mode, active mode, and scanning mode.

Active Mode. During active mode, the MS and BS perform normal operations to exchange the DL/UL traffic transaction between the MS and BS. The MS can perform the fast network reentry procedures after handover. While in handover, the MS maintains any IEEE 802.16-specific IDs required for handover as well as its IP address in accordance with upper-layer protocols. Without going through the access state, the MS may remain in the connected state with target BS, resulting is fast network reentry in an optimized hard handover.

Sleep Mode. During sleep mode, the MS may enable power-saving techniques. An MS in active mode transitions to sleep mode through sleep mode MAC signaling management messages (MOB_SLP-REQ/RSP message). The MS does not transmit and receive any traffic to/from its BS during the sleep interval. An MS can receive an indication message (MOB_TRF-IND message) during the listening interval and then, based on the message content, decide whether it should transit to active mode or to stay in sleep mode. During the sleep interval, the MS may choose to transit to active mode.

Scanning Mode. During scanning mode, the MS performs scanning of neighboring BS's preambles as specified by the serving BS, during which it may be temporarily unavailable to the serving BS. While in active mode, the MS transitions to scanning mode via explicit MAC signaling (MOB_SCN-REQ/RSP message) or implicitly without scanning management messages generation.

2.3.5   Handover

The IEEE 802.16e standard and WiMAX in general support wide-area mobility at high speed during idle, sleep, and connected modes. In connected mode, the MS performs seamless handover from the serving BS to a target BS to maintain its existing connections or receive higher signal quality or QoS. Such handover procedures can be initiated by the BS/network or the MS.

The normal hard handover process in IEEE 802.16 consists of five different stages: scanning and cell reselection, handover decision and initiation, downlink synchronization to target BS, ranging and uplink synchronization, and release of resources at the serving BS.

The cell reselection is the stage in which the MS acquires information about BSs in the network. The information is used in evaluation of the possibility to perform a handover. To be able to perform a cell reselection, the MS needs to acquire information about the radio network topology. The serving BS provides the MS information about its neighboring BSs and their key parameters through a MOB-NBR-ADV message periodically broadcast by the BS. This information will simplify the MS synchronization with a new BS since there is no need for the MS to listen to the target BS's DCD/UCD messages. The MS requests scanning intervals from the serving BS and may start scanning when permitted by the serving BS. During the scanning interval, the MS performs scanning of neighbor BSs with or without the optional association procedure. The goal of the association is to enable the MS to collect and store information about BSs. The gathered information is saved during a reasonable period of time and it can help the MS further in decisions regarding handovers.

The initiation of a handover is the decision to migrate the MS from the serving BS to a target BS. This decision can be triggered in the MS or by the BS/network. To commence the handover, the requesting party sends a handover request that will trigger a sequence of handover-specific messages to be sent between the MS and BS. To establish such communication with the target BS, the MS needs to synchronize to its downlink channel. During this phase, the MS receives downlink and uplink transmission parameters. If the MS previously received information about this BS (through the network topology acquisition), the length of this process can be shortened.

When the MS is synchronized to the channel, it needs to perform initial ranging or handover ranging. Ranging is a procedure whereby the MS receives the correct transmission parameters, for example, time offset and power level.

Termination of services at the serving BS is the last step in the handover process whereby the serving BS terminates all connections associated with the MS and removes all information in queues, counters, and so on.

The baseline handover in mobile WiMAX is an optimized hard handover whereby the target BS obtains information about the MS through the backbone signaling and the MS is provided with extended information about the target BS such that some of handover messaging can be skipped. The optimized handover can reduce handover latency to less than 50 msec.

The IEEE 802.16 standards define two additional modes of handover, namely, macro diversity handover (MDHO) and fast BS switching (FBSS). These modes are not part of system profile release 1.0 and may not be widely deployed.

2.3.6   Service Flows and QoS Support

The WiMAX air interface and IEEE 802.16 MAC are connection oriented, with each connection assigned a service class based on the quality of service (QoS) that is required by the MS and application. The MAC layer manages the radio resources to efficiently support the QoS for each connection established by the BS.

This connection-oriented QoS mechanism is based on service flows (SF), which are unidirectional transport mechanisms from the MAC through the PHY. Each connection is mapped to a SF identified by a service flow ID (SFID) with specific QoS attributes.

The IEEE 802.16e standard has specified five types of SFs, each appropriate for a certain class (see Table 2.10) of services.

- Unsolicited Grant Service (UGS). Supports flows transporting fixed-size data packets on a periodic basis at a constant bit rate (CBR), for example, wireless T1 or wireless fractional T1. Also, UGS SF is appropriate for voice services using fixed-rate vocoders with no silence suppression.
- Real-Time Polling Service (rtPS). Supports real-time services with variable-size data packets issued on a periodic basis, for example, MPEG video.
- Nonreal-Time Polling Service (nrtPS). Supports delay-tolerant data traffic that requires a minimum guaranteed rate, for example, FTP, movie upload/download.
- Best Effort (BE). Supports data services, for example, web browsing, e-mail.
- Enhanced Real-Time Polling Service (ertPS). Supports latency-sensitive real-time applications, for example, VoIP with silence suppression.

Table 2.10. Mobile WiMAX data delivery services and applications

Service class	Real time	Data rate	Application
Unsolicited grant service (UGS)	Yes	Fixed	VoIP
Real-time polling service (rtPS)	Yes	Variable	Streaming video
Nonreal-time polling service (nrtPS)	No	Variable	FTP
Best effort (BE)	No	Variable	Web Surfing
Enhanced real-time polling service (ertPS)	Yes	Variable	VoIP

The QoS of the SFs are maintained through an appropriate bandwidth allocation mechanism among the SFs. On the DL, the bandwidth allocation is performed by the air-link scheduler, and on the UL it is done by bandwidth request from the MSS. Bandwidth assignment is done by the air-link scheduler at the base station and the scheduler at the MS.

The scheduling decision is determined based on the SF's QoS needs, the SF's buffer lengths, and the RF conditions of different MSs. The SF framework specified by IEEE 802.16e provides an excellent QoS granularity on the air interface and also inter-SF isolation. Each service flow is characterized by a set of QoS parameters such as maximum sustained traffic rate (MSTR), minimum reserved traffic rate (MRTR), maximum latency, maximum jitter, traffic priority, service data unit type, ARQ type, and so on. These QoS parameters can be selected appropriately to deliver the desired QoS for each SF.

In IEEE 802.16, the service flows are managed through a DSx messaging procedure that is used to create, change, and delete a service flow for each MS. Typically, a complete procedure includes a DSA (dynamic service addition) procedure, such as request and response, for creating a service flow, DSC (dynamic service change) procedure for modifying a service flow; and DSD (dynamic service flow deletion) procedure to delete a service flow.

During a service flow initiation, the base station sends a DSA-REQ message to the MS. The DSx messages carry important service flow information such as QoS, SFID (service flow identifier), MCID (multicast connection identifier) if the network decides to multicast the data, or a CID (channel identifier) if the network decides to unicast the data in the air to the MS. The MS will then send a response message to the BS, which will acknowledge the success of service flow creation. The service flow state diagram is shown in Figure 2.17.

WiMAX allows the network to establish a virtually unlimited number of SFs with flexible combination of different types and different QoS attributes for each subscriber.

2.4 WiMAX FORUM SYSTEM AND CERTIFICATION PROFILES IN RELEASE 1.0

In this section, a summary of the Mobile WiMAX Release 1.0 air interface system profile feature set as well as applicable certification profile and band classes are

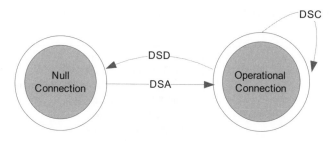

Figure 2.17. Dynamic service flow management.

provided. Tables 2.11 and 2.12 provide the list of main PHY and MAC features, respectively. Note that the list is not comprehensive; please refer to WiMAX Forum Release 1.0 Mobile System Profile [3] for more details.

The Mobile WiMAX Release 1.0 Certification Profile is provided in Table 2.13. The WiMAX Forum Certification Profile contains the list of band classes that are targets for WiMAX enablement.

2.5 SUMMARY

Mobile WiMAX air interface system profiles and interoperability and testing specifications are defined based on IEEE 802.16 standards. System Profile Release 1.0 in the WiMAX Forum is based on IEEE 802.16e (2005) with select features updated based on IEEE 802.16-2009. This chapter provided an overview of the IEEE 802.16 air interface standards with focus on the scalable OFDMA PHY mode and all MAC

Table 2.11. Mobile WiMAX Release 1.0 air interface system profile PHY features

PHY feature	Description
Modulation	QPSK, 16QAM, 64QAM (optional for UL)
Channel coding	CC (1/2, 2/3, 3/4) and CTC (1/2, 2/3, 3/4, 5/6)
Subcarrier allocation mode	Distributed: PUSC and FUSC (DL only)
	Adjacent: AMC
H-ARQ	Chase combining
Power control	Closed and open loop power control
CINR measurement	Physical CINR
	Effective CINR
Fast feedback	
MIMO*	DL: Adaptive MIMO switching, STBC, and SM
	UL: Collaborative MIMO
Beamforming*	Adaptive beamforming

*IO features are those features that are optionally certifiable in base stations to enable service providers' high-performance deployment models without imposing burdens on cost-effective deployment by other operators. Typically, support for IO features is mandatory for mobile stations.

Table 2.12. Mobile WiMAX Release 1.0 air interface system profile MAC features

MAC feature	Description
Convergence Sublayer	IPv4 and IPv6 with PHS
	IPv4 and IPv6 with ROHC
	802.3/Ethernet*, IPv4 over 802.3/Ethernet, and IPv6 over* 802.3/Ethernet*
QoS	BS-initiated service flow
	MS-initiated service flow
ARQ	MAC ARQ
Data delivery services	UGS, BE, ERT-VR
	RT-VR, NRT-VR
Request grant mechanism	Request grant mechanism
Handover initiation	HO initiated by MS
	HO initiated by BS
Handover	Neighbor advertisement
	Scanning and association
	HO optimization
	CID and SAID update
Power-saving modes	Sleep
	Idle
Supported cryptographic suites	No data encryption, no data authentication, 3-DES, 128
	CCM-Mode 128-bit AES, CCM-Mode, AES Key Wrap with 128-bit key
Security	PKMv2
	CMAC
MBS*	Multi-BS MBS

*IO features are those features that are optionally certifiable in base stations to enable service providers' high-performance deployment models without imposing burdens on cost-effective deployment by other operators. Typically, support for IO features is mandatory for mobile stations.

feature and protocols that are included in the system profile. This first major release of Mobile WiMAX focuses initially on TDD implementation of the standard and includes all features needed to support security, QoS, and mobility expected in wide-area cellular networks. Extensions to FDD are made in the Release 1.5 System Profile along with other MAC layer enhancements that are described in Chapter 3.

Table 2.13. Release 1.0 Certification Profile

Band class	Spectrum range (GHz) BW (MHz)	Bandwidth certification group code (BCG)
1	2.3–2.4	
	8.75	1.A
	5 and 10	1.B
2	2.305–2.320, 2.345–2.360	
	3.5 and 5 and 10	2.D
3	2.496–2.69	
	5 and 10	3.A
4	3.3–3.4	
	5	4.A
	7	4.B
	10	4.C
5	3.4–3.8	
	5	5.A
	7	5.B
	10	5.C
7	0.698–0.862	
	5 and 7 and 10	7.A
	8 MHz	7.F

REFERENCES

[1] IEEE Standard 802.16-2009, Local and Metropolitan Area Networks, Part 16: Air Interface for Broadband, Wireless Access Systems.

[2] IEEE Standard 802.16e-2005, Amendment to IEEE Standard for Local and Metropolitan Area Networks—Part 16: Air Interface for Fixed Broadband Wireless Access Systems—Physical and Medium Access Control Layers for Combined Fixed and Mobile Operation in Licensed Bands.

[3] WiMAX Forum™ Mobile System Profile, Release 1.0, Version 09, 2009.

[4] WiMAX Forum™ Mobile System Profile, Release 1.5 Common Part, Version 1, 2009.

[5] WiMAX Forum™ Mobile System Profile, Release 1.5 FDD Specific, Version 1, 2009.

[6] WiMAX Forum™ Mobile System Profile, Release 1.5 TDD Specific, Version 1, 2009.

[7] IEEE Standard 802.16-2004, Local and Metropolitan Area Networks—Part 16: Air Interface.

CHAPTER 3

WiMAX AIR INTERFACE ENHANCEMENTS IN RELEASE 1.5

KAMRAN ETEMAD AND HASSAN YAGHOOBI

3.1 INTRODUCTION

Following the completion of System Profile Release 1.0 and based on new requirements for advanced network services and potential market opportunities in the FDD spectrums, the WiMAX Forum started the development of an interim System Profile Release 1.5 profile. The System Profile Release 1.5 is an extension of System Profile Release 1.0 [2], without requiring major changes in the underlying IEEE 802.16 standards developed to address the following high-level goals:

- To enable WiMAX deployments in the paired spectrum air-interface through support of FDD mode of operation
- To add new system profile features and improvements needed to support advanced network services, such and location-based services and improved multicast/broadcast services in line with network Release 1.5 specifications.
- To improve MAC-layer efficiency with a focus on lowering MAP overhead, especially for VoIP traffic, and reducing latencies
- To include other enhancements, such as closed-loop MIMO, to further improve coverage and capacity for low-mobility usage

The following sections provide a high-level overview of key enhancements in Sys-

WiMAX Technology and Network Evolution. Edited by Kamran Etemad and Ming-Yee Lai
Copyright © 2010 the Institute of Electrical and Electronics Engineers, Inc.

tem Profile Release 1.5. A summary of features and certification band classes introduced in Release 1.5 is included in the last section.

3.2 SUPPORT FOR FREQUENCY DIVISION DUPLEXING (FDD/HFDD)

The Mobile WiMAX Release 1.5 air interface supports frequency division duplexing (FDD), with which the base stations transmit and receive in different frequencies simultaneously. Following the base station configuration, on the Mobile Stations uplink and downlink channels are naturally transmitted in different frequency channels but not simultaneously necessarily. When mobile stations have the capability of transmitting and receiving simultaneously, they are called full-duplex FDD (FD-FDD) devices. To trade off design cost and complexity, mobile stations may be designed in half-duplex FDD (HD-FDD) mode, in which they transmit and receive in different spectrum segments of a paired allocation but not simultaneously. Please note that this does not equate with lost of spectrum efficiency necessarily, as the Base Station operates in full-duplex mode and the system fully utilizes the uplink and downlink spectra at all times.

More specifically, in an FDD system, unlike in a TDD system in which frames are split into uplink and downlink frames, the uplink and downlink channels located on separate frequencies are dedicated to uplink and downlink subframes. A fixed-duration frame is used for both uplink and downlink transmissions. The fact that the uplink and downlink channels utilize a fixed-duration frame simplifies the bandwidth allocation algorithms and facilitates the use of different modulation types in downlink and uplink. It also allows interoperation of use of FD-FDD and HF-FDD mobile stations supported by the same base station. In this case, the scheduler does not allocate uplink bandwidth for HD-FDD users at the same time that it is expected to receive data on the downlink channel, including allowance for the propagation delay, mobile Station transmit/receive transition, and receive/transmit transition gaps. A FD-FDD is capable of continuously listening to the downlink channel, whereas a HD-FDD user can listen to the downlink channel only when it is not transmitting in the uplink channel.

The FDD base station must support both FD-FDD and HD-FDD mobile station types concurrently. Therefore, in its most general scenario, the FDD frame structure supports bandwidth allocations to both FD-FDD and HD-FDD mobile station types.

The FDD frame structure supports a coordinated transmission of two groups (Group-1 and Group-2) of HD-FDD mobile stations that share every downlink frame using nonoverlapping partitioning of the downlink frame. Similarly, every uplink frame is shared by the two groups using nonoverlapping partitioning of the uplink frame. Figure 3.1 shows the frame structure of the FDD system that supports the concurrent operation of HD-FDD and FD-FDD mobile stations.

The downlink frame contains two subframes. Downlink Subframe 1 comprises a preamble symbol, a MAP region (MAP1), and data symbols (DL1). Downlink Subframe 2 comprises a MAP region (MAP2) and data symbols (DL2). The uplink

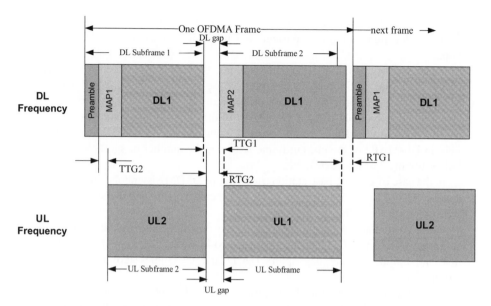

Figure 3.1. OFDMA FDD frame structure supporting HD-FDD mobile stations in two groups.

frame contains two subframes, UL2 and UL1, in the order relevant to the DL1 and DL2 subframes.

Figure 3.1 shows the timing relationship of the uplink subframes UL1 and UL2 relative to the downlink subframes DL1 and DL2. The four parameters TTG1, TTG2, RTG1, and RTG2 are announced in the downlink channel descriptor (DCD) messages and they must be sufficiently large to accommodate the HD-FDD mobile station's transmit-to-receive and receive-to-transmit switching times plus the relevant round-trip propagation delay.

Group-1 HD-FDD mobile stations receive downlink Subframe 1 and transmit during uplink subframe UL1. Group-2 HD-FDD Mobile Stations receive downlink Subframe 2 and transmit during uplink subframe UL2. No HD-FDD mobile station transmits during the preamble transmission in downlink. All FD-FDD mobile stations may transmit during the preamble transmission. The MAP regions (MAP1 and MAP2) are independent from each other and include a frame control header (FCH) and downlink and uplink MAPs.

HD-FDD mobile stations always use Group 1 for all purposes for initial network entry and reentry.

In FDD operation, for the purpose of load balancing between the two groups and other possible applications, the base station must be able to switch HD-FDD mobile stations from Group-1 to Group-2, or vice versa. The group switching can be realized through usage of a group switch information element (IE) or various subburst IEs.

When a mobile station is instructed to switch groups, any existing periodic CQICH allocations and any persistent allocations are deallocated by the base station.

When supporting HD-FDD mobile stations, the base station needs to properly define the group partitioning in the frame. Base stations uses DL-MAP and UL-MAP in frame n to provide partition information for frame $n + 1$ and $n + 2$. The "No. OFDMA Symbols" field in DL-MAP1 of the current frame indicates the number of symbols in Subframe 1 of the current frame and the "No. OFDMA Symbols" field in DL-MAP2 of the current frame indicates the number of symbols in downlink Subframe 2 of the next frame.

In the case of UL, the base station uses "No. of OFDMA symbols" in UL-MAP1 and UL-MAP2 to indicate the size of UL1 of the next frame and the size of UL2 of the next-next ($n + 2$) frame, respectively.

For a particular configuration change, the downlink and uplink channel descriptors (DCD and UCD message) transmitted in Group-1 are equal to the downlink and uplink channel descriptors (DCD and UCD message) transmitted in Group-1.

When frames are partitioned into groups, FD-FDD mobile stations can be supported using two alternative solutions. In the first solution, the base station allocates resources to FD-FDD mobile stations in both Group-1 and Group-2 using the FDD paired allocation IE. As a second solution, FD-FDD mobile stations may negotiate aggregated HARQ channels with the Base Station. If the Mobile Station and Base Station negotiate aggregated HARQ channels, then these HARQ channels will be treated as a paired set of HARQ channels for transmission and reception of bursts and, therefore, improve the latency performance of FD-FDD users.

3.3 MIMO ENHANCEMENTS

There are many enhancements in support of multiple antenna techniques in the Mobile WiMAX Release 1.5 air interface. Mobile WiMAX Release 1.0 air interface supports 2 × 2 open-loop STC/SM adaptive MIMO in downlink and 2 × 1 collaborative SM in uplink in PUSC mode. Mobile WiMAX Release 1.5 air interface supports four types of enhancements—two in downlink and two in uplink:

1. Support for open-loop STC/SM adaptive MIMO in downlink AMC mode
2. Support for closed-loop MIMO in downlink AMC mode
3. Support for collaborative SM for two mobile stations with a single transmit antenna in uplink in AMC mode
4. Support for open-loop STC/SM adaptive MIMO in uplink in PUSC and AMC modes.

For more details information on MIMO solutions, please refer to Chapter 4.

3.4 MAC ENHANCEMENTS

3.4.1 MAP Efficiency and Persistent Assignments

One of the distinctive features of MAC design in IEEE 802.16 is the flexibility of MAP messages for dynamic resource assignments. However, with this flexibility comes considerable overhead that for some applications can result is lower system capacity. For example, for applications such as VoIP, in which the packet arrival rate is somewhat predictable and mostly periodic, a persistent resource assignment may be used to reduce the system overhead. In persistent assignment, transmitting an initial assignment message is used to allocate a radio resource for a sequence of future frames according to some periodicity. Figure 3.2 illustrates the high-level concept of persistent assignment in contrast to dynamic assignment.

As shown in Figure 3.2, the base station needs to make time-frequency resource assignments in frame N for frame $N + P$, $N + 2P$, and so on, where P is the periodicity of the allocation. When dynamic scheduling is used, the base station explicitly assigns resources in frame N, $N + P$, and so on, at any location in the frame and can change the modulation and coding from frame to frame. In persistent assignment, the base station assigns an initial time-frequency resource in frame N that is valid in frame $N + P$, $N + 2P$, and so on. Since the base station uses a single assignment for multiple frames, the MAP overhead is reduced, assuming that frame-by-frame change in assignments attributes is often not required.

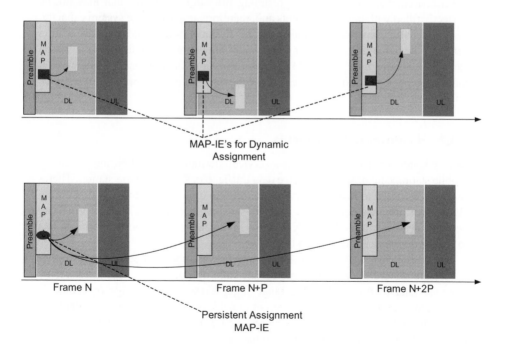

Figure 3.2. Comparing dynamic and persistent assignment concepts.

To enable this type of MAP overhead reduction, IEEE 802.16-2009 and Mobile WiMAX Release 1.5 air interface introduce the concept of persistent allocation using a set of new MAP-IEs. The design of enhanced IEs in 802.16-2009 allows allocation of both persistent and nonpersistent allocations in the same message.

These new IEs are structured such that they reduce the MAP overhead for both persistent and nonpersistent assignments. The design of IEs utilizes the inherent redundancy among resource assignment attributes and parameters, such as MCS level, burst size, and periodicity in multiple dynamic assignments in a single compress message to reduce the overhead. This simple compression technique applied to both persistent and nonpersistent allocations helps with improving overhead for typical VoIP and data mixed-traffic cases. Each persistent assignment has an associated period. For a series of packets, the persistent assignment is only valid for the first HARQ transmission of each packet, that is, follow up retransmissions are dynamically scheduled.

Whereas at high levels, persistent scheduling is a straightforward mechanism for supporting VoIP-type traffic, several secondary issues must be addressed to ensure its reliability and efficiency. For example a special error handling mechanism in needed to avoid error propagation due to lost assignment messages and quick recovery from such errors, which otherwise could result in excessive overhead. Therefore, in addition to new MAP-IEs the standard has defined detailed error handling procedures.

Another important issue is to minimize or avoid holes or fragmented resources in the time-frequency resource space resulting from deassignments that may not be perfectly filled with new assignments and, therefore, impacting the system capacity. The base station has two mechanisms to fill resource holes. First, the base station can fill a resource hole with a new persistent allocation. Second, the base station can fill a resource hole with a HARQ retransmission. In addition, in some cases it may be necessary to relocate some users in the time-frequency space to fill resource holes

3.4.2 Handover Enhancements

During handover (HO), there exists a service interruption period during which the mobile station cannot send or receive data traffic to or from any base station. Therefore, it is essential to keep the service interruption time short enough so that during HO the performance degradation of delay-sensitive applications such as VoIP can be unnoticeable. Mobile WiMAX Release 1.0 air interface supports optimized HO to reduce the service interruption time.

The Mobile WiMAX Release 1.5 air interface supports a mechanism to further reduce the service interruption time for the fully optimized seamless handover by allowing data to flow before the range request/response exchange procedure is completed at target base station. In order to achieve this, the connection identifier allocation update and security procedures should be performed other ways and in advance of ranging request/response exchange.

3.4.3   Load Balancing

When a base station cannot support a mobile station attempt for initial network entry, handover, or reentry from idle mode, it has to move the mobile station to another base station that can provide the support. In the Mobile WiMAX Release 1.5 air interface, the support for load balancing is enabled through the following three protocol elements:

1. Load balancing using preamble index and/or downlink frequency override
2. Load balancing using a ranging abort timer
3. Load balancing using BS-initiated handover

Using methods 1 and 3, serving a base station can designate a specific target base station to the mobile station. Support of method 2 enables the serving base station to prohibit the mobile station for a certain period of time to come back to this base station.

3.4.4   Location-Based Services

A new feature in Release 1.5 air interface supports three LBS methods:

1. GPS-based method
2. Assisted GPS (A-GPS) method to speed up the GPS receiver "cold startup" procedure that requires integrated GPS receiver and network components
3. Non-GPS-based method that works based on the role of the serving and neighboring base stations

For LBS support in Mobile WiMAX Release 1.5 air interface, an LBS advertising message is required. This periodic broadcast message provides the mobile station with the base station coordinates. Based on the serving base station coordinates, the mobile station can estimate its location.

The base station can additionally provide precise time-of-day information and base station frequency accuracy information to assist a GPS module if supported in the mobile station.

3.4.5   Enhanced Multicast and Broadcast Services

As mentioned in the previous chapter, multicast and broadcast services (MBS) is one of the MAC features in IEEE 802.16 that is supported in the Release 1.0 air interface. During the development of Mobile WiMAX Release 1.5 air interface, one of the objectives was to review and align the end-to-end service and network requirements of Mobile WiMAX with the MBS definitions and protocols in the IEEE 802.16 air interface . During this process, the MBS text in IEEE 802.16-2009 was significantly refined and many clarifications, corrections, and enhancements were made. Although most clarifications and corrections were incorporated in the revised

System Profile Release 1.0 as summarized in previous chapter, additional enhancements were made that are supported as part of System Profile Release 1.5. The key MBS enhancements beyond Release 1.0 air interface include the following:

- A compressed group DSx supported as a means to reduce overhead and signaling latency involved in concurrently setting up multiple MBS flows associated with one MBS zone. The Group DSx is a new message that utilizes the redundancies among service flow attributes for several multicast connections and is used to create multiple such connections using a unified and optimized MAC message.
- An inter-MBS continuity feature that involves optional TLVs broadcast by the base station, allowing the serving MBS zone to provide configuration information about friendly neighbor MBS zones. This a priori information about the MBS configuration in the target zone enables the mobile station to continue receiving the service without requiring any mobile station-initiated requests or interaction with the target BS that may result in temporary disruption of data reception.

The end-to-end description of Mobile WiMAX MBS is described in Chapter 12. There are also other clarifications on layer 2 encryption and logical channel IDs that were included in IEEE 802.16-2009 specifications, but they are not part of Mobile WiMAX Release 1.5. For more details on those aspects, readers are referred to IEEE 802.16-2009 specifications [1].

3.4.6 WiMAX/WiFi/Bluetooth Coexistence

Mobile handsets are rapidly evolving into multimedia devices with multiple connectivity capabilities such as (1) WiFi enabling Internet access and access to peripherals like severs and printers, (2) WiMAX enabling WAN access and cellular-type applications, and (3) Bluetooth to provide short-range connectivity to headsets, laptops, and peripherals.

It is important to devise a solution that addresses all three protocols (WiMAX, Bluetooth, and WiFi) in order to broaden the use case and types of potential devices. Interference between WiMAX, Bluetooth, and WiFi arises due to a number of factors. The first is due to the proximity of operating frequency bands where both Bluetooth and WiFi are operating in the 2.4–2.5 GHz band and WiMAX on both sides—2.3–2.4 GHz and 2.5–2.7 GHz. The second is due to colocation of respective antennas in devices. The third is due to the fact that the three protocols are completely uncoordinated in nature.

WiMAX devices have to offer the same connectivity for the end user as other cellular devices to be competitive with other standards.

There are two alternative directions for solutions to this problem: (1) RF domain (through filtering), which is usually suboptimal, very costly, and may need additional design area; and (2) time domain (MAC coordination) through protocol coordination.

Mobile WiMAX Release 1.5 supports the following two modes for co-existence:

1. Colocated coexistence Mode 1, a power-saving class-based generic mode for WiFi and some Bluetooth
2. Colocated coexistence Mode 2, which provides more optimization for WiMAX gaps for coexistence with Bluetooth

Other protocol elements are supported to further optimize WiMAX throughput in case of colocation with WiFi and Bluetooth. Mobile WiMAX Release 1.5 also supports combined uplink-band AMC operation with colocated coexistence to reduce interference relevant to Bluetooth and WiFi by proper AMC band selection.

3.5. SYSTEM PROFILE AND CERTIFICATION PROFILES IN RELEASE 1.5

In this section, a summary of the main Mobile WiMAX Release 1.5 air interface system profile feature set as well as applicable certification profiles (band classes) are provided. Table 3.1 provides the list of main PHY and MAC features.

Please note that the list in Table 3.1 does not include baseline Mobile WiMAX

Table 3.1. Mobile WiMAX Release 1.5 air interface

Feature	Description
FDD mode	Support for both full-duplex and half-duplex FDD
MIMO enhancements	Downlink open-loop MIMO in AMC
	Downlink closed-loop MIMO in AMC
	Uplink-collaborative SM for two mobile stations with single transmit antenna in AMC
	Uplink open-loop STC/SM MIMO in AMC and PUSC
	Cyclic delay diversity
MAC efficiency enhancements	Downlink and uplink persistent allocation IEs to reduce MAP overhead for both persistent and nonpersistent traffic
Handover enhancements	Support for seamless handover
Load balancing	Load balancing using preamble index and/or downlink frequency override
	Load balancing using ranging abort timer
	Load balancing using BS-initiated handover
Location-based services (LBS)	GPS-based method
	Assisted GPS (A-GPS) method
	Non-GPS-based method
Enhanced multicast and broadcast (MBS) services	Optimizations/clarification to MBS procedures such as group DSx and inter-MBS zone continuity messages
WiMAX–WiFi–Bluetooth coexistence	Colocated coexistence Mode 1
	Colocated coexistence Mode 2
	Combined uplink band AMC operation with colocated coexistence

Release 1.0 air interface features. Also, note that the list is not comprehensive. Please refer to WiMAX Forum Release 1.5 Mobile System Profile [3–5] for more details.

A list of Mobile WiMAX Release 1.5 certification profiles is provided in Table 3.2. The WiMAX Forum certification profile contains the list of band classes that are targeted for WiMAX enablement.

Table 3.2. New certification profiles in Release 1.5

Band class	Spectrum range (GHz) BW (MHz)	Duplexing mode BS	Duplexing mode MS	MS transmit band (MHz)	BS transmit band (MHz)	Bandwidth certification group code (BCG)
2	2 × 35 AND 2 × 5 AND 2 × 10	FDD	HFDD	2345–2360	2305–2320	2.E
	5 UL, 10 DL	FDD	HFDD	2345–2360	2305–2320	2.F
3	2 × 5 AND 2 × 10	FDD	HFDD	2496–2572	2614–2690	3.B
5	2 × 5 AND 2 × 7 AND 2 × 10	FDD	HFDD	3400–3500	3500–3600	5.D
6	2 × 5 AND 2 × 10	FDD	HFDD	1710–1770	2110–2170	6.A
	2 × 5 AND 2 × 10 AND 2 × 20 MHz (optional)	FDD	HFDD	1920–1980	2110–2170	6.B
7	2 × 5 AND 2 × 10	FDD	HFDD	776–787	746–757	7.B
	2 × 5	FDD	HFDD	788–793 AND 793–798	758–763 AND 763–768	7.C
	2 × 10	FDD	HFDD	788–798	758–768	7.D
	5 AND 7 AND 10 (TDD), 2 × 5 AND 2 × 7 AND 2 × 10 (H-FDD)	TDD or FDD	Dual Mode TDD/HFDD	698–862	698–862	7.E
8	5 AND 10	TDD	TDD	1785–1805, 1880–1920, 1910–1930, 2010–2025	1785–1805, 1880–1920, 1910–1930, 2010–2025	8.A

3.6 SUMMARY

Mobile WiMAX Release 1.5 air interface is an interim release that has been developed to support FDD mode of operation, provide some MAC-layer enhancements as well as features needed to enable advanced network services, namely LBS and MBS. Mobile WiMAX Release 1.5 is based on the IEEE 802.16-2009 standard.

3.7 REFERENCES

[1] IEEE Standard 802.16-2009, Local and Metropolitan Area Networks, Part 16: Air Interface for Broadband, Wireless Access Systems.

[2] WiMAX Forum™ Mobile System Profile, Release 1.0, 2006.

[3] WiMAX Forum™ Mobile System Profile, Release 1.5 Common Part, 2009.

[4] WiMAX Forum™ Mobile System Profile, Release 1.5 FDD Specific, 2009.

[5] WiMAX Forum™ Mobile System Profile, Release 1.5 TDD Specific, 2009.

CHAPTER 4

MIMO TECHNOLOGIES IN WiMAX

QINGHUA LI, JIANZHONG (CHARLIE) ZHANG, PEIYING ZHU, WONIL ROH, AND XINTIAN EDDIE LIN

4.1 INTRODUCTION

MIMO techniques have been extensively adopted in the IEEE 802.16d/e/j standards to improve the cell throughput, coverage, and average user experience. Examples of the MIMO techniques include single-user MIMO (SU-MIMO), in which data is sent to a single user via multiple antennas; multiuser MIMO (MU-MIMO), in which data is sent to multiple users via multiple antennas simultaneously; open-loop MIMO, in which the transmitter does not require channel state information; closed-loop MIMO, in which the transmitter requires channel state information; collaborative (cooperative) MIMO, in which multiple devices jointly send data using MIMO techniques, and so on. These new techniques add an additional dimension of resource— the spatial dimension. For example, a BS (base station) employing SDMA (spatial division multiple access) can send multiple data streams to multiple MSs (mobile stations) simultaneously on the same time-frequency resource, and multiple BSs/RSs (relay stations) can cooperatively perform space-time coding to send data packets to one MS. However, each MIMO technique is optimized for only a limited set of application scenarios. For example, closed-loop MIMO requires channel state information (CSI) at the transmitter and, thus, does not perform well in high-mobility situations. The support of MIMO techniques also brings additional requirements and constraints in system design and integration. Additional pilots for channel training, for example, are required for the multiple transmit antennas. In addition, the adoption of MIMO techniques often requires a

WiMAX Technology and Network Evolution. Edited by Kamran Etemad and Ming-Yee Lai
Copyright © 2010 the Institute of Electrical and Electronics Engineers, Inc.

tight design integration of PHY, MAC, and higher layers. For example, MU-MIMO requires that the MAC schedules simultaneous MSs with both approximately orthogonal channel vectors and comparable channel qualities (CQI) on the PHY subject to MAC fairness constraints. Besides technical issues, complexity, deployment constraints, and time to market play important roles for wider market penetration. Lower cost or simpler solutions such as antenna selection and collaborative MIMO may be appealing to certain markets.

Not all the MIMO features in the IEEE 802.16e standard are incorporated into the WiMAX profile and products. The Mobile WiMAX System Profile [8] that defines the actual product specification was first released by the WiMAX Forum in early 2006. This release was known as Release 1.0, and was based on the published IEEE 802.16e standard. The MIMO features included in Release 1.0 were chosen after lengthy deliberations on many engineering issues including implementation complexity, operators' limitation at their antenna sites, cost of system and devices, time to market, and so on. The WiMAX Forum just finished a short-term evolution of Release 1.0, known as Release 1.5 [9], which includes slightly more advanced MIMO techniques, mostly in the closed loop. In addition, Release 1.5 is based on the ongoing IEEE 802.16 REV2 standard, which can be viewed as an evolution of 802.16e and was completed in Dec. 2008. Table 4.1 summarizes some of the key MIMO features in Release 1.0 and Release1.5.

In this chapter, we present a survey of the MIMO techniques in WiMAX standards with a focus on the MIMO techniques adopted in WiMAX profile and engineering design issues. We start with the theoretical foundations and engineering considerations of the SU-MIMO techniques adopted in the IEEE 802.16 d/e standards and we then depict the MU-MIMO in IEEE 802.16e and the cooperative MIMO in IEEE 802.16j.

Table 4.1. MIMO techniques adopted by the WiMAX Forum

Key MIMO techniques in 802.16e	WiMAX Release 1.0 [8] (TDD only)	WiMAX Release 1.5 [9]
Open-loop transmit diversity in DL	Included	Included
Open-loop spatial multiplexing in DL	Included	Included
Open-loop transmit diversity in UL	Not included	Included
Open-loop spatial multiplexing in UL	Not included	Included
Collaborative spatial multiplexing in UL (UL MU-MIMO)	Included	Included
Adaptive beam forming, including DL SDMA (DL MU-MIMO)	Not-included	Included
Closed-loop antenna grouping/selection	Not included	Not included
Closed-loop codebook-based precoding	Not included	Not included for TDD Included for FDD
Rank adaptation	Included	Included
Dedicated pilot	Included	Included

4.2 SINGLE-USER MIMO

An SU-MIMO system is illustrated in Figure 4.1. The transmitter and receiver are equipped with N and M antennas, respectively. The channel between each transmit/receive antenna pair is assumed to be a Rayleigh or Rician fading channel. In this chapter, the CSI is assumed to be always available at the receiver. The MIMO system is called closed-loop if full or partial CSI is also available at the transmitter, and is called open-loop otherwise. The increased channel dimension of the MIMO system can be utilized to send multiple data streams or to increase the reliability of a single data stream reception. The increases of data rate and reliability are quantified by multiplexing rate and diversity gain, respectively. In a high signal-to-noise ratio (SNR) regime, the diversity gain indicates how fast the packet error rate (PER) reduces as the noise power approaches zero. The higher the diversity gain, the more reliable the system is. For a narrow-band system with slow fading, the $N \times M$ MIMO and single-input/single-output system (SISO) have the maximum diversity gains $M \times N$ and one, respectively. Meanwhile, the multiplexing rate indicates how many distinctive data streams can be received correctly and simultaneously. The higher the multiplexing rate, the higher the data rate that can be achieved. The maximum multiplexing rate is the minimum of N and M for an $N \times M$ MIMO system, and is one for a SISO system. One can make a trade-off between the multiplexing rate and diversity [3]. For example, a 4×3 system with four transmit and three receive antennas can operate in a reliable mode with multiplexing rate 1 and diversity gain 6, or in a high-rate mode with multiplexing rate 2 and diversity gain 2. For systems with wideband multipath fading or fast fading, additional diversity gain can be collected from the frequency domain or time domain, respectively.

4.2.1 Open-Loop SU-MIMO

IEEE 802.16e features the support of mobility of up to 120 km/h. At such a high speed, the channel variation prevents the transmitter from obtaining accurate CSI and, as a result, open-loop MIMO systems are more desirable in mobility applications because they do not require CSI. Due to its flexible tradeoff between diversity

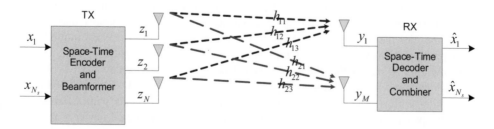

Figure 4.1. An illustrative 3 × 2 MIMO system. The channel between the jth transmit and ith receive antenna is denoted by h_{ij} and the number of transmitted streams is denoted by N_s.

gain and multiplexing rate, IEEE 802.16e adopts space-time coding (STC) as the main open-loop MIMO scheme. IEEE 802.16e provides much more comprehensive STC support than IEEE 802.16d [1], which is primarily designed for fixed wireless channels. For example, in IEEE 802.16e, STC is supported for both the uplink and the downlink, for up to four transmit antennas, and for multiplexing rate up to 4.

WiMAX Release 1.0 mandates the adaptive MIMO implementation with two transmit antennas to achieve two objectives: (1) increased degree of diversity to enhance the signal reliability for the same data throughput, or (2) increased data throughput at the same channel quality condition. The first objective can be achieved by space-time transmit diversity (STTD, referred by Matrix A), and the second one can be achieved by spatial multiplexing (referred by Matrix B) that creates several parallel, independent transmission streams. A rule of thumb to apply the two modes is as follows. When the MS's channel matrix is well conditioned and the received signal is strong (SNR greater than 4 dB), the spatial multiplexing mode can be applied to increase the throughput. On the other hand, when the channel matrix is ill conditioned or the received signal is weak (at the cell edge or line of sight), the STTD mode can be applied to increase the reception robustness for lower packet error rate and retransmission rate. The selection between the two modes is fed back to the BS by the MS jointly with channel quality information (CQI) via the CQICH feedback channel, which decides the final MIMO transmission mode together with the modulation and coding scheme, as shown in Figure 4.2.

Although some STC options in IEEE 802.16e are optimized for high-complexity decoding, for example, the maximum likelihood decoding, the ones in Release 1.0 are of low decoding complexity. An example of Release 1.0 STC is described for two transmit antennas with multiplexing rate 1:

$$\mathbf{A}_1 = \begin{bmatrix} S_1 & -S_2^* \\ S_2 & S_1^* \end{bmatrix} \quad \begin{matrix} \text{Antenna 1} \\ \text{Antenna 2} \end{matrix}$$

$$\downarrow \quad \downarrow$$
$$t=1 \quad t=2$$

(1)

where one information symbol is effectively sent per channel use and the codeword symbols are sent over the same subcarrier on two OFDM symbols adjacent in time. For the other IEEE 802.16e STCs not in Release 1.0, there are multiple mapping patterns that are across time and frequency, and these patterns are alternately applied across the allocated subcarriers per the MS. Besides STC, cyclic delay diversity (CDD) is another open-loop scheme discussed in IEEE 802.16d/e working groups. The additional transmit antennas sending signals with different delays provide a diversity gain at the receiver whose total received power within the whole bandwidth is stabilized. Note that CDD is not mentioned in the IEEE 802.16e specification since the BS can transparently apply CDD without the receiver awareness.

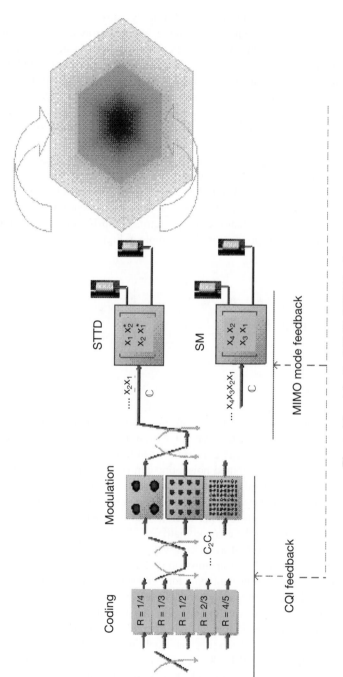

Figure 4.2. Adaptive MIMO-OFDMA scheme.

4.2.2 Closed-Loop SU-MIMO

Exploiting CSI at the transmitter, closed-loop systems can achieve better performance than open-loop systems. For example, antenna array can form a directional radiation pattern (i.e., beam) to enhance transmit (or receive) signal energy in a desired direction, where the signal power is increased by constructive addition of the transmitted (or received) signals. This is called transmit (or receive) beam forming. Transmit and receive beam forming can jointly maximize the received signal power for coverage range extension. Beam forming is capable of increasing the signal strength by a factor of about $\sqrt{M-1} + \sqrt{N}$ for large numbers of antennas [5], where M and N are the numbers of transmit and receive antennas, respectively. The maximum beam forming gain is obtained when the transmitter sends a single spatial stream ino a preferable direction and the receiver receives the signal using receive beam forming. For a 4×4 MIMO, this gain is about 9 dB with respect to SISO, on average. If a BS sends multiple streams to an MS, each stream will require its own beam forming vector. The beam forming vectors comprise a beam forming matrix at the transmitter. Corresponding to the example in Figure 4.1, the signal model is

$$
\begin{bmatrix} y_1 \\ y_2 \end{bmatrix} = \underbrace{\begin{bmatrix} h_{11} & h_{12} & h_{13} \\ h_{21} & h_{22} & h_{23} \end{bmatrix}}_{H} \underbrace{\overbrace{\begin{bmatrix} v_{11} & v_{12} \\ v_{21} & v_{22} \\ v_{31} & v_{32} \end{bmatrix}}^{V} \begin{bmatrix} x_1 \\ x_2 \end{bmatrix}}_{\begin{bmatrix} z_1 \\ z_2 \\ z_3 \end{bmatrix}} + \begin{bmatrix} n_1 \\ n_2 \end{bmatrix} \tag{2}
$$

where the channel matrix and beam forming matrix are denoted by H and V, respectively; z_j is the transmitted signal at the jth transmit antenna; and n_i is the noise at the ith receive antenna. Denote the number of transmitted streams by N_s. For independent, identically distributed noise, one can compute the optimal beam forming matrix from the singular value decomposition (SVD) of H [4]. The first N_s columns of the right unitary matrix of the SVD comprise the optimal beam forming matrix. Note that the beam forming operation at the transmitter only requires the beam forming matrix, not the channel matrix.

The BS can obtain the CSI in several ways. CSI for downlink may be obtained from the reverse uplink (UL) transmission if channel reciprocity is available. However, channel reciprocity does not exist for FDD systems because downlink and uplink operate on different frequency bands. Even for TDD systems, the possession of channel reciprocity requires calibrating the transmit and receive radio frequency (RF) components and transmitting signals over all receive antennas. This results in undesirable cost increases. Therefore, feedback from the MS is usually required for the BS to acquire the CSI. IEEE 802.16e supports six schemes for CSI acquisition: channel matrix feedback, antenna selection feedback, antenna grouping feedback, codebook-based precoding feedback, analog feedback, and uplink channel sounding. Except for analog feedback and uplink channel sounding, the receiver sends digitized CSI to the transmitter.

Medium to High Mobility Support

Supporting medium- to high-speed MSs in closed-loop MIMO requires frequent and timely feedbacks. However, the WiMAX system inherently has two limitations in the feedback channel: low feedback bandwidth and long feedback delay. The feedback delay is 2 frames, which typically corresponds to 10 ms. The frame structure is illustrated in Figure 4.3. As the MS speed increases, the actual beam forming matrix at the BS becomes obsolete and inaccurate due to the temporal channel variation during the 10 ms feedback delay. In IEEE 802.16e, the negative impact of feedback delay is mitigated by two countermeasures: the channel prediction method and the long-term beam forming method. In the first method, the MS predicts the channel response at the time of the actual beam forming and computes the feedback index based on the predicted channel. The beam forming accuracy is thus increased by the virtue of prediction. In the second method, the high-mobility MS only feeds back to the BS the long-term beam forming direction that delivers some level of beam forming gain over a long time period. The long-term beam forming direction corresponds to channel components that vary slowly as compared to the instantaneous CSI, for example, the line-of-sight component of a Ricean fading channel. Although the quantization methods for both the long-term and the instantaneous beam forming directions are the same, the feedback for the long term is much infrequent. Note that the channel prediction is not specified in 802.16e and the long-term beam forming is in IEEE 802.16e but not in WiMAX profiles yet.

Low Mobility Support

Instantaneous CSI feedback is desirable for slow fading situations, where feedback delay is not a major concern. In IEEE 802.16d fixed wireless system, the channel matrix $\mathbf{H}$ is quantized entry by entry, and the quantization indices are fed back. The quantization complexity is low but the feedback overhead is not. In the IEEE 802.16e mobile wireless system, low feedback rate and low storage complexity are required as the feedback is often more than that of the fixed wireless system. Therefore, in IEEE 802.16e the vector quantization approach is supported, which quantizes the beam forming matrix $\mathbf{V}$ in equation (2) as a whole using a codebook. It should be noticed that $\mathbf{V}$ contains about half of the information in $\mathbf{H}$ and, thus, requires half of the feedback bits. In the IEEE 802.16e system, the quantization codebooks are predefined for each pair of transmit antenna number and multiplexing rate, and are stored in both the BS and each MS [6]. Since the ideal beam forming vectors are equally likely and comprise the surface of a M-dimensional complex sphere, the quantization codewords of the vector codebook should be uniformly distributed on the sphere. Similarly, the ideal beam forming matrices comprise the set of all $M \times N_s$ orthogonal matrices (a Grassmann manifold), and the codeword matrix should be uniformly distributed on the surface of the manifold. Here, the uniformity of the codebook helps in reducing the quantization error. Finally, since the

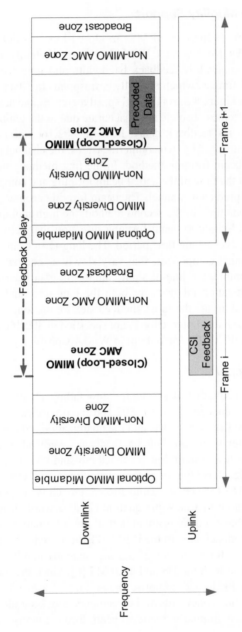

Figure 4.3. Illustration of feedback-based, closed-loop MIMO in FDD.

codebook for each feasible pair of M and N_s needs to be stored in each mobile device, the storage complexity is a concern. For storage reduction, in IEEE 802.16e only vector codebooks of small sizes are stored and large vector codebooks can be computed from a few parameters. Moreover, all matrix codebooks are computed from one or two component vector codebooks

The codebook-based transmit beam forming is referred to codebook-based precoding in WiMAX. It is usually applied in the so-called band AMC zone as shown in Figure 4.3, where the whole bandwidth is divided into subbands, each containing about kHz of contiguous subcarriers. Besides band AMC, the precoding can be applied to wideband permutations such as PUSC. Each participant MS feeds back channel quality information (CQI), a precoding matrix index (PMI), the desired number of data streams (rank), and the index of the subband (band index), for at least the best subband. The feedback of CQI/PMI is usually more often than that of the rank and band index. The feedback can be sent via a dedicated, fast feedback channel called CQICH, or via regular UL data traffic in the form of feedback headers or subheaders. The BS collects the CSI from the MSs and schedules those MSs with strong channel and high-priority traffic over the selected subbands using transmit beam forming.

There are two more methods supporting low-mobility MSs in IEEE 801.16e: antenna selection and antenna grouping. Antenna selection can be viewed as approximating the ideal beam forming matrices using a set of permutation matrices with entries equal to zero and one. Equivalently, the MS selects the best subset of the BS transmit antennas and feeds back the selection. For example, an MS selects two antennas out of four BS transmit antennas and asks the BS to send data over the two selected antennas. In this example, three feedback bits are required to cover the

$$\binom{4}{2} = 6$$

possible selections for each subband. Extending the selection over antennas to also include frequency, the antenna grouping is defined. The MS selects not only the antennas but also the subcarriers to send the STC codeword symbols. For each subband, the feedback bits select one of the mapping patterns specified for the STC. For the example of a four-antenna rate-one STC, there are three time-frequency-space mappings of the STC symbols: A_1, A_1, and A_3. While in the open-loop mode, these mappings are cyclically assigned across frequency; in the antenna grouping, the MS chooses one out of the three using a two-bit feedback. The feedbacks of antenna selection and antenna grouping are sent via CQICH.

Other Feedback Methods in IEEE 802.16e

Besides the quantized feedback approach, analog feedback and uplink channel sounding are supported by IEEE 802.16e. In the analog feedback scheme, the MS essentially sends the time-domain channel responses between each BS transmit antenna and each MS receive antenna for the selected subband. One OFDM subcarrier is modulated by one complex channel response using analog modulation and is scrambled by an uplink sounding symbol. Compared to codebook-based feedback, analog feedback does not suffers from quantization error but suffers from severe feedback error at the cell edge due to the lack of channel coding protection. It is worth noting that IEEE 802.16e also supports the acquisition of downlink CSI using uplink channel sounding for TDD systems for which channel reciprocity can be established. In the uplink channel sounding, the MS sends training symbols over a subband of 18 consecutive subcarriers in a sounding zone. Based on the received training symbols, the BS infers the CSI for the downlink channel. Using the dynamically estimated downlink channel, the BS adaptively applies the computed beam forming weights for beam forming the MS's downlink data. This is a typical scenario for applying beam forming and is referred to as adaptive beam forming in Table 4.1.

Trade-off between Performance and Feedback Overhead

The IEEE 802.16e MIMO systems provide flexible trade-offs between system performance and feedback resource. In the codebook-based precoding, six-bit and three-bit codebooks are available for most antenna configurations. As the codebook size decreases, the beam forming accuracy decreases. The performance of STC, antenna selection, and codebook-based precoding are compared in Figure 4.4. The system has four transmit and two receive antennas. The channel tested is ITU Pedestrian B channel model with 3 km/h mobile speed. QPSK, convolutional code with code rate ½, 512-bit packet is employed. Two spatial streams are sent with equal power. Channel prediction and MMSE detection are used at the receiver. Antenna selection and codebook-based precoding outperform STC by 2 and 3 dBs, respectively, at the cost of three- and six-bit feedback overheads per frame per AMC

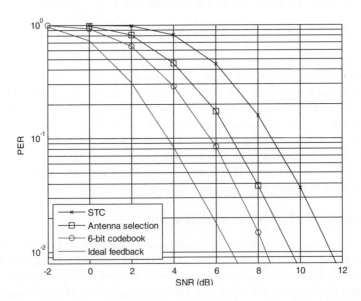

Figure 4.4. PER performances of STC, antenna selection, codebook-based precoding, and ideal SVD precoding.

band. The quantization error and frequency variation within the subband cause a 2 dB loss in codebook-based precoding compared to ideal feedback results.

4.3 MULTIUSER MIMO

MU-MIMO allows multiple MSs to spatially share the same time-frequency resource for improving both the cell spectral efficiency and average user experience. In the past few years, several MU-MIMO schemes have been adopted by industrial standards including WiMAX and LTE. The popularity of MU-MIMO is primarily attributed to its robustness; an MS at the cell edge can benefit from MU-MIMO just as much as an MS at the cell center. MU-MIMO exploits the antennas of multiple devices to form a MIMO system and, therefore, is applicable to scenarios that prohibit SU-MIMO applications. For example, in a cell with highly correlated antennas, the channel rank seen by each MS is one and, therefore, conventional SU-MIMO cannot apply spatial multiplexing to increase data throughput. In addition, MU-MIMO has an edge from a complexity and cost perspective because the burden of complicated MIMO processing is shifted to the BS that is less sensitive to the cost and power consumption of signal processing.

IEEE 802.16e supports basic MU-MIMO operations in both downlink and uplink, as illustrated in Figure 4.5. The downlink MU-MIMO, also known as SDMA in IEEE 802.16e, can be viewed as overlaying up to four single-user beam forming transmissions in the same time-frequency resource. The BS first gathers the downlink CSI of all active MSs in the cell, via either CQI/PMI feedback or uplink channel sounding. After selecting a subset of MSs, the BS computes the beam forming vector for each selected MS. The subset selection and beam forming operation are used to capture the multiuser diversity and to minimize the inter-beam/MS interference. A dedicated downlink pilot is embedded in and beam formed together with each data stream, enabling each selected MS to estimate its beam formed channel and demodulate its intended signal. Note that all the ingredients required to support the downlink MU-MIMO are already present for supporting single-user beam forming in the specification, though the operation of MU-MIMO is not clear in the specifications. It is left to the operator's discretion whether to support and how to support the downlink MU-MIMO.

In the uplink, only two MSs can be allocated the same time-frequency resource for MU-MIMO operation, also known as collaborative spatial multiplexing (C-SM) in IEEE 802.16e. Similar to the downlink, most of the required ingredients are already present in the specification for supporting the single-user uplink transmission, including ranging operation that allows all MSs tightly synchronized with the BS in both frequency and time. The UL MU-MIMO can be operated in either transparent or nontransparent fashion. In the nontransparent mode, the BS distinguishes the two pilot sequences from the two MSs by assigning two orthogonal sets of resources to the two MSs' pilots. Note that WiMAX profile currently only supports single-antenna MS in UL MU-MMO, though IEEE 802.16e allows up to two antennas at each uplink MS in C-SM. On the other hand, in transparent UL MU-MIMO mode,

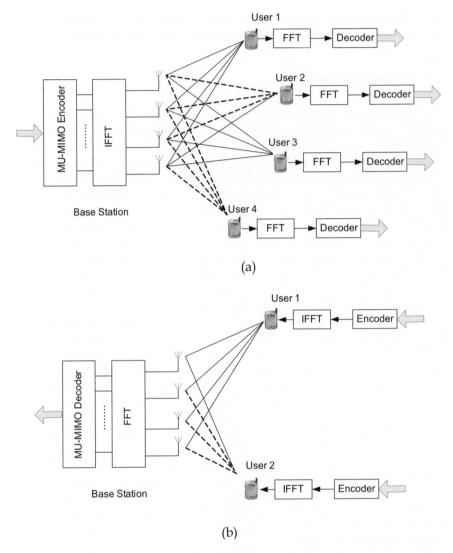

Figure 4.5. (a) Illustration of downlink MU-MIMO in IEEE 802.16e. (b) Illustration of uplink MU-MIMO (C-SM).

the BS simply allocates overlapping frequency-time resources to two MSs with the same pilot sequence, and relies on the receive beam forming capability at the BS to distinguish signals from the two MSs. For example, the defined single-stream beam forming for SU-MIMO can be transparently applied to MU-MIMO as follows. The BS simply schedules two MSs in the same set of frequency-time resources, and receives two sets of precoded data to each of them, respectively. The precoded data comes along with precoded (or dedicated) pilots to estimate each MS's precoded channel. The interference between the two MSs' signals depends on the BS. Namely, the BS needs to select two MSs that have two receive beams near orthogonal in order to achieve a low level of interference.

4.4 DISTRIBUTED MIMO AND RELAY IN IEEE 802.16j

Relay technology is introduced to WiMAX through IEEE 802.16j [7], where the RS is an infrastructure station rather than a MS. For backward compatibility, an IEEE 802.16e MS can benefit from the relay operation without the awareness of the RS presence. Besides the conventional relay operation, IEEE 802.16j also allows the BS and RS to forward the signal in an opportunistic or cooperative fashion. Among BS, RS(s), and MS, a distributed MIMO system is formed with all the transmit antennas of the BS and RS on one side and the receive antennas of the destination MS on the other side.

The opportunistic forwarding is also called RS-assisted H-ARQ in IEEE 802.16j, where the BS and RS jointly transmit to the MS using a switched diversity scheme. As shown in Figure 4.6(a), the BS sends a packet and asks both the MS and RS to decode it. If the RS decoded the packet but the MS missed it, the BS asks the RS to send an H-ARQ packet to the MS so that the MS can combine the initial packet and the H-ARQ packet to decode. This is better than if the BS sends the H-ARQ packet by itself. The reason is that if the channel from the BS to the MS is in deep fading, which causes the initial reception failure, retransmission over the same channel is likely to fail again. Therefore, the retransmission from the RS, going through a different channel, enhances the reliability. However, since three active links need to be maintained, MAC overhead becomes burdensome for densely populated networks. In the cooperative forwarding, STC is performed by multiple devices jointly, each of which may be a BS or an RS. The STC symbols are first dis-

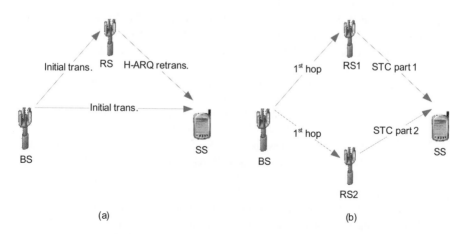

Figure 4.6. Opportunistic (a) and cooperative relays (b).

tributed (or multicast) to the RSs and then sent simultaneously. The distributed sub-sets of STC symbols can be the same, disjoint, or partially overlapped across RSs, for which cases the scheme is referred to as cooperative source diversity, cooperative transmit diversity, and cooperative hybrid diversity. For the example in Figure 4.6(b), RS1 and RS2 jointly conduct the STC in the first part using two transmit antennas each. RS1 sends STC symbols intended for (effective) transmit antennas 1 and 2, while RS2 sends STC symbols intended for (effective) transmit antennas 3 and 4. This is an example of cooperative transmit diversity because the two symbol subsets are disjoint. Finally, the MAC overhead is less than that of the opportunistic forwarding as there is minimum coordination between the first and second hops.

4.5   CONCLUSIONS

WiMAX is the first cellular standard that employs orthogonal frequency division multiple access (OFDMA) technology and provides true integrated services for both fixed and mobile broadband access. Among the many new technologies adopted in WiMAX, MIMO technology plays an essential role in delivering fast, rich-content mobile broadband service reliably over extended coverage areas. In this chapter, we provide a survey on the state of art of the MIMO technologies in current WiMAX standards with an emphasis on practical engineering considerations.

4.6   REFERENCES

[1]   EEE P802.16d, IEEE Standard for Local and Metropolitan Area Networks, Part 16: Air Interface for Fixed Broadband Wireless Access Systems. October 2004.

[2]   IEEE P802.16e, IEEE Standard for Local and Metropolitan Area Networks, Part 16: Air Interface for Fixed and Mobile Broadband Wireless Access Systems, P802.16Rev2/D2. December 2007.

[3]   L. Zheng and D. N. C. Tse, Diversity and multiplexing: A Fundamental Tradeoff in Multiple Antenna Channels, *IEEE Trans. Inform. Theory,* vol. 49, May 2003.

[4]   I. E. Telatar, Capacity of Multi-antenna Gaussian Channels, *AT&T Bell Laboratories Technical Memo,* 1995.

[5]   I. Johnstone, On the Distribution of the Largest Eigenvalue in Principal Components Analysis, *The Annals of Statistics,* vol. 29, no. 2, pp. 295-327, 2001.

[6]   Q. Li, X. E. Lin, and J. C. Zhang, MIMO Precoding in 802.16e WiMAX, *Journal of Communications and Networks,* vol. 9, no. 2, pp.141-149, June 2007.

[7]   IEEE P802.16j, IEEE Standard for Local and Metropolitan Area Networks, Part 16: Air Interface for Fixed and Mobile Broadband Wireless Access Systems: Multihop Relay Specification. P802.16j/D3, February 2008.

[8]   WiMAX Forum™ Mobile System Profile, Release 1.0 (Revision 1.4.0: 2007-05-02), available at http://www.wimaxforum.org/resources/documents/technical/release1.

[9]   WiMAX Forum™ Mobile System Profile, Release 1.5 (Revision 0.2.1: 2009-02-02), available at http://www.wimaxforum.org/resources/documents/technical/release1.5.

CHAPTER 5

OVERVIEW OF IEEE 802.16m RADIO ACCESS TECHNOLOGY

SASSAN AHMADI

5.1 INTRODUCTION TO IEEE 802.16m

The growing demand for mobile Internet and wireless multimedia applications has motivated development of broadband radio access technologies in recent years. Mobile WiMAX was the first mobile broadband wireless access solution, based on the IEEE 802.16-2009 standard [1], that enabled convergence of mobile and fixed broadband networks through a common wide-area radio access technology and flexible network architecture. The mobile WiMAX air interface utilizes orthogonal frequency division multiple access (OFDMA) as the preferred multiple access method in the downlink and uplink for improved multipath performance and bandwidth scalability. The IEEE 802.16 Working Group began development of a new amendment of the IEEE 802.16 standard (i.e., IEEE 802.16m) in January 2007 as an advanced air interface to meet the requirements of ITU-R/IMT-Advanced for fourth-generation of cellular systems.

Depending on the available bandwidth and multiantenna technique used, the next-generation mobile WiMAX will be capable of over-the-air data transfer rates in excess of 1 Gbps and be able to support a wide range of high-quality and high-capacity IP-based services and applications while maintaining full backward compatibility with the existing mobile WiMAX systems in order to preserve investments and provide continuing support for the first-generation products. There are distinctive features and advantages such as flexibility and extensibility of its physical and medium-access layer protocols that make mobile WiMAX and its evolution more attractive and more suitable for the realization of ubiquitous mobile Internet access. This chapter briefly describes the salient technical features of IEEE 802.16m and potentials for successful deployment of the next generation of mobile WiMAX in 2012 and beyond.

The next-generation mobile WiMAX will build on the success of the existing WiMAX technology and its time-to-market advantage over other mobile broadband wireless access technologies. In fact, all OFDM-based mobile broadband radio access technologies that have been developed in recent years were inspired by, exploited, enhanced, or extended the fundamental technical concepts that were originally utilized in mobile WiMAX.

IEEE 802.16m will be suitable for both green-field and mixed deployments with legacy mobile stations (MS) and base stations (BS). The backward compatibility feature will allow smooth upgrades and an evolution path for the existing deployments. It will enable roaming and seamless connectivity across IMT-Advanced and IMT-2000 systems through the use of appropriate interworking functions. The IEEE 802.16m systems can further utilize multihop relay architectures for improved coverage and performance.

5.2   IEEE 802.16m SYSTEM REQUIREMENTS AND EVALUATION METHODOLOGY

Full backward compatibility and interoperability with the reference system is required for IEEE 802.16m systems, [2]. The network operators have the ability to disable legacy support in green-field deployments. The reference system is defined as a system compliant with a subset of the IEEE 802.16-2009 [1] features as specified by the WiMAX Forum Mobile System Profile, Release 1.0 [5]. The following are backward-compatibility requirements for IEEE 802.16m systems [2]:

- An IEEE 802.16m MS shall be able to operate with a legacy BS, at a level of performance equivalent to that of a legacy MS.
- Systems based on IEEE 802.16m and the reference system shall be able to operate on the same RF carrier, with the same channel bandwidth; and should be able to operate on the same RF carrier with different channel bandwidths.
- An IEEE 802.16m BS shall support a mix of IEEE 802.16m and legacy terminals when both are operating on the same RF carrier. The system performance with such a mix should improve with the fraction of IEEE 802.16m terminals attached to the BS.
- An IEEE 802.16m BS shall support handover of a legacy MS to and from a legacy BS and to and from IEEE 802.16m BS, at a level of performance equivalent to handover between two legacy BSs.
- An IEEE 802.16m BS shall be able to support a legacy MS while also supporting IEEE 802.16m terminals on the same RF carrier, at a level of performance equivalent to that a legacy BS provides to a legacy MS.

Consideration and implementation of the above requirements ensure a smooth migration from legacy to new systems without any significant impact on the performance of the legacy systems as long as they exist. The requirements for IEEE 802.16m were selected to ensure competitiveness with the emerging fourth-generation radio access technologies while extending and significantly improving functionalities/features of the reference system such as the layer-2 control/signaling mechanism; coverage of control and traffic channels at the cell edge and downlink/uplink link budgets; air-link access latency; MS power consumption, including uplink (UL) peak to average power ratio (PAPR) reduction; downlink/uplink medium access protocol coding and processing; scan latency and network entry/re-entry procedures; downlink and uplink symbol structure and subchannelization schemes; MAC management messages; MAC headers; and support of FDD duplex mode (this has been already addressed in the latest revision of the IEEE 802.16-2009 standard [1] that is amended by IEEE 802.16m).

The IMT-Advanced requirements defined and approved by ITU-R/Working Party 5D and published as Report ITU-R M.2134 [8] are referred to as target requirements in the IEEE 802.16m system requirement document and are evaluated based on the methodology and guidelines specified by Report ITU-R M.2135 [7]. The baseline performance requirements will be evaluated according to the IEEE 802.16m evaluation methodology document [3]. A careful examination of the IMT-Advanced requirements reveals that they are a subset of and less stringent than the IEEE 802.16m system requirements; therefore, IEEE 802.16m standard is qualified as an IMT-Advanced technology. Table 5.1 summarizes the IEEE 802.16m baseline system requirements and the corresponding requirements specified by Report ITU-R M.2134 [8]. In the next sections, we briefly discuss how these requirements can be met or exceeded through extension and enhancements of the reference system functional features.

In Table 5.1, the cell spectral efficiency is defined as the aggregate throughput—the number of correctly received bits delivered at the data link layer over a certain period of time, of all users, divided by the product of the effective bandwidth, the frequency reuse factor, and the number of cells. The cell spectral efficiency is measured in bit/sec/Hz/cell.

The peak spectral efficiency is the highest theoretical data rate normalized by bandwidth (assuming error-free conditions) assignable to a single mobile station when all available radio resources for the corresponding link are utilized, excluding radio resources that are used for physical layer synchronization, reference signals, guard bands, and guard times (collectively known as layer-1 overhead). Bandwidth scalability is the ability to operate with different bandwidth allocations using single or multiple RF carriers. The normalized user throughput is defined as the number of bits correctly received by a user at the data link layer over a certain period of time, divided by the total spectrum. The cell-edge user spectral efficiency is defined as the 5% point of cumulative distribution function (CDF) of the normalized user throughput.

Control-plane (C-plane) latency is defined as the transition time from idle state to connected state. The transition time (assuming that downlink paging latency and core

Table 5.1. IEEE 802.16m and IMT-Advanced system requirements

Requirements	IMT-Advanced [8]	IEEE 802.16m [2]
Peak data rate (b/s/Hz)	DL: 15 (4 × 4) UL: 6.75 (2 × 4)	DL: 8.0/15.0 (2 × 2/4 × 4) UL: 2.8/6.75 (1 × 2/2 × 4)
Cell spectral efficiency (b/s/Hz/sector)	DL (4 × 2) = 2.2 UL (2 × 4) = 1.4 (base coverage urban)	DL (2 × 2) = 2.6 UL (1 × 2) = 1.3 (mixed mobility)
Cell-edge user spectral efficiency (b/s/Hz)	DL (4 × 2) = 0.06 UL (2 × 4) = 0.03 (base coverage urban)	DL (2 × 2) = 0.09 UL (1 × 2) = 0.05 (mixed mobility)
Latency	C-plane: 100 ms (idle to active)	C-plane: 100 ms (idle to active transition time)
U-plane: 10 ms	U-plane: 10 ms	
Mobility b/s/Hz at km/h	0.55 at 120 km/h 0.25 at 350 km/h (link-level)	Optimal performance up to 10 km/h Graceful degradation up to 120 km/h Connectivity up to 350 km/h Up to 500 km/h depending on operating frequency
Handover interruption time (ms)	Intrafrequency: 27.5 Interfrequency: 40 (in a band) 60 (between bands)	Intrafrequency: 27.5 Interfrequency: 40 (in a band) 60 (between bands)
VoIP capacity (active users/sector/MHz)	40 (4 × 2 and 2 × 4) (base coverage urban)	60 (DL 2 × 2 and UL 1 × 2)
Antenna configuration	Not specified	DL: 2 × 2 (baseline), 2 × 4, 4 × 2, 4 × 4, 8 × 8 UL: 1 × 2 (baseline), 1 × 4, 2 × 4, 4 × 4
Cell Range and coverage	Not specified	Up to 100 km optimal performance up to 5 km
Multicast and broadcast service (MBS)	Not specified	4 bps/Hz for ISD 0.5 km 2 bps/Hz for ISD 1.5 km
MBS channel reselection interruption time	Not specified	1.0 s (intrafrequency) 1.5 s (interfrequency)
Location-based services (LBS)	Not specified	Location determination latency <30 s MS-based position determination accuracy <50 m Network-based position determination accuracy <100 m
Operating bandwidth	Up to 40 MHz (with band aggregation)	5 to 20 MHz (up to 100 MHz through band aggregation)
Duplex scheme	Not specified	TDD, FDD (support for H-FDD terminals)

Table 5.1. IEEE 802.16m and IMT-Advanced system requirements

Requirements	IMT-Advanced [8]	IEEE 802.16m [2]
Operating frequencies (MHz)	IMT bands	IMT bands
	450–470	450–470
	698–960	698–960
	1710–2025	1710–2025
	2110–2200	2110–2200
	2300–2400	2300–2400
	2500–2690	2500–2690
	3400–3600	3400–3600

network signaling delay are excluded) of less than 100 ms is required for IEEE 802.16m systems. The user-plane (U-plane) latency, also known as transport delay, is defined as the one-way transit time between a packet being available at the IP layer of the origin (user terminal in the uplink or base station in the downlink) and the availability of this packet at the IP layer of the destination (base station in the uplink or user terminal in the downlink). User-plane packet delay includes delay introduced by associated protocols and signaling, assuming that the user terminal is in the active mode. The IEEE 802.16m systems are required to achieve a transport delay of less than 10 ms in unloaded conditions (i.e., single user with single data stream) for small IP packets (e.g., 0 byte payload + IP header) for both downlink and uplink.

For the VoIP capacity, a 12.2 kbps codec with a 50% speech activity factor is assumed such that the percentage of users in outage is less than 2%, where a user is defined to have experienced outage if less than 98% of the VoIP packets have been delivered successfully to the user within a one-way radio-access delay bound of 50 ms [2]. It should be noted that the VoIP capacity is the minimum of the capacities calculated for the downlink and uplink divided by the effective bandwidth in the respective link direction. The effective bandwidth is the operating bandwidth appropriately normalized by the DL/UL radio. Minimum performance requirements for enhanced multicast and broadcast are expressed in terms of spectral efficiency over 95% of the coverage area. The performance requirements apply to a wide-area multicell multicast broadcast single-frequency network.

The IEEE 802.16m evaluation methodology document [3] provides simulation parameters and guidelines for evaluation of the candidate proposals against the IEEE 802.16m system requirements [2]. The evaluation scenarios and associated parameters specified in the IEEE 802.16m evaluation methodology document are mainly based on those that were used for evaluation of the reference system [10] in order to benchmark the relative improvements. There are some similarities and differences between evaluation guidelines, test scenarios, and configuration parameters specified by Report ITU-R M.2135 [7] and the IEEE 802.16m evaluation methodology document [3]. The compliance with Report ITU-R M.2134 requirements in at least three test environments is required.

Table 5.2 shows service requirements for IMT-Advanced systems, for which compliance of proposed radio interfaces are verified through inspection [7]. There

Table 5.2. IMT-Advanced service requirements [7]

User experience class	Service class	Service parameters (numerical values)	
Conversational	Basic conversational service	Throughput	20 kbps
		Delay	50 ms
	Rich conversational service	Throughput	5 Mbps
		Delay	20 ms
	Conversational low delay	Throughput	150 kbps
		Delay	10 ms
Streaming	Streaming live	Throughput	2–50 Mbps
		Delay	100 ms
	Streaming nonlive	Throughput	2–50 Mbps
		Delay	1 s
Interactive	Interactive high delay	Throughput	500 kbps
		Delay	200 ms
	Interactive low delay	Throughput	500 kbps
		Delay	20 ms
Background	Background	Throughput	5–50 Mbps
		Delay	< 2s

are four user-experience classes, each of which is further divided into a number of service classes based on the intrinsic characteristics of corresponding services, such as required throughput and latency, to ensure satisfaction of the QoS requirements for each class. When the service class requirements are translated into system requirements, a limited number of QoS attributes, such as data throughput, packet delay and/or delay variations (often referred to as jitter), bit/packet error rate, and other similar aspects can only be considered.

IEEE 802.16m was submitted to ITU-R/Working Party 5D as a candidate for IMT-Advanced in October 2009. The self-evaluation results (i.e., the analytical and simulation results generated by the IEEE 802.16 Working Group members) are shown in Table 5.3. It is evident that IEEE 802.16m exceeds all requirements in all test environments specified by Report ITU-R M.2134. In Table 5.3, the acronyms InH, UMi, UMa, and RMa denote indoor hotspot (indoor), microcellular (urban microcell), Base coverage urban (urban macrocell), High-speed (rural macrocell) test environments, respectively. It must be noted that these performance results were obtained based on the parameters and configurations specified by Report ITU-R M.2135 [7].

5.3   IEEE 802.16m REFERENCE MODEL AND PROTOCOL STRUCTURE

The WiMAX Network Architecture Release 1.5 [6] specifies a nonhierarchical end-to-end network reference model (shown in Figure 5.1) for mobile WiMAX that can

Table 5.3. Performance results for IEEE 802.16m based on Report ITU-R M.2135 configurations [7]

Requirements	Duplex scheme	DL/UL	Test environments			
			InH	UMi	UMa	RMa
Cell spectral efficiency (bit/s/Hz/cell)	TDD	DL	6.93	3.22	2.41	3.23
ITU-R requirement			3.0	2.6	2.2	1.1
Cell-edge spectral efficiency (bit/s/Hz/cell)			0.260	0.092	0.069	0.093
ITU-R requirement			0.1	0.075	0.06	0.04
Cell spectral efficiency (bit/s/Hz/cell)	FDD		6.87	3.27	2.41	3.15
ITU-R requirement			3.0	2.6	2.2	1.1
Cell-edge spectral efficiency (bit/s/Hz/cell)			0.253	0.097	0.069	0.091
ITU-R requirement			0.1	0.075	0.06	0.04
Cell spectral efficiency (bit/s/Hz/cell)	TDD	UL	5.99	2.58	2.57	2.66
ITU-R requirement			2.25	1.8	1.4	0.7
Cell-edge spectral efficiency (bit/s/Hz/cell)			0.426	0.111	0.109	0.119
ITU-R requirement			0.07	0.05	0.03	0.015
Cell spectral efficiency (bit/s/Hz/cell)	FDD		6.23	2.72	2.69	2.77
ITU-R requirement			2.25	1.8	1.4	0.7
Cell-edge spectral efficiency (bit/s/Hz/cell)			0.444	0.119	0.114	0.124
ITU-R requirement			0.07	0.05	0.03	0.015
VoIP capacity (users/dector/MHz)	TDD	DL/UL	140	82	74	89
ITU-R requirement			50	40	40	30
VoIP capacity (users/dector/MHz)	FDD		139	77	72	90
ITU-R requirement			50	40	40	30

be expanded to further include optional relay entities (to be specified by IEEE 802.16m standard) for coverage and performance enhancement. It is expected that the future releases of WiMAX network architecture will specify the reference points between the base station (BS) and relay station (RS) and between two relay stations in a multihop network. The reference points R_i in Figure 5.1 are specified in [6].

The IEEE 802.16-2009 standard [1] describes medium access control (MAC) and physical layer (PHY) protocols for fixed and mobile broadband wireless access systems. The MAC and PHY functions can be classified into three categories: data plane, control plane, and management plane. The data plane comprises functions in the data processing path such as header compression as well as MAC and PHY data packet processing functions. A set of layer-2 (L2) control functions is needed to support various radio resource configuration, coordination, signaling, and management functions. This set of functions is collectively referred to as control plane functions. A management plane is also defined for external management and system configuration. Therefore, all management entities fall into the management plane category.

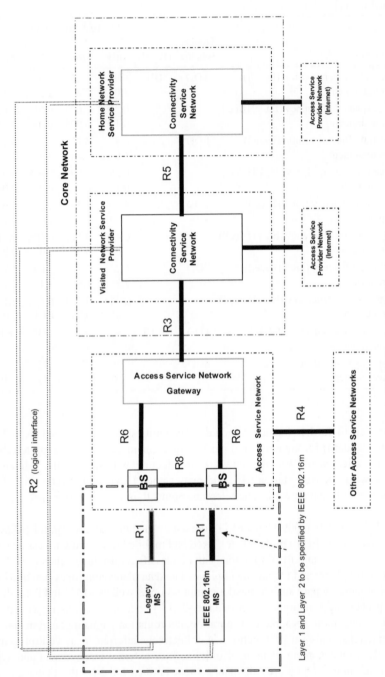

Figure 5.1. WiMAX network architecture [6].

The IEEE Std 802.16-2009 MAC layer is composed of two sublayers: convergence sublayer (CS) and MAC common part sublayer (MAC CPS) [1]. For convenience, we logically classify MAC CPS functions into two groups based on their characteristics, as shown in Figure 5.2.

The upper and lower classes are referred to as resource control and management functional group and medium access control functional group, respectively. The control plane functions and data plane functions are also separately classified. This would allow a more organized, efficient, structured method for specifying the MAC services in the IEEE 802.16m standard specification. As shown in Figure 5.2, the radio resource control and management functional group comprises several functional blocks including:

- **Radio resource management.** This block adjusts radio network parameters related to the traffic load, and also includes the functions of load control (load balancing), admission control, and interference control.
- **Mobility management.** This block scans neighbor BSs and decides whether the MS should perform a handover operation.
- **Network-entry management.** This block controls initialization and access procedures and generates management messages during initialization and access procedures.
- **Location management.** This block supports location-based service (LBS), generates messages including the LBS information, and manages location update operation during idle mode.
- **Idle mode management.** This block controls idle mode operation and generates the paging advertisement message based on paging message from the paging controller in the core network.
- **Security management.** This block performs key management for secure communication. Using a managed key, traffic encryption/decryption and authentication are performed.
- **System configuration management.** This block manages system configuration parameters, and generates broadcast control messages such as superframe header.
- **Multicast and broadcast service (MBS).** This block controls and generates management messages and data associated with MBS.
- **Service flow and connection management.** This block allocates station identifier (STID) and flow identifiers (FIDs) during access/handover service flow-creation procedures.

The medium access control functional group includes functional blocks that are related to physical layer and link controls such as:

- **PHY control.** This block performs PHY signaling such as ranging, channel quality measurement/feedback (CQI), and HARQ ACK or NACK signaling.

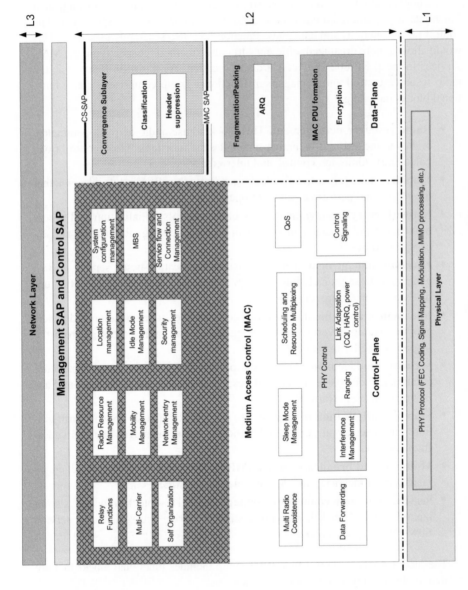

Figure 5.2. IEEE 802.16m protocol stack [4].

- **Control signaling.** This block generates resource allocation messages such as an advanced medium access protocol (A-MAP) as well as specific control signaling messages.

- **Sleep mode management.** This block handles sleep mode operation and generates management messages related to sleep operation. It may communicate with the scheduler block in order to operate properly according to sleep period.

- **Quality-of-service (QoS).** This block performs rate control based on QoS input parameters from the connection management function for each connection.

- **Scheduling and resource and multiplexing.** This block schedules and multiplexes packets based on properties of connections.

The data plane includes functional blocks such as:

- **Fragmentation/packing.** This block performs fragmentation or packing of MAC service data units (MSDU) based on input from the scheduling and resource multiplexing block.

- **Automatic repeat request (ARQ).** This block performs the MAC ARQ function. For ARQ-enabled connections, a logical ARQ block is generated from fragmented or packed MSDUs of the same flow and sequentially numbered.

- **MAC protocol data unit (PDU) formation.** This block constructs the MAC PDU (MPDU) such that the BS/MS can transmit user traffic or management messages into PHY channels.

The IEEE 802.16m protocol structure is similar to that of IEEE 802.16, with some additional functional blocks for new features, including the following:

- **Relay functions.** Relay functionalities and packet routing in relay networks.

- **Self organization and self-optimization functions.** This block performs the self-configuration and self-optimization procedures based on MS measurement reports.

- **Multicarrier functions.** These enable control and operation of a number of adjacent or nonadjacent RF carriers (virtual wideband operation) where the RF carriers can be assigned to unicast and/or multicast and broadcast services. A single MAC instantiation will be used to control several physical layers. The mobile terminal is not required to support multiple carriers. However, if it does support multicarrier operation it may receive control and signaling, broadcast, and synchronization channels through a primary carrier, and traffic assignments (or services) may be made on the secondary carriers.

  A generalization of the protocol structure to multicarrier support using a single MAC instantiation is shown in Figure 5.3. The load balancing functions and RF carrier mapping and control are performed via radio resource

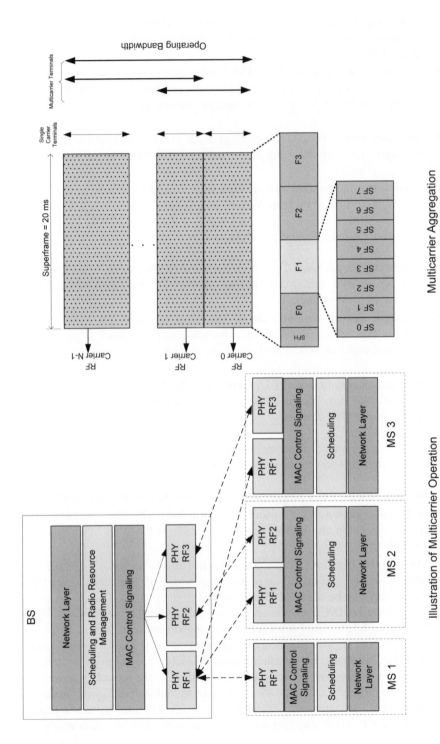

Figure 5.3. IEEE 802.16m multicarrier operation concept [4, 13].

control and management functional class. The carriers utilized in a multicarrier system, from the perspective of a mobile station, can be divided into two categories:

1. A **primary RF carrier** is the carrier that is used by the BS and the MS to exchange traffic and full PHY/MAC control information. The primary carrier delivers control information for proper MS operation, such as network entry. Each mobile station acquires only one primary carrier in a cell.
2. A **secondary RF carrier** is an additional carrier that the BS may use for traffic allocations for mobile stations capable of multicarrier support. The secondary carrier may also include dedicated control signaling to support multicarrier operation.

Based on the primary and/or secondary usage, the carriers of a multicarrier system may be configured differently as follows:

○ **Fully configured carrier.** A carrier for which all control channels including synchronization, broadcast, multicast, and unicast control signaling are configured. Further, information and parameters regarding multicarrier operation and the other carriers can also be included in the control channels. A primary carrier is fully configured, whereas a secondary carrier may be fully or partially configured depending on usage and deployment model.

○ **Partially configured carrier.** A carrier with only essential control channel configuration to support traffic exchanges during multicarrier operation.

In the event that the user terminal RF front end and/or its baseband is not capable of processing more than one RF carrier simultaneously, the user terminal may be allowed, in certain intervals, to monitor secondary RF carriers and to resume monitoring of the primary carrier prior to transmission of the synchronization, broadcast, and nonuser-specific control channels.

● **Multiradio Coexistence Functions.** IEEE 802.16m provides protocols for the multiradio coexistence functional blocks of MS and BS to communicate with each other via air interface. MS generates management messages to report the information about its colocated radio activities obtained from interradio interface, and BS generates management messages to respond with the corresponding actions to support multiradio coexistence operation. Furthermore, the multiradio coexistence functional block at BS communicates with the scheduler functional block to operate properly according to the reported colocated coexistence activities. The multiradio coexistence function can be used independently from sleep mode operation to enable optimal power efficiency with a high level of coexistence support. However, when sleep mode provides sufficient colocated coexistence support, the multiradio coexistence function may not be used (see Figure 5.4).

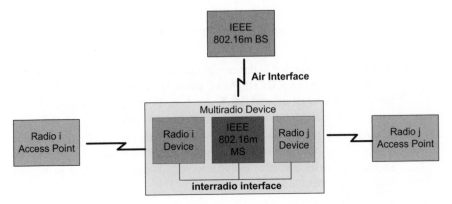

Figure 5.4. A Generic multiradio coexistence model [4, 13].

5.4 IEEE 802.16m MOBILE STATION STATE DIAGRAM

A mobile state diagram (i.e., a finite set of states and procedures between which the mobile station transits when operating in the system to receive and transmit data) for the reference system based on a common understanding of its behavior can be established as follows (see Figure 5.5):

- **Initialization State.** A state in which a mobile station without any connection performs cell selection by scanning and synchronizing to a BS preamble and acquires the system configuration information through the superframe header.
- **Access State.** A state in which the mobile station performs network entry to the selected base station. The mobile station performs the initial ranging process in order to obtain uplink synchronization. Then the MS performs basic capability negotiation with the BS. The MS later performs the authentication and authorization procedure. Next, the MS performs the registration process. The mobile station receives IEEE 802.16m-specific user identification as part of access state procedures. The IP address assignment may follow using appropriate procedures.
- **Connected State.** A state consisting of the following modes: sleep mode, active mode, and scanning mode. During the connected state, the MS maintains at least one transport connection and two management connections as established during access state, while the MS and BS may establish additional transport connections. Moreover, in order to reduce power consumption of the MS, the MS or BS can request a transition to sleep mode. Also, the MS can scan neighbor base stations to reselect a cell that provides more robust and reliable services.
- **Idle State.** A state comprising two separate modes: paging available mode and paging unavailable mode. During idle state, the MS may save power by switching between paging available mode and paging unavailable mode. In the paging available mode, the MS may be paged by the BS. If the MS is

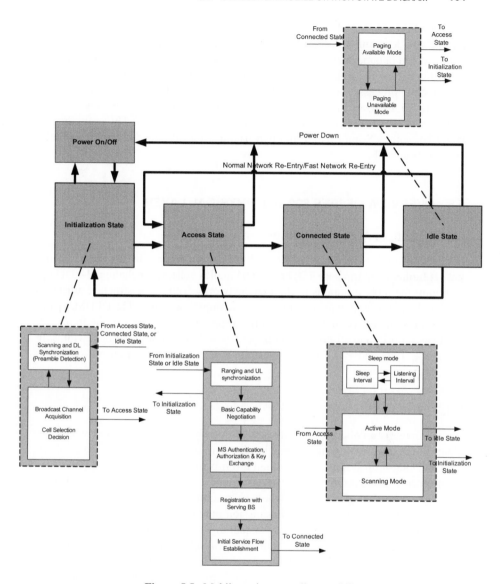

Figure 5.5. Mobile station state diagram [4].

paged, it transitions to the access state for its network reentry. The MS performs a location update procedure during idle state.

The MS state diagram for the IEEE 802.16m is similar to that of the reference system with the exception of the initialization state that has been simplified to reduce the scan latency and to enable fast cell selection or reselection. The location of the essential system configuration information is fixed so that upon successful DL synchronization, the essential system configuration information can be acquired.

This would enable the MS to make a decision for attachment to the BS without acquiring and decoding MAC management messages and waiting for the acquisition of the system parameters, resulting in power saving in the MS due to shortening and simplification of the initialization procedure. Although both normal and fast network reentry processes are shown as transition from the idle state to the access state in Figure 5.5, there are differences that differentiate the two processes. The network reentry is similar to network entry, except that it may be shortened by providing the target BS with MS information through paging controller or other network entity over the backhaul.

5.5 OVERVIEW OF IEEE 802.16m PHYSICAL LAYER

5.5.1 Multiple Access Schemes

IEEE 802.16m uses OFDMA as the multiple-access scheme in downlink and uplink. It further supports both TDD and FDD duplex schemes including H-FDD operation of the mobile stations in the FDD networks. The frame structure attributes and baseband processing are common for both duplex schemes. The OFDMA parameters are summarized in Table 5.4. Tone dropping at both edges of the frequency band based on 10 and 20 MHz systems can be used to support other bandwidths. In Table 5.4, TTG and RTG denote transmit/receive and receive/transmit transition gaps, respectively.

5.5.2 Frame Structure

A superframe is a collection of consecutive equally sized radio frames whose beginning is marked with a superframe header. The superframe header carries short-term and long-term system configuration information.

In order to decrease the air-link access latency, the radio frames are further divided into a number of subframes where each subframe is comprised of an integer number of OFDM symbols. The transmission time interval is defined as the transmission latency over the air link and is equal to a multiple of subframe length (default is one subframe). There are four types of subframes: (1) type-1 subframe, which consists of six OFDM symbols; (2) type-2 subframe, which consists of seven OFDM symbols; (3) type-3 subframe which consists of five OFDM symbols; and (4) type-4 subframe, which consists of nine OFDM symbols. In the basic frame structure shown in Figure 5.6, superframe length is 20 ms (comprised of four radio frames), radio frame size is 5 ms (comprised of eight subframes), and subframe length is 0.617 ms. The use of subframe concept with the latter parameter set would reduce the one-way air-link access latency to less than 10 ms.

The concept of time zones applies to both TDD and FDD systems. These time zones are time-division multiplexed across the time domain in the DL to support both new and legacy mobile stations. For UL transmissions, both time and frequency-division multiplexing approaches can be used to support legacy and new termi-

Table 5.4. IEEE 802.16m OFDM parameters [4, 13]

Nominal channel bandwidth (MHz)			5	7	8.75	10	20
Sampling factor			28/25	8/7	8/7	28/25	28/25
Sampling frequency (MHz)			5.6	8	10	11.2	22.4
FFT size			512	1024	1024	1024	2048
Sub-carrier spacing (kHz)			10.94	7.81	9.76	10.94	10.94
Useful symbol time T_u (μs)			91.429	128	102.4	91.429	91.429
		Symbol time T_s (μs)	102.857	144	115.2	102.857	102.857
CP $T_g = 1/8\ T_u$	FDD	Number of OFDM symbols per 5 ms frame	48	34	43	48	48
		Idle time (μs)	62.857	104	46.40	62.857	62.857
	TDD	Number of OFDM symbols per 5 ms frame	47	33	42	47	47
		TTG + RTG (μs)	165.714	248	161.6	165.714	165.714
		Symbol time T_s (μs)	97.143	136	108.8	97.143	97.143
CP $T_g = 1/16\ T_u$	FDD	Number of OFDM symbols per 5 ms frame	51	36	45	51	51
		Idle time (μs)	45.71	104	104	45.71	45.71
	TDD	Number of OFDM symbols per 5 ms frame	50	35	44	50	50
		TTG + RTG (μs)	142.853	240	212.8	142.853	142.853
		Symbol Time T_s (μs)	114.286	160	128	114.286	114.286
CP $T_g = 1/4\ T_u$	FDD	Number of OFDM symbols per 5 ms frame	43	31	39	43	43
		Idle time (μs)	85.694	40	8	85.694	85.694
	TDD	Number of OFDM symbols per 5 ms frame	42	30	37	42	42
		TTG + RTG (μs)	199.98	200	264	199.98	199.98

nals. The nonbackward-compatible improvements and features are restricted to the new zones. All backward-compatible features and functions are used in the legacy zones. In the absence of any legacy system, the legacy zones will disappear and the entire frame will be allocated to the new zones.

Coexistence between IEEE 802.16m and E-UTRA in TDD mode may be facilitated by inserting either idle symbols within the IEEE 802.16m frame or idle subframes, for certain E-UTRA TDD configurations [12]. An operator-configurable delay or offset between the beginning of an IEEE 802.16m frame and an E-UTRA TDD frame can be applied in some configurations to minimize the time allocated to idle symbols or idle subframes.

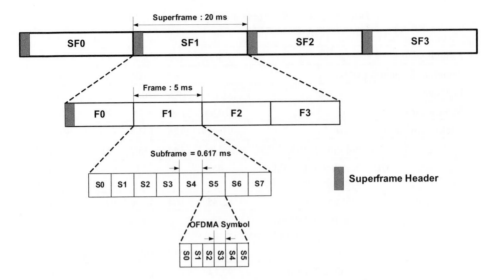

Figure 5.6. IEEE basic frame structure [4, 13].

The legacy and IEEE 802.16m frames are offset by a fixed number of subframes to accommodate new features such as the IEEE 802.16m synchronization channels (preambles), superframe header (system configuration information), and control channels, as shown in Figure 5.7. The "frame offset" is a time offset between the start of the legacy frame and the start of the IEEE 802.16m frame carrying the superframe header, defined in a unit of subframes. In the case where an IEEE 802.16m BS coexists with a legacy BSs, two switching points are inserted in each TDD radio frame.

The support for multiple RF carriers can be accommodated with the same frame structure used for single-carrier operation. All RF carriers are time aligned at the frame, subframe, and symbol levels (see Figure 5.3). Alternative frame structures for CP = 1/16 and CP = 1/4 are used that incorporate different numbers of OFDM symbols in certain subframes or different numbers of subframes per frame [4, 13].

5.5.3 Physical and Logical Resource Blocks

The downlink/uplink subframes are divided into a number of frequency partitions, where each partition consists of a set of physical resource units over the available number of OFDMA symbols in the subframe. Each frequency partition can include localized and/or distributed physical resource units. This concept is different than the legacy system, in which each zone can only accommodate localized or distributed subchannels. Frequency partitions can be used for different purposes such as fractional frequency reuse (FFR). The downlink/uplink resource petitioning and mapping is illustrated in Figure 5.8.

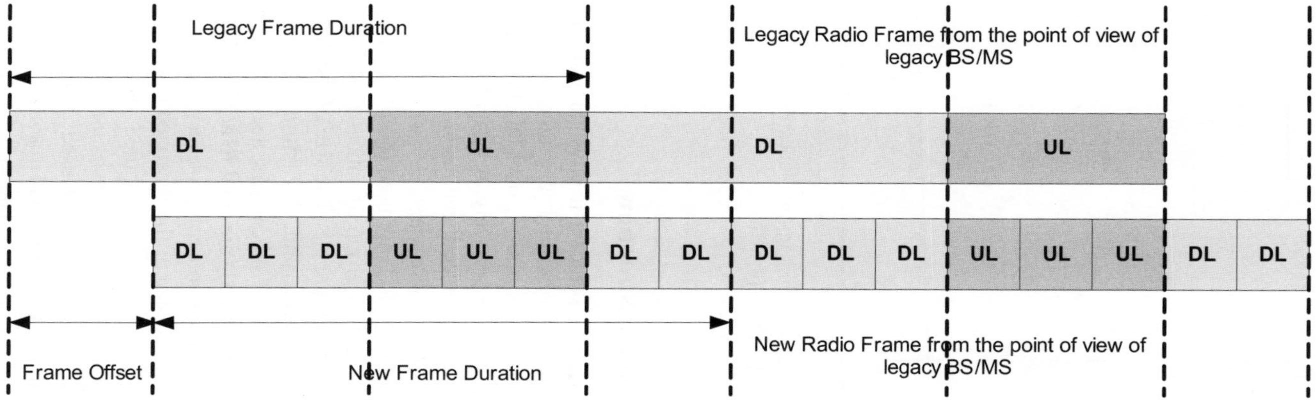

Figure 5.7. Relative position of the legacy and new frames (TDD duplex scheme DL/UL ratio 5:3) [4, 13].

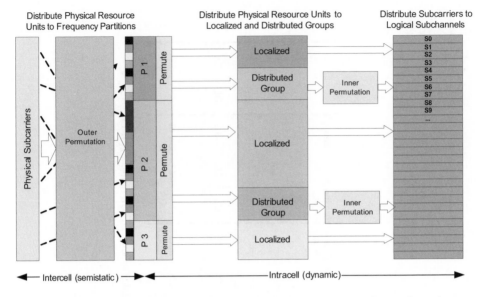

Figure 5.8. Illustration of the downlink/uplink subcarrier-to-resource unit mapping, where P_i denotes the ith frequency partition [4, 13].

5.5.3.1 Downlink Symbol Structure

A physical resource unit (PRU) is the basic physical unit for resource allocation that comprises P_{sc} contiguous subcarriers by N_{sym} contiguous OFDM symbols, where P_{sc} is 18 subcarriers and N_{sym} is 6, 7, 5, or 9 OFDM symbols for type-1, type-2, type-3, and type-4 subframes, respectively. A logical resource unit (LRU) is the basic logical unit for distributed and localized resource allocations. A logical resource unit is comprised of $P_{sc} \times N_{sym}$ subcarriers.

Distributed resource units (DRU) are used to achieve frequency diversity. A distributed resource unit contains a group of subcarriers that are spread across a frequency partition as shown in Figure 5.8. The size of the distributed resource units is equal to that of physical resource unit. The minimum unit for forming a DRU is equal to one subcarrier or a pair of subcarriers, also known as a tone-pair. Localized or contiguous resource units (CRU) are used to achieve frequency-selective scheduling gain. A localized resource unit comprises a group of subcarriers that are contiguous across frequency. The size of the localized resource units is equal to that of the physical resource units. To form distributed and localized resource units, the subcarriers over the OFDM symbols of a subframe are partitioned into guard and used subcarriers. The DC subcarrier is not used. The used subcarriers are divided into physical resource units. Each physical resource unit contains pilot and data subcarriers. The number of pilot and data subcarriers depends on MIMO mode, rank, and number of multiplexed mobile stations, as well as the number of OFDM symbols within a subframe.

The downlink subcarrier to resource unit mapping process is defined as follows and illustrated in Figure 5.8:

1. An outer permutation is applied to the PRUs in the units of N_1 and N_2 PRUs where $N_1 = 4$ and $N_2 = 1$.
2. The reordered PRUs are distributed into frequency partitions.
3. The frequency partition is divided into localized and/or distributed resource allocations. The sizes of the distributed/localized groups are flexibly configured per sector. Adjacent sectors are not required to have the same configuration for the localized and distributed groups.
4. The localized and distributed resource units are further mapped into LRUs by direct mapping for CRUs and by subcarrier permutation for DRUs.

The subcarrier permutation defined for the downlink distributed resource allocations spreads the subcarriers of the DRU across all the distributed resource allocations within a frequency partition. After mapping all pilots, the remaining used subcarriers are used to define the DRUs. To allocate the LRUs, the remaining subcarriers are paired into contiguous subcarrier pairs. Each LRU consists of a group of subcarrier pairs.

Let us assume that there are N_{RU} DRUs. A permutation sequence, P, for the distributed group is provided and the subchannelization for downlink distributed resource allocations is performed using the following procedure for every kth OFDMA symbol in the subframe:

1. Let n_k denote the number of pilot tones in the kth OFDM symbol within a PRU. n_k pilots in the kth OFDMA symbol within each PRU are allocated.
2. Let N_{RU} denote the number of DRUs within the frequency partition. The remaining $N_{RU}(P_{sc} - n_k)$ data subcarriers of the DRUs are renumbered in order from 0 to $N_{RU}(P_{sc} - n_k) - 1$ subcarriers.
3. The contiguous and logically renumbered subcarriers are grouped into N_{RU} $(P_{sc} - n_k)/2$ pairs and are renumbered from 0 to $N_{RU}(P_{sc} - n_k)/2 - 1$.
4. A subcarrier permutation formula is applied with the permutation sequence P or data subcarrier pairs.
5. Each set of logically contiguous $(P_{sc} - n_k)$ subcarriers is mapped into distributed LRUs and form a total of N_{RU} distributed LRUs.

5.5.3.2 Uplink Symbol Structure

Similar to the downlink symbol structure, a physical resource unit is defined as the basic physical unit for resource allocation that comprises P_{sc} consecutive subcarriers by N_{sym} consecutive OFDM symbols. P_{sc} is 18 subcarriers and N_{sym} is 6, 7, and 5 OFDM symbols for type-1, type-2, and type-3 subframes, respectively. A logical resource unit is the basic logical unit for distributed and localized resource allocations and its size is $P_{sc} \times N_{sym}$ subcarriers for data transmission. The subcarriers of an OFDM symbol are partitioned into $N_{g,left}$ left guard subcarriers, $N_{g,right}$ right guard subcarriers, and N_{used} used subcarriers. The DC subcarrier is not used. The N_{used} subcarriers over a subframe are divided into PRUs. Each PRU contains pilot and data subcarriers. The number of used pilot and data subcarriers depends on MIMO mode, rank, and number of mobile stations multiplexed, and the type of resource allocation; that is, distributed or localized resource allocations as well as the type of the subframe.

The uplink distributed units comprise a group of subcarriers that are spread over a frequency partition. The size of distributed unit is equal to logical resource blocks. The minimum unit for constructing a distributed resource unit is a tile. The uplink tile sizes are 6 subcarriers by 6 OFDM symbols. The localized resource unit is used to achieve frequency-selective scheduling gain. A localized resource unit contains a group of subcarriers that are contiguous across frequency. The size of the localized resource unit equals the size of the logical resource units for localized allocations, that is 18 subcarriers by 6 OFDM symbols.

The PRUs are first subdivided into subbands and minibands. A subband comprises N_1 adjacent PRUs and a miniband consists of N_2 adjacent PRUs, where $N_1 = 4$ and $N_2 = 1$. The subbands are suitable for frequency-selective allocations, whereas the minibands are suitable for frequency-diversity allocation and are permuted in frequency. The main features of resource mapping include support of CRUs and DRUs by frequency-division multiplexing, DRUs comprising multiple tiles spread across the distributed resource allocations to obtain frequency-diversity gain, and FFR may be applied in the uplink. The uplink resource unit mapping process is illustrated in Figure 5.8 and defined as follows:

1. An outer permutation is applied to the PRUs in the units of N_1 and N_2 PRUs.
2. The reordered PRUs are distributed into frequency partitions.
3. A frequency partition is divided into localized and/or distributed resource allocations. Sector-specific permutation can be applied and direct mapping of the resources can be supported for localized resources. The sizes of the distributed/localized groups are flexibly configured per sector. Adjacent sec-

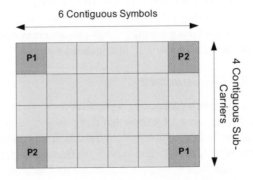

Figure 5.9. New uplink PUSC structure for two streams [4, 13].

tors do not need to have same configuration of localized and diversity resources.

4. The subcarriers in the localized and distributed resource allocations are further mapped into LRUs by direct mapping for CRUs and by tile permutation for DRUs.

An inner permutation is defined for the uplink distributed resource allocations; it spreads DRU tiles across all distributed resource allocations within a frequency partition. Each DRU in an uplink frequency partition is divided into three tiles of six adjacent subcarriers over N_{sym} symbols. The tiles within a frequency partition are collectively tile-permuted to obtain frequency diversity gain across the allocated resources.

The IEEE 802.16m uplink physical structure supports both frequency- and time-division multiplexing of the new and legacy system. If the legacy system uses PUSC or AMC subchannelization schemes, then the type of multiplexing can be either FDM or TDM. When the legacy system operates in the PUSC mode, a symbol structure based on the new PUSC subchannelization [4, 13] should be used in order to provide FDM-based legacy support. In that case, unlike the distributed structure, the new PUSC resource unit contains six tiles of four subcarriers by six symbols (shown in Figure 5.9).

5.5.4   Modulation and Coding

The performance of adaptive modulation generally suffers from the power inefficiencies of multilevel modulation schemes. This is due to the variations in bit reliabilities caused by the bitmapping onto the signal constellation. To overcome this issue, a constellation-rearrangement scheme is utilized whereby signal constellation of QAM signals between retransmissions is rearranged; that is, the mapping of the bits into the complex-valued symbols between successive HARQ retransmissions is changed, resulting in averaging bit reliabilities over several retransmissions and lower packet error rates. The mapping of bits to the constellation point depends on the constellation-rearrangement type used for HARQ retransmissions and may also depend on the MIMO scheme. The complex-valued modulated symbols are mapped to the input of the MIMO encoder. Incremental redundancy HARQ is used in determining the starting position of the bit selection for HARQ retransmissions.

Figure 5.10 shows the channel coding and modulation procedures. A cyclic re-

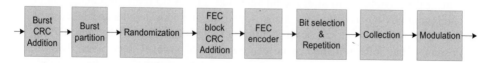

Figure 5.10. Coding and modulation procedures.

dundancy check (CRC) is appended to a burst (i.e., a physical layer data unit) prior to partitioning. The 16-bit CRC is calculated over all the bits in the burst. If the burst size including burst CRC exceeds the maximum FEC block size, the burst is partitioned into K_{FB} FEC blocks, each of which is encoded separately. If a burst is partitioned into more than one forward error correction (FEC) blocks, a FEC block CRC is appended to each FEC block before the FEC encoding. The FEC block CRC of a FEC block is calculated based on all of the bits in that FEC block. Each partitioned FEC block including the 16-bit FEC block CRC has the same length. The maximum FEC block size is 4800 bits. Concatenation rules are based on the number of information bits and do not depend on the structure of the resource allocation (number of logical resource units and their size). IEEE 802.16m utilizes the convolutional turbo code (CTC) with code rate of 1/3 as defined in [1]. The CTC scheme is extended to support additional FEC block sizes. Furthermore, the FEC block sizes can be regularly increased with predetermined block size resolutions. The FEC block sizes that are multiple of seven are removed for the tail-biting encoding structure. The encoder block depicted in Figure 5.10 includes the interleaver.

Bit selection and repetition are used in IEEE 802.16m to achieve rate matching. Bit selection adapts the number of coded bits to the size of the resource allocation, which may vary depending on the resource unit size and subframe type. The total subcarriers in the allocated resource unit are segmented to each FEC block. The total number of information and parity bits generated by the FEC encoder are considered to be the maximum size of circular buffer. Repetition is performed when the number of transmitted bits is larger than the number of selected bits. The selection of coded bits is done cyclically over the buffer. Table 5.5 summarizes the coding and modulation schemes used in IEEE 802.16m. The mother-code bits, the total number of information and parity bits generated by the FEC encoder, are considered to be the maximum size of the circular buffer. If the size of the circular buffer N_{buffer} is smaller than the number of mother-code bits, the first N_{buffer} bits of mother-code bits are considered to be selected bits.

Modulation constellations of QPSK, 16QAM, and 64QAM are supported as defined in IEEE 802.16-2009 standard. The mapping of bits to the constellation point depends on the constellation-rearrangement (CoRe) version used for HARQ retransmission as described and further depends on the MIMO scheme. The QAM symbols are mapped into the input of the MIMO encoder. Only the burst sizes N_{DB} listed in Table 5.5 are supported in the physical layer. The sizes include the addition of CRCs (per burst and per FEC block), if applicable. Other sizes require padding to the next burst size. The code rate and modulation depend on the burst size and the resource allocation.

Incremental redundancy HARQ (HARQ-IR) is used in IEEE 802.16m by determining the starting position of the bit selection for HARQ retransmissions. Chase combining HARQ (HARQ-CC) is also supported and considered a special case of HARQ-IR. The two-bit subpacket identifier (SPID) is used to identify the starting position. The CoRe scheme can be expressed by a bit-level interleaver. The resource allocation and transmission formats in each retransmission in the downlink can be adapted with control signaling. The resource allocation in each retransmis-

Table 5.5. Supported burst sizes in IEEE 802.16m

Index	N_{DB} (Byte)	K_{FB}	Index	N_{DB} (Byte)	K_{FB}	Index	N_{DB} (Byte)	K_{FB}
1	6	1	23	90	1	45	1200	2
2	8	1	24	100	1	46	1416	3
3	9	1	25	114	1	47	1584	3
4	10	1	26	128	1	48	1800	3
5	11	1	27	145	1	49	1888	4
6	12	1	28	164	1	50	2112	4
7	13	1	29	181	1	51	2400	4
8	15	1	30	205	1	52	2640	5
9	17	1	31	233	1	53	3000	5
10	19	1	32	262	1	54	3600	6
11	22	1	33	291	1	55	4200	7
12	25	1	34	328	1	56	4800	8
13	27	1	35	368	1	57	5400	9
14	31	1	36	416	1	58	6000	10
15	36	1	37	472	1	59	6600	11
16	40	1	38	528	1	60	7200	12
17	44	1	39	600	1	61	7800	13
18	50	1	40	656	2	62	8400	14
19	57	1	41	736	2	63	9600	16
20	64	1	42	832	2	64	10800	18
21	71	1	43	944	2	65	12000	20
22	80	1	44	1056	2	66	14400	24

sion in the uplink can be fixed or adaptive according to control signaling. In HARQ retransmissions, the bits or symbols can be transmitted in a different order to exploit the frequency diversity of the channel.

For HARQ retransmission, the mapping of bits or modulated symbols to spatial streams may be applied to exploit spatial diversity with a given mapping pattern, depending on the type of HARQ-IR. In this case, the predefined set of mapping patterns should be known to the transmitter and receiver. In the downlink HARQ, the BS may transmit coded bits exceeding the currently available soft buffer capacity.

5.5.5 Pilot Structure

5.5.5.1 Downlink Pilot Structure

Transmission of pilot subcarriers in the downlink/uplink is necessary to allow channel estimation, channel quality measurement (e.g., CQI), frequency offset estimation, and so on. To optimize the system performance in different propagation environments, IEEE 802.16m supports both common and dedicated pilot structures. The classification of pilots into common and dedicated is done based on their usage. The common pilots can be used by all mobile stations. Dedicated pilots can be used with both localized and distributed allocations. Pilot subcarriers that can be used only by a group of mobile stations within an FFR group are a special case of common pilots. The dedicated pilots are associated with a specific FFR group and can be only used by the mobile stations assigned to that group; therefore, they can be precoded or beam-formed similar to the data subcarriers. The pilot structure is defined for up to eight spatial streams and there is a unified design for common and dedicated pilots. There is equal pilot density per spatial stream; however, there is not necessarily equal pilot density per OFDMA symbol. There are the same number of pilots for each physical resource unit allocated to a particular mobile station (see Figure 5.11).

For the subframe consisting of five or seven OFDM symbols, one of the OFDMA symbols is deleted or repeated. To overcome the effects of pilot interference among the neighboring sectors or base stations, an interlaced pilot structure is utilized by cyclically shifting the base pilot pattern such that the pilots of neighboring cells do not overlap.

E-MBS-zone-specific pilots are transmitted for multicell multicast broadcast, single-frequency network (SFN) transmissions. An E-MBS zone is a group of base stations involved in an SFN transmission. The E-MBS-zone-specific pilots are common within an E-MBS zone but different between neighboring E-MBS zones. Synchronous transmissions of the same content with a common pilot pattern from multiple base stations in an E-MBS zone would result in improved channel estimation. The E-MBS-zone-specific pilot patterns depend on the maximum number of transmit streams. Pilot structures up to two transmit streams are supported.

5.5.5.2 Uplink Pilot Structure

The uplink pilots are dedicated to localized and distributed resource units and are precoded using the same precoding as the data subcarriers of the resource allocation. The pilot structure is defined for up to four spatial streams with orthogonal patterns. The pilot pattern may support variable power boosting. When pilots are power-boosted, each data subcarrier should have the same transmission power across all OFDMA symbols in a resource block. The 18×6 uplink resource blocks use the same pilot patterns as the downlink counterpart for up to four spatial streams. The pilot pattern for the 6×6 tile structure is different and is shown in Figure 5.11.

5.5.6 Control Channels

Downlink control channels carry essential information for system operation. Depending on the type of control signaling, information is transmitted over different time intervals (i.e., from superframe to subframe intervals). The system configuration parameters are transmitted at the superframe intervals, whereas control signaling related to user data allocations is transmitted at the frame/subframe intervals. In mixed-mode operation, an IEEE 802.16m mobile station can access the system

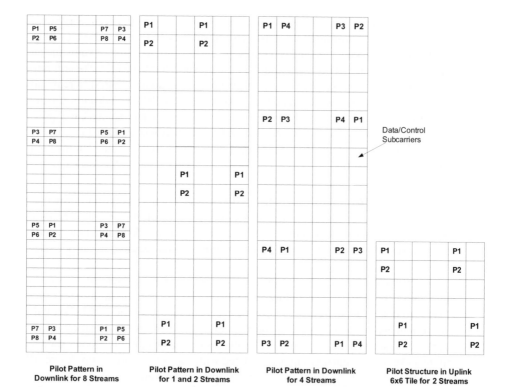

Figure 5.11. Downlink/uplink pilot structures for 1, 2, 4, and 8 streams for the type-1 subframe.

without decoding legacy frame control header and legacy DL/UL-MAP messages [1, 4].

5.5.6.1 Downlink Control Channels

Superframe Header. The superframe header carries essential system parameters and configuration information. The content of superframe header is divided into two segments: primary and secondary superframe headers. The information transmitted in the secondary superframe header is further divided into different subpackets. The primary superframe header is transmitted every superframe, whereas the secondary superframe header is transmitted over one or more superframes. The primary and secondary superframe headers are located in the first subframe within a superframe and are time-division-multiplexed with the advanced preamble. The superframe header occupies narrower bandwidth relative to the system bandwidth (i.e., 5 MHz bandwidth). The primary superframe header is transmitted using a predetermined modulation and coding scheme. The secondary superframe header is transmitted using a predetermined modulation scheme and its repetition coding factor is signaled in the primary superframe header. The primary and secondary superframe headers are transmitted using two spatial streams and space-frequency block coding to improve coverage and reliability. The MS is not required to know the antenna configuration prior to decoding the primary superframe header. The information transmitted in the secondary superframe header is divided into different subpackets. The secondary superframe header subpacket 1 (SP1) includes information needed for network reentry. The secondary superframe header subpacket 2 (SP2) contains information for initial network entry. The secondary superframe header subpacket 3 (SP3) contains the remaining system information for maintaining communication with the BS.

Advanced MAP (A-MAP). The advanced MAP (i.e., unicast control information) consists of both user-specific and nonuser-specific control information. Nonuser-specific control information includes information that is not dedicated to a specific user or a specific group of users. It contains information required to decode user-specific control signaling.

User-specific control information consists of information intended for one or more users. It includes scheduling assignment, power control information, and HARQ feedback. Resources can be allocated persistently to the mobile stations. The periodicity of the allocation is configurable. Group control information is used to allocate resources and/or configure resources to one or multiple mobile stations within a user group. Each group is associated with a set of resources. Voice over IP (VoIP) is an example of the class of services that can take advantage of group messages. Within a subframe, control and data channels are frequency-division-multiplexed. Both control and data channels are transmitted on logical resource units that span over all OFDMA symbols within a subframe [4, 13].

Each DL subframe contains a control region including both nonuser-specific and user-specific control information. All advanced MAPs share a physical time-frequency region known as the A-MAP region. The control regions are located in every subframe. The corresponding UL allocations occurs L subframes later, where L is determined by A-MAP relevance. The coding rate of the control blocks is known to the MS through group size indication in nonuser-specific control information in order to reduce the complexity of blind detection by the MS.

An advanced MAP (A-MAP) allocation information element (IE) is defined as the basic element of unicast service control. A unicast control IE may be addressed to one user using a unicast identifier or to multiple users using a multicast/broadcast identifier. The identifier is masked with CRC in the advanced MAP allocation information element. It may contain information related to resource allocation, HARQ, MIMO transmission mode, and so on. Each unicast control information element is coded separately. Note that this method is different from the legacy system control mechanism in which the information elements of all users are jointly coded. Nonuser-specific control information is encoded separately from the user-specific control information. The transmission format of nonuser-specific control information is predetermined. In the DL subframes, each frequency partition may contain an A-MAP region. An A-MAP region occupies the first few distributed resource units in a frequency partition. The structure of an A-MAP region is illustrated in Figure 5.12. The resource occupied by each A-MAP physical channel may vary depending on the system configuration and scheduler operation. There are different types of A-MAPs as follows:

- **Assignment A-MAP** (A-A-MAP) contains resource assignment information that is categorized into multiple types of resource assignment IEs. Each assignment A-MAP IE is coded separately and carries information for one or a group of users. The size of the assignment A-MAP is indicated by nonuser-specific A-MAP.
- **HARQ Feedback A-MAP** (HF-A-MAP) contains HARQ ACK/NACK information for UL data transmission.
- **Power Control A-MAP** includes fast power control command to mobile stations.

There are different A-A-MAP IE types that distinguish between DL/UL, persistent/nonpersistent, single user/group resource allocation, and basic/extended IE scenarios.

5.5.6.2   Uplink Control Channels

MIMO Feedback. MIMO feedback provides information obtained through MS measurements on the wideband and/or narrowband spatial characteristics of the channel that are required for MIMO operation. The MIMO mode, precoding matrix index, rank adaptation information, channel covariance matrix elements, power loading factor, eigenvectors, and channel sounding are examples of MIMO feedback information.

HARQ Feedback. HARQ feedback (ACK/NACK) is used to acknowledge downlink data transmissions. The uplink HARQ feedback channel starts at a predetermined offset with respect to the corresponding downlink transmission. The HARQ feedback channel is frequency-division-multiplexed with other control and data channels. Orthogonal codes are used to multiplex multiple HARQ feedback channels. The HARQ feedback channel comprises three distributed minitiles.

Bandwidth Request. Bandwidth requests are used to indicate the amount of bandwidth required by a user. Bandwidth requests are transmitted through indicators or messages. A bandwidth request indicator notifies the base station of an uplink grant request by the mobile station that is sending the indicator. Bandwidth request messages can include information about the status of queued traffic at the mobile station, such as buffer size and quality of service parameters.

Contention- or noncontention-based random access is used to transmit bandwidth request information on this control channel. The contention-based bandwidth

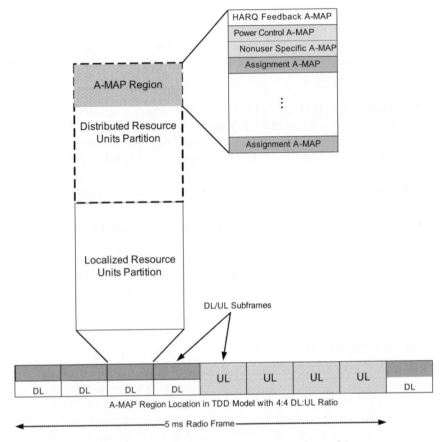

Figure 5.12. A-MAP location and structure (example).

request procedure is illustrated in Figure 5.13. A five-step regular procedure or an optional three-step quick access procedure is utilized. Steps 2 and 3 can be skipped in a quick access procedure. In step 1, the MS sends a bandwidth request (BW-REQ) indicator for quick access that may indicate information such as MS addressing, and/or request size and/or uplink transmit power report, and/or QoS parameters, and the BS may allocate an uplink grant based on a certain policy. The MS may piggyback additional BW-REQ information along with user data during uplink transmission (step 5). In steps 2 and 4, the BS acknowledges receipt of the messages (see Figure 5.13).

The bandwidth request channel starts at a configurable location with the configuration defined in a downlink broadcast control message. The bandwidth request channel is frequency-division-multiplexed with other uplink control and data channels. A BW-REQ tile is defined as six subcarriers by six OFDM symbols. Each BW-REQ channel consists of three distributed BW-REQ tiles. Multiple bandwidth

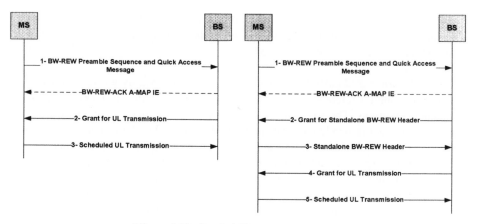

Figure 5.13. Bandwidth request procedure.

request indicators can be transmitted on the same BW-REQ channel using code-division multiplexing.

Channel Quality Indicators. Channel quality feedback provides information about channel conditions as seen by the user. This information is used by the base station for link adaptation, resource allocation, power control, and so on. The channel quality measurement includes both narrowband and wideband measurements. The CQI feedback overhead can be reduced through differential feedback or other compression techniques. Examples of CQI include physical carrier to interference plus noise ratio (CINR), effective CINR, and band selection.

The default subframe size for transmission of uplink control information is six symbols. The fast feedback channel carries channel quality feedback and MIMO feedback. There are two types of uplink fast feedback control channels: primary and secondary fast feedback channels. The primary fast feedback channel provides wideband feedback information, including channel quality and MIMO feedback. The secondary fast feedback control channel carries narrowband CQI and MIMO feedback information. The secondary fast feedback channel can be used to support CQI reporting at higher code rates and, thus, more CQI information bits. The fast feedback channel is frequency-division-multiplexed with other uplink control and data channels.

The fast feedback channel starts at a predetermined location, with the size defined in a downlink broadcast control message. Fast feedback allocations to a mobile station can be periodic and the allocations are configurable. For periodic allocations, the specific type of feedback information carried on each fast feedback opportunity can be different. The secondary fast feedback channel can be allocated in a nonperiodic manner based on traffic or channel conditions. The number of bits carried in the fast feedback channel can be adaptive. For efficient transmission of feedback channels, a minitile is defined comprising two subcarriers by six OFDM symbols. One logical resource unit consists of nine minitiles and can be shared by multiple fast feedback channels [4, 13].

Uplink Sounding Channel. The sounding channel is used by a user terminal to transmit sounding reference signals to enable the base station to measure uplink channel conditions. The sounding channel may occupy either specific uplink subbands or the entire bandwidth over an OFDM symbol. The base station can configure a mobile station to transmit the uplink sounding signal over predefined subcarriers within specific subbands or the entire bandwidth. The sounding channel is orthogonally multiplexed (in time or frequency) with other control and data channels. Furthermore, the base station can configure multiple user terminals to transmit sounding signals on the corresponding sounding channels using code-, frequency-, or time-division multiplexing. Power control for the sounding channel can be utilized to adjust the sounding quality. The transmit power from each mobile terminal may be separately controlled according to certain CINR target values [4, 13].

Ranging Channel. The ranging channel is used for uplink synchronization. The ranging channel can be further classified into ranging for nonsynchronized and synchronized mobile stations. A random access procedure, which can be contention- or noncontention-based, is used for ranging. Contention-based random access is used for initial ranging. Noncontention based random access is used for periodic ranging and handover. The ranging channel for nonsynchronized mobile stations starts at a configurable location with the configuration signaled in a downlink broadcast control message. The ranging channel for nonsynchronized mobile stations is frequency-division multiplexed with other uplink control and data channels.

The ranging channel for nonsynchronized mobile stations consists of three fields: (1) ranging cyclic prefix (RCP), (2) ranging preamble (RP), and (3) guard time (GT). The length of RCP is longer than the sum of the maximum delay spread and round-trip delay of supported cell size. The length of GT is chosen longer than the round-trip delay of the supported cell size. The length of ranging preamble is chosen to be equal to or longer than the length of RCP. To support large cell sizes, the ranging channel for nonsynchronized user terminals can span multiple concatenated subframes. Figure 5.14 shows the default ranging channel structure over one subframe. A single preamble can be used by different nonsynchronized users for increasing ranging opportunities. When the preamble is repeated as a single opportunity, the second RCP can be omitted for coverage extension. A number of guard subcarriers are reserved at the edges of the nonsynchronized ranging channel physical resource.

When multiantenna transmission is supported by a mobile station, it can be used to increase the ranging opportunity by using spatial multiplexing. The ranging channel for synchronized mobile stations is used for periodic ranging. The ranging channel for synchronized user terminals starts at a configurable location with the configuration signaled in a downlink broadcast control message. The ranging channel for synchronized mobile stations is frequency-division-multiplexed with other uplink control and data channels.

Power Control. The power control mechanism is supported for downlink and uplink. The base station controls the transmit power per subframe and per user. Using downlink power control, user-specific information is received by the terminal with the controlled power level. The downlink advanced MAPs are power-controlled based on the terminal uplink channel quality feedback. The per-pilot-subcarrier power and the per-data-subcarrier power can jointly be adjusted for adaptive downlink power control. In the case of dedicated pilots, this is done on a per-user basis and in the case of common pilots this is done jointly for the users sharing the pilots.

The uplink power control is supported to compensate the path loss, shadowing, fast fading, and implementation loss, as well as to mitigate intercell and intracell interference. The uplink power control includes open-loop and closed-loop power control mechanisms. The base station can transmit necessary information through the control channel or message to terminals to support uplink power control. The parameters of the power control algorithm are optimized on system-wide basis by the base station and broadcast periodically or trigged by certain events.

In high-mobility scenarios, the power control scheme may not be able to compensate for the fast fading channel effect because of the variations of the channel impulse response. As a result, the power control is used to compensate for the distance-dependent path loss, shadowing, and implementation loss only.

The channel variations and implementation loss are compensated via open-loop power control without frequently interacting with the base station. The terminal can determine the transmit power based on the transmission parameters sent by the serving base station, uplink channel transmission quality, downlink channel state information, and interference knowledge obtained from the downlink. Open-loop power control provides a coarse initial power setting of the terminal when an initial connection is established.

The dynamic channel variations are compensated through closed-loop power control with power control commands from the serving base station. The base station measures the uplink channel state and interference information using uplink

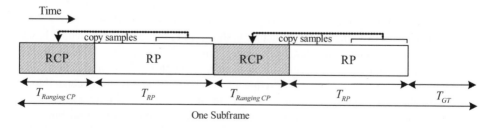

Figure 5.14. Ranging channel structure for nonsynchronized access [4, 13].

data and/or control channel transmissions and sends power control commands to the terminal. The terminal adjusts its transmission power based on the power control commands from the base station.

5.5.7   Downlink Synchronization (Advanced Preamble)

IEEE 802.16m utilizes a new hierarchical structure for the DL synchronization in which two sets of preambles at superframe and frame intervals are transmitted (Figure 5.15). The first set of preamble sequences mark the beginning of the superframe and are common to a group of sectors or cells. The primary advanced preamble carries information about system bandwidth and carrier configuration. The primary advanced preamble has a fixed bandwidth of 5 MHz and can be used to facilitate location-based services. A frequency reuse of one is applied to the primary advanced preamble in the frequency domain. The second set of advanced preamble sequences (secondary advanced preamble) is repeated every frame, spans the entire system bandwidth, and carries the cell ID. A frequency reuse of three is used for this set of sequences to mitigate inter-cell interference. The secondary advanced preambles carry 768 distinct cell IDs. Secondary advanced preamble sequences are partitioned and each partition is dedicated to a specific BS type such as macro BS, femto BS, and so on. The partition information is broadcast in the secondary superframe header.

5.5.8   Multiantenna Techniques in IEEE 802.16m

5.5.8.1   Downlink MIMO Structure

IEEE 802.16m supports several advanced multiantenna techniques including single and multiuser MIMO (spatial multiplexing and beamforming) as well as a number of transmit diversity schemes. In the single-user MIMO (SU-MIMO) scheme, only one user can be scheduled in one (time, frequency, space) resource unit. In multiuser MIMO (MU-MIMO), on the other hand, multiple users can be scheduled in one resource unit. Vertical encoding (or single codeword) utilizes one encoder block (or layer), whereas horizontal encoding (or multicodeword) uses multiple encoders (or multiple layers). A layer is defined as a coding and modulation input path to the MIMO encoder. A stream is defined as the output of the MIMO encoder that is further processed through the precoder block [11]. For spatial multiplexing, the rank is defined as the number of streams to be used for the user. Each SU-MIMO or MU-MIMO open-loop or closed-loop schemes is defined as a MIMO mode.

The DL MIMO transmitter structure is shown in Figure 5.16. The encoder block contains the channel encoder, interleaving, rate-matching, and modulating blocks per layer. The resource mapping block maps the complex-valued modulation symbols to the corresponding time-frequency resources. The MIMO encoder block maps the layers onto the streams, which are further processed through the beamforming or precoder block. The precoder block maps the streams to antennas by generating the antenna-specific data symbols according to the selected MIMO

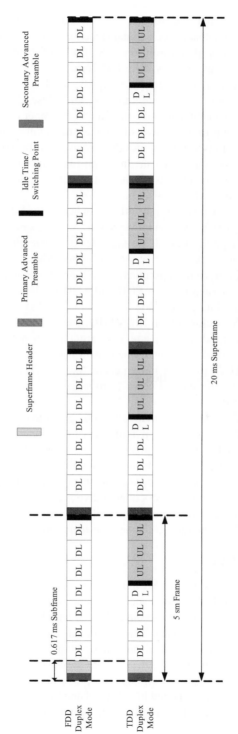

Figure 5.15. Structure of the IEEE 802.16m advanced preambles.

mode. The OFDM symbol construction block maps antenna-specific data to the OFDM symbols. The feedback block contains feedback information such as CQI or channel state information (CSI) from the MS. Table 5.6 contains information on various MIMO modes supported by IEEE 802.16m.

The minimum antenna configuration in the DL and UL is 2×2 and 1×2, respectively. For open-loop spatial multiplexing and closed-loop SU-MIMO, the number of streams is constrained to the minimum of number of transmit or receive antennas. For open-loop transmit diversity modes, the number of streams depends on the space-time coding (STC) schemes that are used by the MIMO encoder. The MU-MIMO can support up to two streams with two transmit antennas and up to four streams for four and eight transmit antennas. Table 5.7 summarized DL MIMO parameters for various downlink MIMO modes.

For SU-MIMO, vertical encoding is utilized, whereas for MU-MIMO horizontal encoding is employed at the BS and only one stream is transmitted to each MS. The stream-to-antenna mapping depends on the MIMO scheme that is used. CQI and rank feedback are transmitted to assist the BS in rank adaptation, mode switching, and rate adaptation. For spatial multiplexing, the rank is defined as the number of streams to be used for each user. In FDD and TDD systems, a unitary codebook based precoding is used for closed-loop SU-MIMO. An MS may feedback some information to the BS in closed-loop SU-MIMO, such as rank, subband selection, CQI, precoding matrix index (PMI), and long-term channel state information.

The MU-MIMO transmission with one stream per user is supported. The MU-MIMO schemes include two transmit antennas for up to two users, and four and eight transmit antennas for up to 4 users. Both unitary and nonunitary MU-MIMO schemes are supported.

If the columns of the precoding matrix are orthogonal to each other, it is defined as unitary MU-MIMO. Otherwise, it is defined as nonunitary MU-MIMO [11]. Beam forming is enabled with this precoding mechanism. IEEE 802.16m has the capability to adapt between SU-MIMO and MU-MIMO in a predefined and flexible manner. Multi-BS MIMO techniques are also supported for improving sector and cell-edge throughput using multi-BS collaborative precoding, network coordinated beamforming, or intercell interference cancellation. Both open-loop and closed-loop multi-BS MIMO techniques are under consideration.

5.5.8.2 Uplink MIMO

The block diagram of the uplink MIMO transmitter is illustrated in Figure 5.17. Note the similarities of MIMO baseband processing in the downlink and uplink.

The BS will schedule users to resource blocks and determines the modulation and coding scheme (MCS) level and MIMO parameters (mode, rank, etc.). The supported antenna configurations include one, two, or four transmit antennas and more than two receive antennas. In the UL, the MS measurements of the channel are based on DL reference signals (e.g., common pilots or a midamble). UL MIMO modes and parameters are contained in Table 5.8 and Table 5.9.

A number of antenna configurations and transmission rates are supported in UL

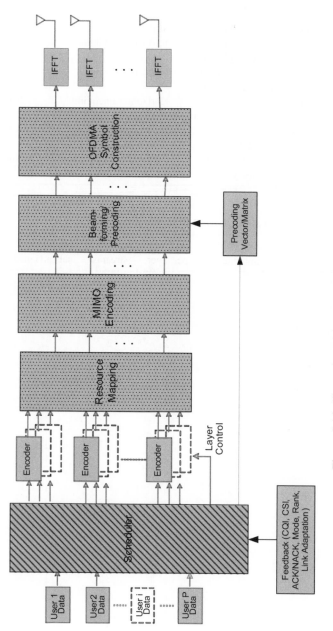

Figure 5.16. Illustration of downlink MIMO structure [4, 13].

Table 5.6. DL MIMO modes [4, 13]

Mode index	Description	MIMO encoding format	MIMO precoding
Mode 0	Open-Loop SU-MIMO	SFBC	Nonadaptive
Mode 1	Open-Loop SU-MIMO (Spatial Multiplexing)	Vertical Encoding	Nonadaptive
Mode 2	Closed-Loop SU-MIMO (Spatial Multiplexing)	Vertical Encoding	Adaptive
Mode 3	Open-Loop MU-MIMO (Spatial Multiplexing)	Horizontal Encoding	Nonadaptive
Mode 4	Closed-Loop MU-MIMO (Spatial Multiplexing)	Horizontal Encoding	Adaptive
Mode 5	Open-Loop SU-MIMO (TX Diversity)	Conjugate Data Repetition	Nonadaptive

Table 5.7. DL MIMO parameters [4, 13]

	Number of transmit antennas	STC rate per layer	Number of streams	Number of subcarriers	Number of layers
MIMO Mode 0	2	1	2	2	1
	4	1	2	2	1
	8	1	2	2	1
MIMO Mode 1 and MIMO Mode 2	2	1	1	1	1
	2	2	2	1	1
	4	1	1	1	1
	4	2	2	1	1
	4	3	3	1	1
	4	4	4	1	1
	8	1	1	1	1
	8	2	2	1	1
	8	3	3	1	1
	8	4	4	1	1
	8	5	5	1	1
	8	6	6	1	1
	8	7	7	1	1
	8	8	8	1	1
MIMO Mode 3 and MIMO Mode 4	2	1	2	1	2
	4	1	2	1	2
	4	1	3	1	3
	4	1	4	1	4
	8	1	2	1	2
	8	1	3	1	3
	8	1	4	1	4
MIMO Mode 5	2	1/2	1	2	1
	4	1/2	1	2	1
	7	1/2	1	2	1

open-loop SU-MIMO, including two and four transmit antennas with rate 1 (i.e., transmit diversity mode) and two and four transmit antennas with rates 2, 3, and 4 (i.e., spatial multiplexing). The supported UL transmit diversity modes include two and four transmit antenna schemes with rate 1 such as space frequency block coding (SFBC) and rank 1 precoder.

The multiplexing modes supported for open-loop single-user MIMO include two and four transmit antenna rate 2 schemes with and without precoding, four transmit antenna rate 3 schemes with precoding, and four transmit antenna rate 4 schemes. In FDD and TDD systems, unitary codebook-based precoding is supported. In this mode, an MS transmits a sounding reference signal in the UL to assist in the UL scheduling and precoder selection in the BS. The BS signals the resource allocation, MCS, rank, preferred precoder index, and packet size to the MS. UL MU-MIMO enables multiple mobile stations to be spatially multiplexed on the same radio resources. Both open-loop and closed-loop MU-MIMO are supported. The mobile stations with a single transmit antenna can operate in open-loop MU-MIMO mode.

5.6 OVERVIEW OF THE IEEE 802.16M MAC LAYER

There are various MAC functionalities and features that are specified by the IEEE 802.16m standard, some of which are extensions of the existing features in the mobile WiMAX [1, 5]. The following sections briefly describe selected MAC features.

5.6.1 MAC Addressing

IEEE 802.16m standard defines global and logical addresses for a mobile station that identify the user and its connections during a session. The MS is identified by the globally unique 48-bit IEEE extended unique identifier assigned by the IEEE Registration Authority [1]. The mobile station is further assigned the following logical identifiers: (1) a station identifier during network entry (or network reentry) that uniquely identifies the MS within the cell, and (2) a flow identifier that uniquely identifies the management connections and transport connections with the MS.

5.6.2 Network Entry

Network entry is the procedure through which a mobile station detects a cellular network and establishes a connection with that network. The network entry has the following steps (see Figure 5.5):

- Synchronization with the BS by acquiring the preambles
- Acquiring necessary system information such as BS and network service provider identifiers for initial network entry and cell selection
- Initial ranging
- Basic capability negotiation

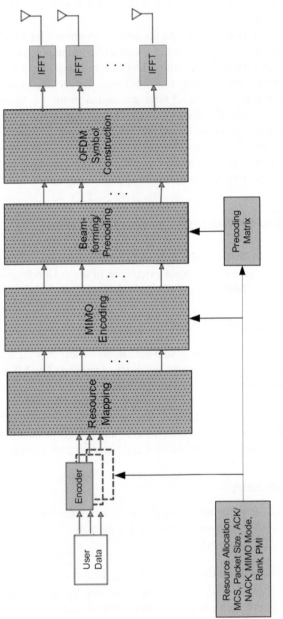

Figure 5.17. Illustration UL MIMO structure [4, 13].

Table 5.8. UL MIMO modes [4, 13]

Mode index	Description	MIMO encoding format	MIMO precoding
Mode 0	Open-Loop SU-MIMO	SFBC	Nonadaptive
Mode 1	Open-Loop SU-MIMO (Spatial Multiplexing)	Vertical Encoding	Nonadaptive
Mode 2	Closed-Loop SU-MIMO (Spatial Multiplexing)	Vertical Encoding	Adaptive
Mode 3	Open-Loop Collaborative Spatial Multiplexing (MU-MIMO)	Vertical Encoding	Nonadaptive
Mode 4	Closed-Loop Collaborative Spatial Multiplexing (MU-MIMO)	Vertical Encoding	Adaptive

- Authentication/authorization and key exchange
- Registration and service flow setup

Neighbor search is based on the same downlink signals as initial network search except that some information can be provided by the serving BS (i.e., neighbor advertisement messages, NBR-ADV). The BS responds to the MS initial ranging code transmission by broadcasting a status indication message in the following predefined downlink frame/subframe.

Table 5.9. UL MIMO parameters

	Number of transmit antennas	STC rate per layer	Number of streams	Number of subcarriers	Number of layers
MIMO Mode 0	2	1	2	2	1
	4	1	2	2	1
MIMO Mode 1 and	2	1	1	1	1
MIMO Mode 2	2	2	2	1	1
	4	1	1	1	1
	4	2	2	1	1
	4	3	3	1	1
	4	4	4	1	1
MIMO Mode 3 and	2	1	1	1	1
MIMO Mode 4	4	1	1	1	1
	4	2	2	1	1
	4	3	3	1	1

5.6.3 Connection Management

Connections are identified by the combination of station identifier and flow identifier. Two types of connections (i.e., management and transport connections) are specified. Management connections are used to carry MAC management messages. Transport connections are used to carry user data including upper layer signaling messages and data-plane signaling such as ARQ feedback. Fragmentation and augmentation of the MAC SDUs are supported on unicast connections.

Management connection is bidirectional and predefined values of flow identifier are reserved for unicast management connections. Management connections are automatically established after a station identifier is assigned to an MS during initial network entry. Transport connections are unidirectional and are established with a unique flow identifier assigned during the service-flow-establishment procedure. Each active service flow is uniquely mapped to a transport connection.

5.6.4 Quality of Service

IEEE 802.16m MAC assigns a unidirectional flow of packets with specific QoS requirements to a service flow. A service flow is mapped to a transport connection with a flow identifier. The QoS parameter set is negotiated between the BS and the MS during the service flow setup/change procedure. The QoS parameters can be used to schedule traffic and allocate radio resources. The uplink traffic may be regulated based on the QoS parameters.

IEEE 802.16m supports adaptation of service flow QoS parameters. One or more sets of QoS parameters are defined for each service flow. The MS and BS negotiate the possible QoS parameter sets during the service flow setup procedure. When QoS requirement/traffic characteristics for uplink traffic change, the BS may adapt the service flow QoS parameters, such as grant/polling interval or grant size, based on predefined rules. In addition, the MS may request the BS to switch the service flow QoS parameter set with explicit signaling. The BS then allocates resources according to the new service flow parameter set. In addition to the scheduling services supported by the legacy system, IEEE 802.16m provides a specific scheduling service and dedicated ranging channel to support real-time, non-periodical applications such as interactive gaming.

5.6.5 MAC Management Messages

To satisfy the latency requirements for network entry, handover, and state transitions IEEE 802.16m supports fast and reliable transmission of MAC management connections. The transmission of unicast MAC management messages is made more reliable using HARQ, whereby retransmissions can be triggered by an unsuccessful outcome from the HARQ entity in the transmitter. If the MAC management message is fragmented into multiple MAC service data units, only unsuccessful fragments are retransmitted.

5.6.6   MAC Header

IEEE 802.16m specifies a number of efficient MAC headers for various applications comprised of fewer fields and shorter size compared to the IEEE 802.16-2009 standard generic MAC header [1]. The advanced generic MAC header consists of extended header indicator, flow identifier, and payload length fields. Other MAC header types include two-byte short-packet fragmentation extended headers for management connections, and multiplexing extended header that is used when data from multiple connections associated with security association is present in the payload of the MAC PDU. The structures of some MAC headers are shown in Figure 5.18.

5.6.7   ARQ and HARQ Functions

An ARQ block is generated from one or multiple MAC service data units (SDUs) or MAC SDU fragment(s). ARQ blocks can be variable in size and are sequentially numbered. When both ARQ and HARQ are applied on a flow, HARQ and ARQ interactions are applied to the corresponding flow. If the HARQ entity in the transmitter determines that the HARQ process was terminated with an unsuccessful outcome, the HARQ entity in the transmitter informs the ARQ entity in the transmitter about the failure of the HARQ burst. The ARQ entity in the transmitter can then initiate retransmission and resegmentation of the appropriate ARQ blocks.

   IEEE 802.16m uses adaptive asynchronous and nonadaptive synchronous HARQ schemes in the downlink and uplink, respectively. The HARQ operation is relying on an N-process (multichannel) stop-and-wait protocol. In adaptive asynchronous HARQ, the resource allocation and transmission format for the HARQ retransmissions may be different from the initial transmission. In case of retransmission, control signaling is required to indicate the resource allocation and transmission format along with other HARQ necessary parameters. A nonadaptive synchronous HARQ scheme is used in the uplink when the parameters and the resource allocation for the retransmission are known a priori.

5.6.8   Mobility Management and Handover

IEEE 802.16m supports both network-controlled and MS-assisted handover (HO). The handover procedure may be initiated by either MS or BS; however, the final handover decision and target BS selection are made by the serving BS or the network. The MS executes the handover as directed by the BS or cancels the procedure through a HO cancellation message. The network reentry procedure with the target BS may be optimized by target BS possession of MS information obtained from the serving BS via the core network. The MS may also maintain communication with the serving BS while performing network reentry at the target BS, as directed by the serving BS. Figure 5.19 illustrates the HO procedure.

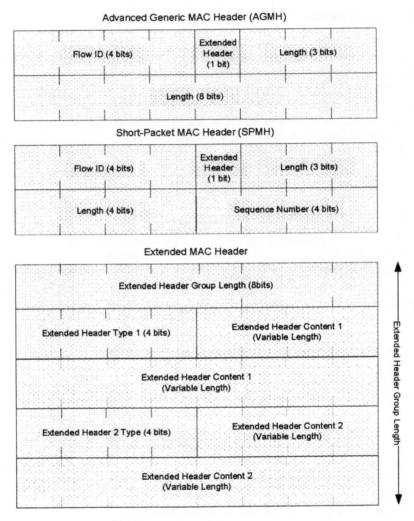

Figure 5.18. Structure of the IEEE 802.16m MAC headers.

The handover procedure is divided into three stages: (1) HO initialization, (2) HO preparation, and (3) HO execution. Upon completion of HO execution, the MS is ready to perform network reentry with the target BS. In addition, the HO cancellation procedure is defined to allow the MS to cancel a HO procedure [4, 13].

The HO preparation is completed when the serving BS informs the MS of its handover decision via a HO control command. The signaling may include dedicated ranging resource allocation and resource preallocations for the MS at the target BS for optimized network reentry. The control signaling includes an action time for the MS to start network reentry at the target BS and an indication whether the MS should maintain communication with the serving BS during network reentry. If the

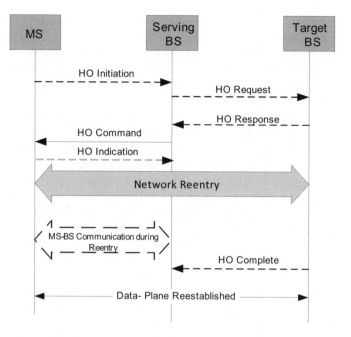

Figure 5.19. General handover call flow [4, 13].

communication cannot be maintained between the MS and the serving BS during network reentry, the serving BS stops allocating resources to the MS for transmission in action time. If directed by the serving BS via a HO control command, the MS performs network reentry with the target BS during action time while it continuously communicates with serving BS. However, the MS stops communication with the serving BS after network reentry with the target BS is completed. In addition, the MS cannot exchange data with the target BS prior to completion of network reentry.

5.6.9 Power Management

IEEE 802.16m provides power management functions, including sleep mode and idle mode, to mitigate power consumption of the mobile station. Sleep mode is a state in which an MS performs prenegotiated periods of absence from the serving BS. The sleep mode may be enacted when an MS is in the connected state. Using the sleep mode, the MS is provided with a series of alternative listening and sleep windows. The listening window is the time interval in which MS is available for transmit/receive of control signaling and data. IEEE 802.16m has the capability of dynamically adjusting the duration of sleep and listening windows within a sleep cycle based on changing traffic patterns and HARQ operations. When the MS is in active mode, sleep parameters are negotiated between the MS and BS. The base station instructs the MS to enter sleep mode. MAC management messages are used for sleep mode request/response. The period of the sleep cycle is measured in units of frames or superframes and is the sum of a sleep and listening windows. During the MS listening window, the BS may transmit the traffic indication message intended for one or multiple mobile stations. The listening window can be extended through explicit or implicit signaling. The maximum length of the extension is to the end of the current sleep cycle [4, 13].

Idle mode allows the MS to become periodically available for downlink broadcast traffic messaging, such as paging messages, without registration with the network. The network assigns mobile stations in the idle mode to a paging group during idle mode entry or location update. The MS monitors the paging message during listening interval. The start of the paging listening interval is calculated based on the paging cycle and paging offset. The paging offset and paging cycle are defined in terms of number of superframes. The serving BS transmits the list of paging group identifiers (PGID) at the predetermined location at the beginning of the paging-available interval. An MS may be assigned to one or more paging groups. If an MS is assigned to multiple paging groups, it may also be assigned multiple paging offsets within a paging cycle where each paging offset corresponds to a separate paging group. The MS is not required to perform location update when it moves within its assigned paging groups. The assignment of multiple paging offsets to an MS allows monitoring paging message at different paging offset when the MS is located in one of its paging groups. When an MS is assigned to more than one paging group, one of the paging groups is called *Primary Paging Group* and others are known as *Secondary Paging Groups*. If an MS is assigned to one paging group, that paging group is considered the *Primary Paging Group*. When different paging offsets are assigned to an MS, the *Primary Paging Offset* is shorter than the *Secondary Paging Offsets*. The distance between two adjacent paging offsets should be long enough so that the MS paged in the first paging offset can inform the network before the next paging offset in the same paging cycle so that the network avoids unnecessary paging of the MS in the next paging offset. An idle state MS (while in paging listening interval) wakes up at its primary paging offset and looks for primary PGIDs information. If the MS does not detect the primary PGID, it will wake up during its secondary paging offset in the same paging cycle. If the MS can find neither primary nor secondary PGIDs, it will perform a location update. The paging indications, if present, are transmitted at the predetermined location. The paging message contains identification of the mobile stations to be notified of pending traffic or location update. The MS determines the start of the paging listening interval based on the paging cycle and paging offset. During the paging-available interval, the MS monitors the superframe header and if there is an indication of any change in system configuration information, the MS will acquire the latest system information at the next instance of superframe header transmission (i.e., the next superframe header). To provide location privacy, the paging controller may assign temporary identifiers to uniquely identify the mobile stations in the idle mode in a particular paging group. The temporary identifiers remain valid as long as the MS stays in the same paging group.

An MS in idle mode performs a location update if any of these conditions are met: paging group location update, timer-based location update, or power-down location update. The MS performs the location update when the MS detects a change in paging group. The MS detects the change of paging group by monitoring the PGIDs, which are transmitted by the BS. The MS periodically performs a location update procedure prior to the expiration of the idle mode timer. At every location update including the paging group update, the idle mode timer is reset.

5.6.10 Security

Security functions provide subscribers with privacy, authentication, and confidentiality across the Mobile WiMAX IEEE 802.16m network. The MAC protocol data units are encrypted over the connections between the MS and BS. Figure 5.20 shows the functional blocks of IEEE 802.16m security architecture.

The security architecture is divided into security management and encryption and integrity logical entities. The security management functions include overall security management and control, EAP encapsulation/deencapsulation, privacy key management (PKM) control, security association management, and identity/location privacy. The encryption and integrity protection entity functions include user data encryption and authentication, management message authentication, and message confidentiality protection [4, 13].

Authorization is a process in which base station and mobile station mutually authenticate the identity of each other. Authentication is performed during initial network entry after security capabilities and policies are negotiated. The basic mobile station capability negotiation is performed prior to authentication and authorization.

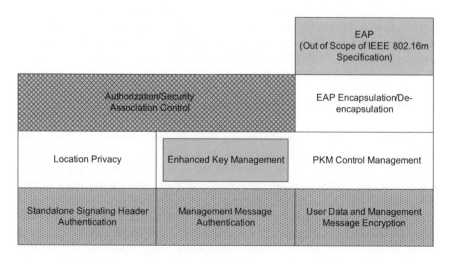

Figure 5.20. Functional blocks of IEEE 802.16m security architecture.

Reauthentication is performed before the authentication credentials expire. Data transmission may continue during the reauthentication process.

In the legacy system, there is no explicit means by which the identity of a user is protected. During initial ranging and certificate exchange, the MS MAC address is transmitted over the air, revealing user identity or user location information that may result in compromising the security aspects of the system. IEEE 802.16m incorporates mechanisms such a pseudoidentity to mitigate this problem.

IEEE 802.16m inherits the key hierarchies of the legacy system. The IEEE 802.16m uses the PKM v3 protocol to transparently exchange authentication and authorization (EAP) messages. The PKM protocol provides mutual and unilateral authentication and establishes confidentiality between the MS and the BS. Some IEEE 802.16m keys are derived and updated by both the BS and MS. The key exchange procedure is controlled by the security key state machine, which defines the allowed operations in the specific states. The key exchange state machine is similar to that of the legacy system.

A security association (SA) is the set of information required for secure communication between the BS and MS. However, the SA is not equally applied to messages within the same flow. According to the value of MAC header fields (e.g., EC, EKS, Flow ID), the SA is selectively applied to the management connections. When a service flow is established between the BS and the group of mobile stations, it is considered as multicast and it is serviced by a group SA.

The mobile station and the base station may support encryption methods and algorithms for secure transmission of MAC PDUs. The advanced encryption standard (AES) is the only supported cryptographic method in IEEE 802.16m [4, 13]. The legacy system does not define confidentiality protection for control-plane signaling. IEEE 802.16m selectively protects the confidentiality of control-plane signaling. The use of MAC message authentication code in legacy systems only proves the originator of messages and ensures integrity of the messages. IEEE 802.16m supports the selective confidentiality protection over MAC management messages. If the selective confidentiality protection is activated, the negotiated keying materials and cipher suites are used to encrypt the management messages. Integrity protection is applied to the stand-alone MAC signaling header.

5.7 PHY AND MAC ASPECTS OF MULTICARRIER OPERATIONS

The carriers of a multicarrier system may be configured differently either as fully configured or partially configured carrier. Fully configured carrier is a standalone carrier for which all control channels including synchronization, broadcast, multicast and unicast control signaling are configured. Fully configured carrier supports both single carrier MS and multicarrier MS. A partially configured carrier is a carrier configured for downlink only transmission in TDD or a downlink carrier without paired UL carrier in FDD mode. The partially configured carriers may be used only in conjunction with a primary carrier and cannot operate standalone to offer IEEE

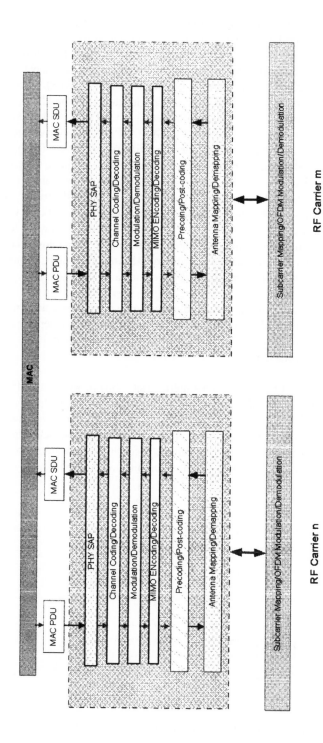

Figure 5.21. Example physical layer processing for multicarrier operation [4].

135

802.16m services to an MS. If a partially configured carrier is used for DL unicast traffic, the required UL feedback channels are provided by the primary carrier. In multicarrier aggregation, the UL control channels corresponding to the secondary partially configured carriers (i.e., DL-only secondary carriers are located in distinct nonoverlapping control regions in the UL of the primary carrier). The UL control regions for the DL only secondary carriers are located after the UL control region for the primary carrier. The location information of the UL control channels for the DL only secondary carriers is signaled via AAI_SCD MAC control message. The MS uses the UL control channels on the primary carrier to feedback HARQ ACK/NACK and channel quality measurements corresponding to transmission over DL only secondary carrier. Note that only an FDD primary carrier may be used to provide UL feedback channels for DL partially configured carriers. A partially configured carrier may be used for broadcast services only in that case it would not need UL feedback channels. A primary carrier is fully configured while a secondary carrier may be fully or partially configured depending on the deployment scenarios. Carrier configuration information is broadcasted through advance primary preamble. The MS does not attempt network entry or handover to partially configured carriers.

A secondary carrier for an MS, if fully configured, may serve as primary carrier for other MS. The mobile stations with different primary RF carriers may also share the same secondary carrier. The following multicarrier operation modes are defined in IEEE 802.16m standard:

- Multicarrier Aggregation: The multicarrier mode in which the MS maintains its physical layer connection and monitors the control signaling on the primary carrier while processing data on the secondary carrier. The resource allocation to an MS may span across a primary and multiple secondary carriers. Link adaptation feedback mechanisms incorporate measurements relevant to both primary and secondary carriers. In this mode the system may assign sec-

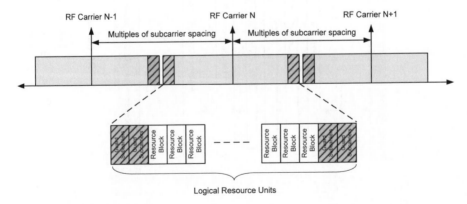

Figure 5.22. Example of data transmission using the guard subcarriers [4, 13].

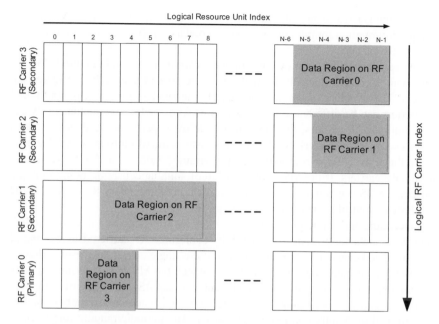

Figure 5.23. Example of data region mapping in multicarrier operation [4].

ondary carriers to an MS in the downlink and/or uplink asymmetrically based on system load, peak data rate, or QoS requirements.

- Multicarrier Switching: The multicarrier mode in which the MS switches its physical layer connection from the primary to the partially configured or fully configured secondary carrier according to base station's instruction to receive E-MBS services on the secondary carriers. The MS connects with the secondary carrier for the specified duration of time and then returns to the primary carrier. When the MS is connected to the secondary carrier, the MS is not required to maintain its transmission or reception through the primary carrier.

The guard subcarriers between contiguous RF carriers in the new zones can be utilized for data transmission if the subcarriers of contiguous RF carriers are frequency aligned. The serving BS and the MS need to negotiate their capabilities to support data transmission over guard subcarriers. The subcarriers at band edges on each contiguous RF carrier can be used as additional data subcarriers. The guard subcarriers are grouped to form resource blocks. The physical resource structure used for guard subcarriers is the same as that of the ordinary resource units described in Section 5.3. The BS provides information regarding the use of guard subcarriers for data channels. Guard subcarriers are not used for control channel transmission. Figure 5.22 illustrates an example of exploiting guard subcarriers for data transmission.

Signaling is used to indicate the allocation of data regions. For each MS, the allocation information for both primary and secondary RF carriers is sent through the primary RF carrier.

An example of allocation over multiple RF carriers is shown in Figure 5.23.

Downlink control channels defined earlier are present over fully configured RF carriers. The structure of control and superframe headers for partially configured RF carriers can be different. For the case in which the MS can simultaneously decode multiple RF carriers, the MS can decode the superframe headers associated with its secondary RF carriers. The BS may instruct the MS through control signaling on the primary RF carrier to decode superframe headers of a specific set of secondary RF carriers. If the MS is not capable of simultaneously decoding multiple carriers, the BS can convey the system information of secondary RF carriers to the MS through control signaling on the primary RF carrier.

Superframe and frame preambles as well as the broadcast and downlink/uplink control channels are present over each fully configured RF carrier. The location of these overhead channels is the same as that of the single carrier described earlier.

The BS configures the set of carriers for which the MS reports channel quality information. The BS may only allocate resource to the MS on a subset of those configured RF carriers. The uplink initial ranging for nonsynchronized mobile stations is performed on a fully configured RF carrier. The periodic ranging for synchronized mobile stations is performed on the primary RF carrier. The serving BS transmits the ranging response on the same RF carrier on which the uplink ranging is received.

The uplink sounding is performed on the primary and secondary carriers. The bandwidth request channel is transmitted only on the primary RF carrier. Depending on the channel correlation between RF carriers, a separate power control channel for each RF carrier may be necessary. Although multiple power control channels are allowed, the power control messages can be sent to the MS through the primary RF carrier. If an MS switches to a secondary RF carrier in carrier switching mode, the secondary RF carrier can carry power control messages if supported by its configuration.

The MAC layer in multicarrier mode operates in the same way as the single carrier that includes MAC addressing, headers, PDU formation and segmentation, and so on. All security procedures between the MS and BS are performed using only the primary RF carrier.

The MS attempts initial ranging and network entry exclusively over a fully configured RF carrier. An MS needs to detect fully configured RF carriers. An MS scans different RF carriers and attempts to detect certain overhead channels such as superframe/frame preambles and superframe header. Once a potential primary RF carrier is determined, the initial network entry procedures are the same as in single carrier mode. The RF carrier on which the MS successfully performs initial network entry becomes the primary carrier for the MS. After successful ranging, the MS follows the capability negotiation and exchange information, such as carrier aggregation or carrier switching capabilities with the base station. The BS may assign secondary carriers to the MS through negotiation with the MS.

Bandwidth requests are transmitted over the designated primary RF carrier using uplink control channels similar to the single carrier mode. Bandwidth request using the piggyback scheme is also allowed over the secondary carriers. The BS may allocate uplink resources that belong to a specific RF carrier or a combination of multiple RF carriers.

The QoS and connection management functions in multicarrier mode are similar to the single carrier scheme. The station ID and all the flow IDs assigned to an MS are unique identifiers for a common MAC and are used over all RF carriers. The connection setup signaling is performed only through the primary RF carrier. The connection is defined for a common MAC entity. The QoS parameters are managed per service flow for each MS, and are applicable to primary carrier and secondary RF carriers. The flow ID is maintained per MS for both primary and secondary RF carriers. The required QoS for a service flow may be one of the parameters considered in order to determine the number of secondary carriers assigned to the MS.

The BS may instruct the MS, through control signaling on the current primary carrier, to switch its primary RF carrier to one of the available fully configured RF carriers within the same BS for load balancing and to avoid congestion on a certain RF carrier. The BS may instruct the MS to perform scanning on other RF carriers that are not used by the MS. The MS performs synchronization with the target RF carrier, if necessary. Primary to secondary RF carrier switching in the multicarrier mode is supported when the secondary RF carrier is partially configured. The RF carrier switching between a primary carrier and a secondary carrier can be periodic or event-triggered, with timing parameters defined by multicarrier switching message on the primary carrier.

An MS in multicarrier operation follows the handover procedures specified for the single carrier mode. The MAC management messages corresponding to handover between an MS and a BS are transmitted over the primary RF carrier. The MS terminates communication with the serving BS on primary or secondary RF carriers following successful completion of network reentry with the target BS. To facilitate scanning of neighbor a BS's fully configured RF carriers, the serving BS may provide the neighbor BS's multicarrier configuration information to the MS. An MS capable of concurrently processing multiple RF carriers may perform scanning with neighbor BSs and signaling with the target BS using one or more of its available RF carriers, while maintaining normal operation on the primary carrier and secondary carriers of the serving BS. The MS may negotiate with its serving BS in advance to prevent allocation over those carriers used for scanning with neighboring BSs and HO signaling with the target BS.

The MS is only assigned to one or more secondary carriers during the active mode. Therefore, the power saving procedures in multicarrier operation are the same as those specified for single carrier operation. All MAC messaging, including idle mode procedures and state transitions, are performed through the primary RF carrier. In active mode, the MS can be directed via signaling on the primary carrier to disable transmission/reception on some secondary carriers to minimize power consumption.

When a mobile station enters sleep mode, the negotiated sleep mode is applied to a common MAC, regardless of multicarrier mode, and all carriers will power down according to the negotiated sleep pattern. During the listening window of the sleep cycle, the traffic indication is transmitted through the primary RF carrier.

During the paging-available interval, the MS monitors paging indication on a fully configured RF carrier. The procedure for paging is the same as that defined for a single carrier. Messages and procedures to enter the idle mode between the MS and the BS are exchanged via the primary RF carrier. The network reentry procedure from idle mode is similar to that of initial network entry. One set of unified idle mode parameters (i.e., paging-available and paging-unavailable interval configuration) is used for an MS regardless of single carrier or multicarrier operation.

5.8 PHY AND MAC ASPECTS OF MULTICAST AND BROADCAST SERVICES

Enhanced multicast and broadcast services are point-to-multipoint communication systems in which data packets are transmitted simultaneously from a single source to multiple destinations. The term broadcast refers to the ability to deliver contents to all users. Multicast, on the other hand, refers to contents that are directed to a specific group of users that have the associated subscription for receiving such services.

The enhanced multicast and broadcast service (E-MBS) functions consist of MAC and PHY protocols that define interactions between the base stations and mobile stations. Although the basic operation principles are consistent with the latest revision of the IEEE 802.16 standard [1], some enhancements and extensions are defined to provide improved functionality and performance. The E-MBS protocol structure is a logical structure that describes MAC operations for E-MBS and complements the IEEE 802.16m protocol structure for unicast transmission.

In the control plane, the E-MBS MAC function (see Figure 5.2) operates in conjunction with other unicast MAC functions. The E-MBS MAC function consists of the following functional blocks:

- **E-MBS Group Configuration.** This function manages the configuration advertisement of E-MBS groups. A BS could belong to multiple E-MBS groups (similar to paging groups).
- **E-MBS Transmission Mode Configuration.** This function describes the transmission mode in which E-MBS is delivered over the air interface, such as single-BS and multi-BS transmission.
- **E-MBS Session Management.** This function manages E-MBS service registration or deregistration, session start, update, or termination.
- **E-MBS Mobility Management.** This block manages the E-MBS group update procedures when an MS crosses the E-MBS group boundary.
- **E-MBS Control Signaling.** This block broadcasts the E-MBS scheduling

and logical-to-physical channel mapping to facilitate E-MBS reception and support power saving.

E-MBS group-specific pilots are transmitted for multicast and broadcast in single frequency network operation. An E-MBS group is a set of BSs that form a single frequency network. The E-MBS group-specific pilot structure, which is common within an E-MBS group but different between neighboring E-MBS groups, is configurable. Synchronous transmissions of the same content with common pilots from multiple base stations in one E-MBS group (i.e., macro diversity) would result in more accurate channel estimation. The E-MBS group-specific pilot structure depends on the maximum number of MIMO streams used in the E-MBS group.

The E-MBS MAPs are classified into cell-specific and non-cell-specific control channels. Within a subframe in which multicast data and control are carried, multicast service control and data channels are frequency-division multiplexed. Within an E-MBS region, control information is transmitted before E-MBS data transmission in order to help decode burst information.

An E-MBS MAP information element (E-MBS MAP-IE) is defined as the basic element of the multicast service control. A multicast service control information element is non-user-specific and is addressed to all users in the cell. Open-loop spatial multiplexing schemes described in Section 5.8 are used for E-MBS. E-MBS data bursts may be transmitted in terms of several subpackets, where these subpackets may be transmitted in different subframes.

Two types of E-MBS access are supported: (1) single-BS access and (2) multi-BS access. Single-BS access is realized over multicast and broadcast transport connections within one BS, whereas multi-BS access is implemented by transmitting service flow(s) over multiple BSs. The MS may support both single-BS and multi-BS access. The E-MBS service may be delivered either via a dedicated carrier or a mixed unicast and broadcast carrier (see Figure 5.24).

The macro-diversity operation for E-MBS consists of a wide-area, multicell multicast broadcast, single frequency network in which E-MBS traffic from multiple cells, with timing errors within the cyclic prefix length, are received at the MS.

A multi-BS E-MBS in which the transmission of data across BSs in an E-MBS group are synchronized at the symbol level allows macro-diversity combining of signals and higher cell-edge performance. It requires the multiple BSs participating in the same multi-BS E-MBS service to be synchronized in the transmission of common multicast and broadcast data. Each BS transmits the same PDUs, using the same transmission format.

The E-MBS traffic can be time-division multiplexed with unicast traffic within a frame. The E-MBS subframes are located contiguously at the end of downlink subframes (i.e., E-MBS region). Both TDM and FDM schemes are supported for the mixed unicast and E-MBS operation.

In a unicast/multicast mixed-RF carrier, E-MBS uses the same frame structure used for a unicast RF carrier. The E-MBS traffic is multiplexed with unicast traffic. The secondary superframe header indicates the location of the E-MBS zone, which may span over multiple subframes (see Figure 5.25).

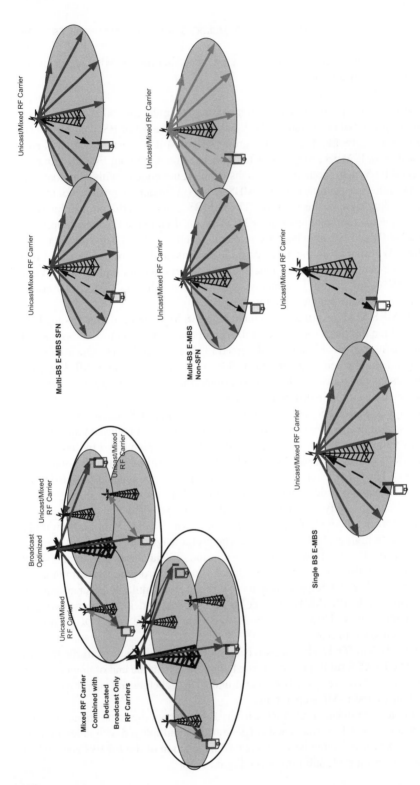

Figure 5.24. E-MBS deployment scenarios.

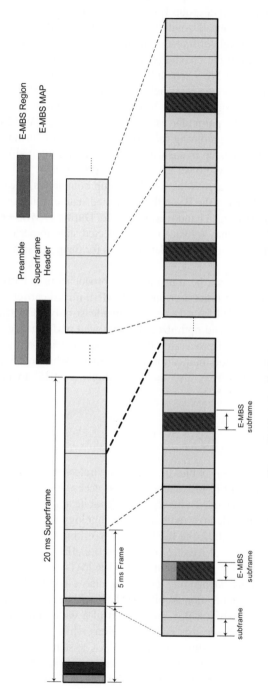

Figure 5.25. E-MBS MAP and resource allocation.

Each E-MBS group has a unique group ID. All the BSs in an E-MBS group broadcast the same E-MBS group ID. If a BS belongs to several E-MBS groups, it broadcasts all of the group IDs with which it is associated. The E-MBS scheduling interval can span several subframes. The length of the E-MBS scheduling interval may be constrained by the channel switching time requirements [2]. For each E-MBS group, there is an MBS scheduling interval that refers to the number of successive frames in which the access network may schedule traffic for the streams associated with the MBS group prior to the start of the interval. The length of this interval depends on the particular use case of the E-MBS. An MS decodes only a specific E-MBS data burst during the time when the user is using a certain E-MBS program. The MS wakes up in each E-MBS scheduling interval in order to decode the data burst. This allows power saving in E-MBS service.

When an MS roams across the E-MBS group boundaries, it can continue to receive E-MBS traffic from the BS in the connected state or idle state. In the connected state, the MS performs a handover procedure. During E-MBS group transition in the idle state, the MS may transit to the connected state to perform handover or it may initiate an E-MBS location update process for the purpose of MBS group transition.

The E-MBS traffic can be transmitted in broadcast-only carrier mode. In this case, a fully configured unicast or unicast/E-MBS mixed carrier is used to provide signaling support needed for service. The broadcast-only RF carrier may be transmitted at higher power and optimized for improved performance. The multicarrier MS, which is capable of processing multiple radio carriers at the same time, may perform normal data communication at one carrier while receiving E-MBS data over another carrier. It may also receive multiple E-MBS streams from multiple carriers simultaneously.

5.9 SUMMARY

The growing demand for mobile Internet and wireless multimedia applications has motivated development of broadband wireless access technologies in recent years. Mobile WiMAX has enabled convergence of mobile and fixed broadband networks through a common wide-area radio-access technology and flexible network architecture. Since January 2007, the IEEE 802.16 Working Group has been developing a new amendment to the IEEE 802.16 standard (i.e., IEEE 802.16m) as an advanced air interface to meet the requirements of ITU-R/IMT-Advanced for fourth-generation cellular systems. Depending on the available bandwidth and multiantenna mode used, the next-generation mobile WiMAX will be capable of over-the-air data throughputs in excess of 1 Gbps and support of a wide range of high-quality and high-capacity IP-based services and applications while maintaining full backward compatibility with the existing mobile WiMAX systems to preserve investments and provide continuing support for the first-generation products. This chapter briefly described the prominent technical features of IEEE 802.16m and potentials for successful deployment of the next generation of mobile WiMAX in 2012 and beyond.

REFERENCES

[1]   IEEE Standard for Local and Metropolitan Area Networks, PART 16: Air Interface For Broadband Wireless Access Systems, May 2009.

[2]   IEEE 802.16m-07/002r9, IEEE 802.16m System Requirements, http://ieee802.org/16/tgm/index.html, September 2009.

[3]   IEEE 802.16m-08/004r5, IEEE 802.16m Evaluation Methodology Document, http://ieee802.org/16/tgm/index.html, January 2009.

[4]   IEEE 802.16m-08/0034r2, IEEE 802.16m System Description Document, http://ieee802.org/16/tgm/index.html, September 2009.

[5]   WiMAX Forum Mobile System Profile, Release 1.0 Approved Specification (Revision 1.7.1: 2008-11-07), http://www.wimaxforum.org/technology/documents.

[6]   WiMAX Forum Network Architecture Release 1.5 Version 1—Stage 2: Architecture Tenets, Reference Model and Reference Points, http://www.wimaxforum.org/resources/documents/technical/release, September 2009.

[7]   Report ITU-R M.2135, Guidelines for Evaluation of Radio Interface Technologies for IMT-Advanced, November 2008.

[8]   Report ITU-R M.2134, Requirements Related to Technical System Performance for IMT-Advanced Radio Interface(s), November 2008.

[9]   IMT-Advanced Submission and Evaluation Process, http://www.itu.int/ITU-R.

[10] WiMAX System Evaluation Methodology, http://www.wimaxforum.org/technology/documents, July 2008.

[11] Andrea Goldsmith, *Wireless Communications,* Cambridge University Press, New York, 2005.

[12] E. Dahlman et al., *3G Evolution: HSPA and LTE for Mobile Broadband,* 2nd Edition, Academic Press, 2008.

[13] P802.16m/D5: IEEE Standard for Local and Metropolitan Area Networks—Part 16: Air Interface for Broadband Wireless Access Systems, Advanced Air Interface, May 2010.

[14] Sassan Ahmadi, An Overview of Next-Generation Mobile WiMAX Technology, *IEEE Communications Magazine,* June 2009.

CHAPTER 6

OVERVIEW OF WiMAX NETWORK ARCHITECTURE AND EVOLUTION

KAMRAN ETEMAD, JICHEOL LEE, AND YONG CHANG

6.1 INTRODUCTION

Extending the scope of interoperability beyond the air interface, the WiMAX Forum defines end-to-end network architectures and protocols [2] based on and consistent with MAC and PHY specifications in IEEE802.16 and the WiMAX air interface System Profile. Stage 1 requirements are defined by the Service Provider's Working Group (SPWG) and stage 2 and 3 specifications are developed by the Network Working Group (NWG) of the WiMAX Forum. These network specification efforts involve interactions with other standard organizations such as IETF, 3GPP, 3GPP2, DSL Forum, and OMA.

The following are some of the key tenets and attributes of WiMAX network design:

- **All IP and functional frameworks reuse IETF-based protocols.** The WiMAX network architecture is designed to meet the requirements while maximizing the use of open standards and IETF protocols in a simple all-IP architecture. These network specifications reuse and integrate many IETF-defined IP protocols for control and bearer interconnectivity. The functional architecture framework allows logical separation of access, connectivity, and application services in WiMAX networks with flexible composition of functions into physical network entities while maintaining interoperable reference points between them.

- **A unified framework for fixed/mobile or greenfield/overlay access networks.** WiMAX network specifications allow simplified configurations for fixed-access deployments all the way to large-scale cellular networks supporting high speed, and macromobility, and roaming. A WiMAX network can be deployed as a Greenfield network without any unnecessary complexity due to legacy circuit-switched system support. It can also be deployed as an overlay to existing fixed or mobile access networks such as 2.5G/3G cellular systems or cable/DSL networks by supporting different levels of interworking to ensure service continuity.

- **Support for different usage models and devices profiles.** The same network can used for a variety of usage models such as wireless backhaul to Wi-Fi hot spots, fixed/nomadic access to CPE's and residential gateways (RGs), and for mobile access to notebooks, smart phones, and next-generation WiMAX-embedded ultramobile devices; see Figure 6.1.

- **Unbundling of access, connectivity, and application services networks.** WiMAX network design functionally separates the radio access network from the connectivity service core network such that the access network architecture becomes almost unaffected by the operator's domain, enabling (1) network access provider's (NAPs) infrastructure sharing by multiple network service providers (NSPs), (2) service offering through multiple radio access infrastructure aggregation, (3) radio access network evolution independent of operator core network.

- **Support for both operator-managed IP and open internet services.** The WiMAX architecture enables open access to web-based applications and enhanced Internet services as well as operator-managed "walled garden" services and IMS interworking in the same network, allowing operators to explore creative service offerings and Internet-friendly business models. The network can support various real-time and nonreal-time, server-client, and peer-to-peer applications, including VoIP, TCP, unicast and multicast video and audio streaming, and interactive gaming.

- **Support for operators agnostic or "retail" device models.** The over-the-air (OTA) activation and dynamic provisioning protocols and associated network conformance testing and certification in the WiMAX Forum are structured to ensure successful network entry and provisioning of a variety of mobile internet devices, including embedded communications devices and consumer electronics distributed through retail channels.

In the following sections we present a high-level overview of the Mobile WiMAX network architecture as defined in WiMAX Forum specifications. The focus of this chapter is on the basic network reference model and functionalities defined in Release 1.0 but it also includes highlights of network evolution to releases 1.5 and 2.0. Detailed description of some end-to-end features in Release 1.0 such as mobility, QoS, security, and accounting, as well advanced features of Release 1.5, including 3G interworking, location-based services, multicast and broadcast services, enhanced Internet services, and roaming, are presented in future chapters.

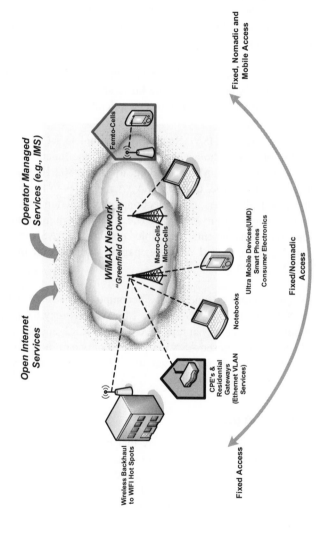

Figure 6.1. Mobile WiMAX enabling a variety of usage/device models in the same network.

149

6.2   WiMAX BASIC NETWORK REFERENCE MODEL

The baseline WiMAX network architecture is logically represented by a network reference model (NRM), which identifies key functional entities and reference points over which the network interoperability specifications are defined [2]. The WiMAX NRM, as illustrated in Figure 6.2, consists of several logical network entities, namely, mobile station (MS), access service network (ASN), and connectivity service network (CSN), and their interactions through reference points R1–R8. Each of the MS, ASN, and CSN represents a logical grouping of functions as described in the following:

Mobile Station (MS). Generalized user equipment set providing wireless connectivity between a single or multiple hosts and the WiMAX network. In this context, the term MS is used more generically to refer to both mobile and fixed-device terminals.

Access Service Network (ASN). Represents a complete set of network functions required to provide radio access to the MS. These functions include Layer-2 connectivity with the MS according to IEEE 802.16 standards and the WiMAX system profile, transfer of AAA messages to Home NSP (H-NSP), preferred NSP discovery and selection, relay functionality for establishing Layer-3 (L3) connectivity to the MS (i.e., IP address allocation), as well as radio resource management. To enable mobility, the ASN may also support ASN and CSN anchored mobility, paging and location management, and ASN-CSN tunneling.

The ASN architecture is functionally decomposed based on what used to be known as ASN Profile C in NWG Release 1.0. An ASN may be implemented in a fashion that only exposes external reference points R1, R3, and R4 and does not expose R6 and R8. A decomposed ASN may consist of one or more base stations (BS) and at least one instance of an ASN gateway (ASN-GW). The BS and ASN-GW functions can be described as follows.

6.2.1   Base Station (BS)

A base station is a logical network entity that primarily consists of radio-related functions of ASN interfacing with MS over the air link according to MAC and PHY specifications in IEEE 802.16, subject to applicable interpretations and parameters defined in the WiMAX Forum system profile. In this definition, each BS is associated with one sector with one frequency assignment but it may incorporate additional implementation-specific functions such as downlink and uplink schedulers.

6.2.2   ASN Gateway (ASN-GW)

An ASN Gateway is a logical entity that represents an aggregation of centralized functions related to QoS, security, and mobility management for all the data con-

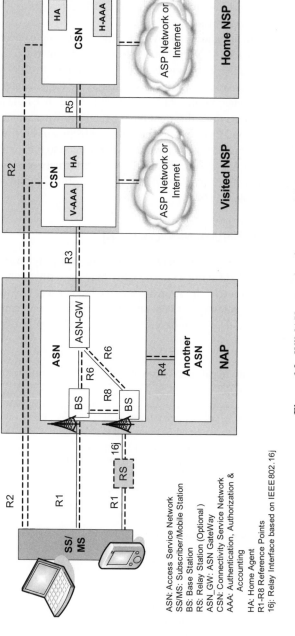

Figure 6.2. WiMAX network reference model.

ASN: Access Service Network
SS/MS: Subscriber/Mobile Station
BS: Base Station
RS: Relay Station (Optional)
ASN_GW: ASN GateWay
CSN: Connectivity Service Network
AAA: Authentication, Authorization &
    Accounting
HA: Home Agent
R1-R8 Reference Points
16j: Relay Interface based on IEEE802.16j

151

nections served by its association with BSs through the R6 reference point. The ASN-GW also hosts functions related to IP layer interactions with the CSN through R3, as well as interactions with other ASNs through R4 in support of mobility.

Typically, multiple BSs may be logically associated with an ASN-GW. Also, a BS may be logically connected to more than one ASN gateways to allow load balancing and redundancy options.

The WiMAX Forum has chosen the ASN Profile C as a single decomposed ASN profile (formerly known as ASN Profile C) with open R6 interface in Release 1.5. The former deployed implementations that have followed the Profile-A and Profile-B architecture can be considered as an ASN that may be implemented in a fashion that only exposes reference points R1, R3, and R4. Note that Release 1.0 of the WiMAX network included three different types of ASN profiles which were known as Profile A, Profile B, and Profile C. Due to down-selection of Profile C as a single ASN profile, Profile A and Profile B have been deprecated to reduce the number of implementation options and create a better framework for network interoperability. The detailed functional decomposition of the chosen Profile C of the ASN architecture is described in Table 6.1.

6.2.3 Connectivity Service Network (CSN)

A connectivity network is a set of network functions that provide IP connectivity services to the WiMAX subscriber(s). The CSN may be further comprised of network elements such as routers, AAA proxy/servers, home agent and user databases, as well as interworking gateways or enhanced network servers to support multicast and broadcast services and location-based services. A CSN may be deployed as part

Table 6.1. Functional decomposition of Profile C ASN architecture

Features	Functions	BS	ASN-GW
Handover management	Data path function	x	x
	Handover control Function	x	
	Context server and client	x	x
	IP mobility (PMIP Client/MIP FA)		x
Security	Authenticator		x
	Authentication relay	x	
	Key distributor		x
	Key receiver	x	
Radio resource management	Radio resource controller	x	
(RRM)	Radio resource agent	x	
Paging	Paging controller		x
	Paging agent	x	
	Location register		x
Quality of service	SF authorization		x
	SF manager	x	
	AAA Client		x

of a Greenfield WiMAX NSP or as part of an incumbent WiMAX NSP. The following are some of the key functions of CSN:

- IP address management
- AAA proxy or server
- QoS policy and admission control based on user subscription profiles
- ASN-CSN tunneling support
- Subscriber billing and interoperator settlement
- Inter-CSN tunneling for roaming
- CSN-anchored inter-ASN mobility
- Connectivity to Internet and managed WiMAX services such as IP multimedia services (IMS), location-based services, peer-to-peer services, broadcast and multicast services
- Over-the-air activation and provisioning of WiMAX devices

The WiMAX network reference model recognizes the network access provider (NAP) and network service provider (NSP) as being two potentially different business entities. The NAP is a business entity that provides and operates a WiMAX radio access network deployed as one or more ASNs, whereas an NSP is the business entity that operates the CSN and provides IP connectivity and WiMAX services to WiMAX subscribers according to some negotiated service-level agreements (SLAs) through one or more NAPs.

The NSP supplies this functionality by implementing a full CSN or CSN elements as part of its infrastructure. The network architecture allows one NSP to have relationships with multiple NAPs in one or different geographical locations. It also enables NAP sharing by multiple NSPs. In some cases, the NSP may be the same business entity as the NAP.

This flexible framework clearly accommodates the conventional 3G mobile operator's business model, in which the NSP is the same as the NAP, that is, where the core network operator owns the RAN by definition.

The model also accommodates MVNOs as well as 3GPP/2 interworking scenarios in which WiMAX ASNs (owned by the cellular operator or by another NAP operator) are connected directly to the cellular 3G core network (which functions as a CSN) and allows sharing of the core network server as well as seamless interworking. This flexibility also introduces some challenges in the network discovery and selection (ND&S) procedure, described later in this chapter, as the WiMAX device must be able to discover the NSP(s) behind the NAP, make a decision as to which one to connect to and notify the NAP about its decision.

The WiMAX network architecture has an end-to-end protocol stack as shown in Figure 6.3. For IP connectivity service to the MS, the BS, ASN-GW, and home agent (HA) perform their respective operations and these network nodes transfer or forward IP packets to each other on top of their corresponding protocols. The MS and the BS are logically connected through the IEEE802.16e over-the-air interface shown as the R1 reference point. The IP packets are transferred via IEEE 802.16

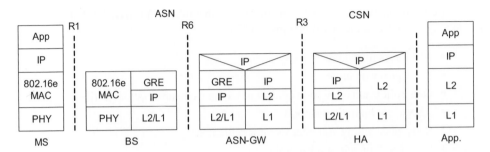

Figure 6.3. End-to-end protocol stack of the mobile WiMAX network.

medium access control on top of the IEEE 802.16 physical layer. The BS and the ASN-GW are logically connected through the IP network and their logical connection is shown as the R6 reference point. This connection delivers the classified IP packets encapsulated by the generic routing protocol to each other. The ASN-GW and the HA are connected through the IP network and its logical connection is shown as the R3 reference point. The IP packets are transferred normally through IP in IP encapsulation or, sometimes, GRE encapsulation on R3.

From the MS's application, in order for an IP layer application to send an IP packet toward the Internet, the application must deliver the packet down to the IEEE 802.16 MAC layer. The IEEE 802.16 MAC layer classifies the received IP packet and may fragment it into multiple MAC PDUs if necessary in order to transfer it over the air with the distinct and logical traffic connection identifier with corresponding QoS parameters and policy rules. The classified IP packet is formatted as an IEEE 802.16e MAC PDU (packet data unit) and is sent over the air as physical layer frames to the BS. After receiving PHY frames over the air, the PHY layer of the BS delivers the MAC PDUs in frames to the MAC layer. The MAC layer reconstructs the IP packet from the MAC-PDUs that were transferred via the logical over-the-air connection. The IP packets are forwarded to the ASN-GW as encapsulated in GRE packets . The ASN-GW decapsulates the received GRE packets and sends the IP packet with IP-in-IP (sometimes GRE) encapsulation to the HA. The HA finally decapsulates the IP-in-IP packet and forwards it to the right application server on the Internet.

In the opposite direction, when the HA receives the IP application packet sent from the application server, it forwards the packet to the ASN-GW over the mobile IP tunnel after looking at the destination IP address of the IP header. The ASN-GW classifies the IP packet into the proper service flow, which is mapped to the GRE tunnel toward the BS by the table of service classification rule known to the MS. The BS receives the packet classified by GRE key and it identifies the distinct and logical traffic connection identifier over the air and delivers it down to the 802.16e MAC layer. The application of the MS handles the IP packets that are being delivered up from IP layer after the MS receives the IP packet over IEEE 802.16e MAC and the PHY.

R1 is a reference point between the MS and the BS. Its protocol stack (see Figure 6.4) includes the IEEE 802.16 physical layer and medium access control layer. The physical layer performs all modulation coding, channel coding, AMC (adaptive modulation coding), H-ARQ (hybrid automatic repeat request), and power control. Moreover, it supports MIMO (multiple input and multiple output) and beam forming as 4G prominent technologies over the underlying OFDMA. The MAC (medium access control) layer is composed of the three sublayers: CS (convergence sublayer), MAC CPS (common part sublayer), and the security sublayer. The CS classifies IP flows to the corresponding service flow as well as the RoHC (robust header compression) and packet header suppression. The MAC CPS includes the functions necessary to handle the packets over the physical layer, such as random access, bandwidth allocation, downlink scheduling, connection management, and ARQ. The security sublayer provides the encryption for the MAD-PDU and it also provides transfer service for EAP packets for user and/or device authentication. The detailed operations of IEEE 802.16e PHY and MAC layers are described in Chapter 2.

The R6 reference point logically connects both the BS and the ASN-GW. There are both control-plane and bearer-plane protocols on the R6. On the control plane of R6, there are the WiMAX Control Protocol, designed by the WiMAX Forum Network Working Group, on top of UDP/IP. The WiMAX Control Protocol defines its own signaling message header, signaling the message body to be applied to all signaling between the BS and the ASN-GW in WiMAX network. The WiMAX Control Protocol Header is composed of function type and message Type to identify the signaling message type, and its required transaction management functions. The message body part is composed of well-known type-length value to include necessary information elements that belong to the specific messages. On the bearer plane of R6, all payloads [e.g., IPv4 packet, IPv6 packet, compressed/suppressed IP packets, or handover data integrity (HODI) packet] are encapsulated in GRE over IP.

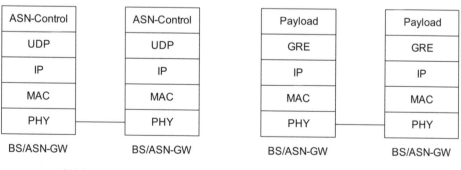

Figure 6.4. Protocol stack of R4, R6, and R8 reference points.

The R4 reference-point logically connects two different ASN-GWs. There are both control-plane and bearer-plane protocols on R4, similar to those of R6.

The R8 reference point logically connects two different BSs. There are only control-plane protocols on R8, similar to that of R6.

The R3 reference point logically connects both the ASN-GW to the CSN entities. It consolidates AAA protocols like RADIUS or DIAMETER interfacing with the AAA server and Mobile IP-based protocol to support IP-based mobility interfacing with the mobile IP home agent (see Figure 6.5).

There are two kinds of AAA protocols on the R3 interface with AAA (see Figure 6.6). One is RADIUS with some extensions specified in the WiMAX Forum to include WiMAX-specific attributes to support its own design requirements of the WiMAX network. This RADIUS-based protocol was designed in Release 1.0. The other is DIAMETER, which has been extended along with RAIDUS in Release 1.5. These two types of AAA protocol have been defined so that operators can choose one of them based on their own purpose.

The IP-based Mobility Protocol is based on Mobile IP which was designed in IETF. The R3 Mobility Protocol leverages the underlying Mobile IP Protocol to support the tunnel between the ASN-GW and the HA. In order to register or revoke binding entry for the given MS, the Mobile IP signaling messages are used on top of UDP/IP. The IP traffic is either IP in IP or GRE/IP tunneled and transferred between the ASN-GW and the HA.

The R2 reference point logically connects between the MS and the CSN. It corresponds to the logical protocol on top of IETF-based protocols, for examples, EAP or Mobile IPv6 known as Client-based Mobile IPv6 (CMIPv6). On R2, there exist the protocols-specific EAP methods such as EAP-TLS, EAP-TTLS, or EAP-AKA running on top of EAP between the MS and the CSN when the MS is to be authenticated, and CMIPv6 for the binding update between the MS and the HA.

The R5 reference point logically connects between the visited CSN and the home CSN, especially in roaming environment. The RADIUS or DIAMETER protocols can be applied to the R5 control protocol.

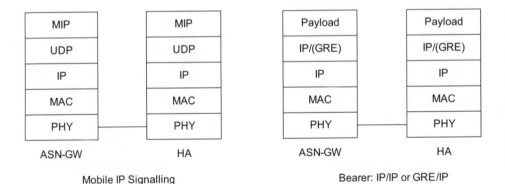

Figure 6.5. Protocol stack of the R3 mobile IP.

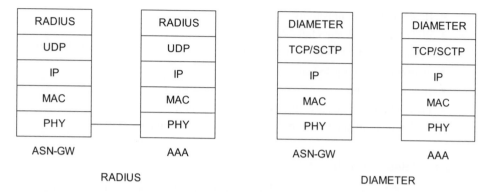

Figure 6.6. Protocol stack of the R3 AAA interface.

6.3   WiMAX NETWORK ROAD MAP: RELEASES 1.0, 1.5, 1.6, AND 2.0

The Mobile WiMAX road map currently shows three releases of the network specifications: 1.0, 1.5, 1.6, and 2 (see Figure 6.7).

The initial release of WiMAX Network Release 1.0 targeted basic mobile Inter-
net services using an all-IP architecture supporting both integrated and decomposed
ASN profiles. This release supports all basic features needed to enable early
WiMAX deployments, including:

- ASN, CSN mobility (for mobility support)
- Paging and location management

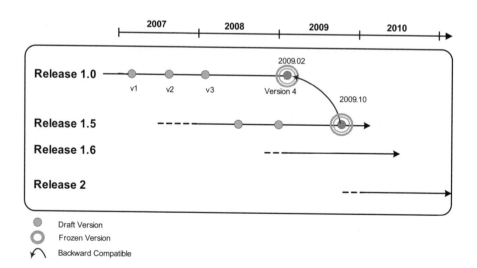

Figure 6.7. Release planning of network specification.

- IPv4 and IPv6 connectivity
- Preprovisioned/static QoS
- Optional radio resource management (RRM)
- Network discovery/selection
- IP/Eth CS support
- Flexible credentials, pre- and postpaid accounting
- Authentication, authorization, accounting, and roaming (RADIUS only)

6.3.1   Release 1.5 Network

The WiMAX Network Release 1.5 development started in 2007 with focus on enabling retail distributed devices, advanced managed IP services, and 3G interworking as well as commercial grade VoIP. The following are some of the key features of network Release 1.5:

- Over-the-air (OTA) activation and provisioning
- Location-based services (LBS)
- Multicast broadcast service (MBS)
- IMS integration
- Dynamic QoS and policy and charging (PCC) compatible with 3GPP Release 7
- Telephony VoIP with emergency call services and lawful interception
- Handover data integrity
- Multihost support
- Ethernet services, VLAN, DSL IWK
- Enhanced open-Internet services
- DIAMETER-based AAA

Release 1.5 also defines the normative R8 reference point between BSs and discontinues the ASN profile A, leaving only one R6 definition in a single decomposed ASN configuration (Profile C).

6.3.2   Release 1.6 Network

The work on WiMAX Network Release 1.6 began at end of 2008 and went full scale in the first quarter of 2009. Five major work areas have been signed up for by network working group members, as follows:

1. Femtocell and self-organization network (SON)
2. Architecture evolution
3. WiFi-WiMAX interworking
4. IPv4–IPv6 transition
5. OTA evolution and DRMD (device reported metrics and diagnostics)

6.3.3   Release 2.0 Network

At the time of this publication, the full scope of Release 2.0 of the WiMAX Network has not been finalized and it is expected to enable E2E features to complement enhancements made in System Profile Release 2.0 based in IEEE802.16m. WiMAX Forum Network Working has approved the following work items for release 2:

- Evolution to support the advanced air interface (IEEE 802.16m)
- WiMAX Voice Over IP Service
- Multiple IP Support
- WiMAX Network Support for Emergency Telecommunications Service Phase 2
- Local Routing
- Inter NAP Handover
- OTA/DRMD Release 2

6.4   OVERVIEW OF MAJOR FEATURES IN RELEASE 1.0

The WiMAX Network has been designed for the users who want to use Internet service over a broadband packet wireless network. WiMAX Network Release 1.0 features are indispensible for achieving this ultimate goal. This section describes the major features on top of the simplest WiMAX network architecture in the world.

When a user tries to connect to the wireless Internet, he wants to select one of the wireless network if there are multiple access providers. On other hand, the operators want to cover as large as possible area for users to use their IP service. The feature called network discovery and selection makes this possible in the WiMAX Network.

Remember that the WiMAX network has two different types of operators. one that provides access service is the network access provider (NAP) and the other one provides connectivity service is the network service provider (NSP). Once a user subscribed to a NSP, when his mobile or laptop turns on, it needs to select one of the NAPs that connect with the NSP he subscribed to, so WiMAX has this feature.

Once the user or his/her mobile selects the NAP and NSP, the mobile or the laptop need to connect with the WiMAX network. The mobile starts random access to the network, negotiates the necessary parameters with the network, is authenticated by the network, is allocated over-the-air resources for the logical over-the-air connection to transfer packets, it finally gets an IP address from the network that provides IP connection to the user. This whole process is called the network entry process.

The user who tries to use the WiMAX network should be authenticated. The WiMAX network provides flexible architecture by using the IETF-designed Extensible Authentication Protocol. Since WiMAX supports the EAP framework, the WiMAX Network can use the existing authentication methods running on top of the

EAP. WiMAX has chosen EAP-TLS for the device authentication and the EAP-TTLS with MSCHAPv2 and EAP-AKA for the subscription authentication.

After authentication is done for the user, the mobile should be given an IP address, of course, to connect to Internet. Although there had been an IETF-designed, IP-based mobility protocol called Client-based mobile IP technology, the IP allocation protocol stack Dynamic Host Configuration Protocol (DHCP) was mainly adopted in the major operating systems as the participants in the WiMAX Forum preferred to have a solution for existing DHCP stack on the mobile side. Proxy Mobile IP is the technology that can provide IP-based mobility without requiring a mobile IP stack in the client.

6.4.1 Network Discovery and Selection (ND&S)

The purpose of ND&S specification is to define how and what information to supply to the MS over R1 so that it is able to identify all the NSPs residing behind the received NAP(s). The ND&S procedures also define how the MS should use the information about the selected NSP and notify the NAP about its decision so it will be able to route the MS traffic to the right NSP.

The important point is that the MS gets its connectivity from the CSN that is owned by the NSP but the only information the MS can receive prior to establishing a connection to the network is trough R1 (R2 is also based on R1 transport for the air-link part), which is the ASN information managed by the NAP.

The ND&S algorithm includes 4 steps:

1. NAP discovery—scanning for available NAPs
2. NSP discovery—identify the NSPs behind the discovered NAPs
3. NSP selection—choosing the NSP to connect to
4. ASN attachment—performing NW-entry with the chosen NSP

In these four steps, the NSP is identified by a unique NSPiD (24 bit) that is mapped into an NSP verbose name and a realm. The NAP is also identified by a unique NAPiD (if the NSP owns the NAP, it will be identical to the NSPiD), which is the 24 most significant bits of the BSiD transmitted as defined by IEEE802.16e (see Figure 6.8).

6.4.1.1 NAP Discovery

The MS detects available NAPs by performing a scan procedure as defined by IEEE802.16e/.16-2009. The MS decodes DL-MAPs/DCD received on scanned channels and uses the 24 most significant bits of the BSiD as the NAPiD.

6.4.1.2 NSP Discovery

The NSP discovery step is where the MS detects all NSPs that are directly connected to all discovered NAPs (see Figure 6.8). The detected NSPs are identified by the 24-bit NSPid and by the NSP verbose name, which can be presented to the user in case of manual selection.

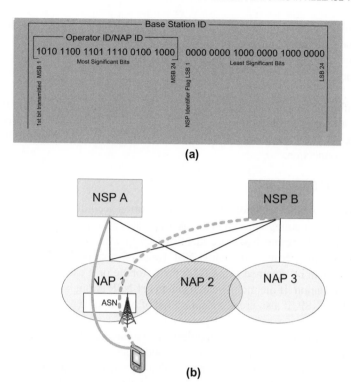

Figure 6.8. ND&S: (a) splitting BS-ID into BS and NAP ID fields, (b) NSP selection concept.

In order to perform the NSP discovery step, the MS needs to have a mapping between NAP and NSP. The source for this mapping can be an advertisement message by the ASN of all NSPs directly connected to the NAP, or an internal configuration file on the MS (obtained by provisioning and by the ASN advertisement). Once MS has the mapping information for a specific NAP, it can create a list of all NSPs connected to this NAP.

After the MS establishes such mapping for all received NAPs, it has a list of all NSPs that can be directly reached through any discovered NAP. For each NSP, the list also includes the list of all discovered NAPs that can be used for connecting to that NSP in prioritized order (the order is based on user/operator policy, PHY parameters, etc.).

NAP provides two methods to advertise NSP information for the MS to establish the NAP-NSP mapping formation as follows:

1. Broadcasting method. BSs belonging to the same NAP broadcast SII-ADV messages periodically (or per request). The SII-ADV message includes the list of all NSP IDs and verbose NSP names directly connected to the NAP.

2. Solicitation method. If the mobile includes a SIQ-TLV in the SBC-REQ to request the same information as broadcast by the SII-ADV, the BS responds it by sending a SBC-RSP including the list of all NSP IDs and verbose names.

NSP information and the change counter of this information must be uniformly set across all base stations within the same NAP so the MS can skip this procedure during a handover while using the most updated information stored internally from the former connection.

6.4.1.3 NSP Selection

In preparation for NSP selection, the NSP list is sorted according to type of discovered NSP (home/roaming/other) and the user/operator policy, which is out of scope for this chapter. Roaming NSPs are identified by internal information on the MS that maps roaming partners of each H-NSP (this information is not advertised by the ASN as it can be obtained using OTA provisioning or it can be preprovisioned in the mobile station. If the mobile can reach the H-NSP through more than one detected NAP or NSP, it also creates a sorted list of all possible V-NSPs for each H-NSP.

The WiMAX Network specification defines two methods for NSP selection: automatic and manual selection. In the manual mode, the list is presented to the upper layer (user or application) and it is expected that a connect command with the NSP chosen by the user will be received. In the automatic mode, the MS automatically selects the first NSP in the sorted list to connect to. Based on the NSP selection, the MS chooses the highest priority NAP linked to this NSP to connect to.

6.4.1.4 ASN Attachment

ASN attachment involves performing network entry with the selected NSP using the selected NAP, which means connecting to one of the serving base stations according to radio/PHY parameters or other criteria.

In order to notify the NAP which selected NSP to connect to, the MS uses the selected NSP's realm in the NAI of the EAP authentication (user@realm). The ASN-GW knows to route the traffic to the correct NSP based on the supplied realm.

In the case of roaming, the NAP is not connected to the target NSP and will not be able to route it directly and, therefore, it also needs to know through which V-NSP to send the traffic. In this case, the MS should append a "routing realm" decoration to the NAI ([routingrealm!]user@realm) that tells the NAP which NSP to use for the first hop (in case there is only one NSP connected to the NAP, the routing realm may be omitted as the NAP will use the only possible NSP).

6.4.2 Mobility Support

The network supports two tiers of mobility architecture. ASN anchored mobility occurs on every base station handover, irrespective of IP subnet/prefix or Mobile IP foreign agent changes. This handover is completely transparent to the core network (CSN) as the ASN functions manage data path changes within or between ASNs to maintain the UL and DL bearer path for the MS. The CSN anchored handover is triggered when the foreign agent or IP subnet/prefix changes. This handover procedure is based on the Client or Proxy Mobile IP for IPv4 or IPv6. For more details on end-to-end mobility in the WiMAX System, see Chapter 8.

6.4.3   IP Address Management

The WIMAX network can support both IPv4 and IPv6 addressing and classification based on the IP convergence Sublayer of IEEE802.16-2009. The IPv4 support may take three forms:

1. Client Mobile IPv4 with mandatory foreign agent support
2. Client DHCP in "simple" IP configuration for nomadic access deployments
3. Client DHCP with Proxy Mobile IPv4, in which the ASN provides mobile node (MN) functionality as a proxy transparent to the client

IPv6 address assignment can be based on stateless address autoconfiguration or DHCPv6, and its mobility is supported via Client Mobile IPv6.

6.4.4   Idle Mode and Paging Operations

In mobile communications systems, when a mobile moves around without using actual data communication, the modem does not have to be awake all the time. It can enter into a power-saving mode such as idle mode. During the idle mode, the terminal has an idle period of time so that it can save battery consumption. After the mobile or the system detects no packet transmission during a certain amount of time, the mobile or the system can initiate entry into the idle mode. During this idle mode, the mobile is supposed to wake up for a short period of time periodically in order to check if the data packet comes to the mobile. While the mobile stays in idle mode, the serving BS does not have to maintain a traffic connection associated with the mobile. The BS stores the entire context for the idle mobile in the central point, which is the ASN-GW (paging controller). These stored contexts and negotiated parameters are restored when the mobile wakes up to communicate with the system. When the mobile is about to send an IP packet to the network in the idle mode state, the mobile has to wake up by sending a ranging message (RNG-REQ). The BS requests the ASN-GW (paging controller) to retrieve all the stored contexts. After retrieval of the context, the mobile and the BS can have a traffic channel and are ready to transfer data packets to each other. When the packet comes toward the mobile during idle mode, the paging controller detects that packet is coming and pages the mobile by advertising this event to one or more BSs with which the mobile was supposed to have stayed recently. The BS pages the mobile during the interval when the mobile is supposed to listen. If the mobile received this paging event successfully, the mobile wakes up and can receive the packet directed to it. WiMAX Network Release 1.0 fully supports all these operations.

6.4.5 Security Support

One of the indispensible features of wireless packet communication is security. Since WiMAX adopted IETF-based specifications as much as possible, the authentication framework is also based on the IETF-based Extensible Authentication Protocol (EAP) defined in RFC3748. Although the EAP framework supports various kinds of EAP methods, the WiMAX Forum further narrows down the choices of EAP methods for interoperability so that operators and terminal vendors can implement it. The WiMAX Forum Network Working Group has decided to use EAP-TLS for the device authentication and has chosen the EAP-TTLS with MSCHAPv2 and EAP-AKA for subscription authentication.

After the mobile is authenticated with the chosen method to enter the network, the mobile and the ASN-GW(authenticator) possesses the primary master key (PMK) as a result of successful authentication. By having the PMK, the mobile is permitted to access the network system. While the mobile possesses the PMK, there are BSs connected to the ASN-GW. In order to protect the other BSs in a situation where one BS is exposed to a security threat, the WiMAX security team has designed levels of key hierarchy so that the BS has one level of key hierarchy down from the authenticator (ASN-GW). In other words, each BS has a unique key named AK which can be derived from the PMK owned by the ASN-GW(authenticator). All the MAC layer management message exchanges are signed by the AK, so we secure the all the MAC management messages.

To secure traffic connection, the IEEE 802.16e over-the-air interface provides security association (SA) and cipher suites. One of them is AES-CCM, which was chosen by the WiMAX Forum. Since AES-CCM provides both encryption and integrity protection, all the traffic over the WiMAX air interface can be protected. For a more detailed end-to-end description of security support in WiMAX networks, see Chapter 9.

6.4.6 Quality of Service (QoS) Support

IEEE802.16e access technology supports various type of scheduling algorithms to provide the categories of service listed in Table 6.2. To initiate and enable the various kinds of over-the-air service flows with different types of QoS, the network-initiated service flow creation and mobile-initiated service flow creation procedures are supported.

As a part of the initial entry procedure, the preprovisioned service flows can be established for each mobile. The policy and QoS information of the preprovisioned service flows are normally provisioned in the AAA server and downloaded into the ASN-GW during the authentication procedure. The MS can make use of these service flows after the initial network entry procedure.

Table 6.2. Quality of service categories (IEEE 802.16e)

IEEE 802.16e Quality of Service	Example Applications
Unsolicited grant service (UGS)	VoIP without silence suppression, EI/T1
Real-time polling service (rtPS)	MPEG video
Extended real-time polling service (ertPS)	VoIP with silence suppression
Non-real-time polling service (nrtPS)	File transfer
Best-effort (BE) service	Various IP data applications including IP-based signaling (e.g., DHCP etc.)

In addition to the preprovisioned service flows, the MS and the network can initiate the dynamic QoS service flows by using the mobile-initiated service flow creation and activation procedure and/or the network-initiated service flow creation and activation procedures. For a more detailed end-to-end description of QoS in WiMAX networks, see Chapter 10.

6.4.7 Radio Resource Management Support

Radio resource management is a function performed by the BS in a WiMAX network to increase the radio resource usage efficiency. If RRM is supported, radio resource control and radio resource agent functions are located in the BS.

Two types of message primitives are supported. One is Per-BS space capacity reporting and the other is Per-BS radio configure update. Per-BS space capacity reporting procedure may be used by a BS to retrieve information about the current load situation of any other BS, in particular neighboring BSs. The Per-BS radio configure update procedure may be used by a BS to report some critical resource configuration update to the serving BS, such as DCD and UCD burst profile changes.

6.4.8 Initial Network Entry Procedure

This section provides an overview of initial network entry procedures to better understand the ASN/CSN architecture. These procedures are described in multiple phases as follows (see Figure 6.9):

- Phase I. Random Access and Ranging Process. After the MS decides to enter the network through the BS, the MS starts random access ranging followed by message ranging. The BS grants the MS access and allocates the basic and primary connection ID to the MS.
- Phase II. Basic Capability Negotiation Procedure. After successfully performing the ranging procedure, the MS performs the basic capability negotiation procedure, in which basic physical parameters can be adjusted and security parameters negotiated.
- Phase III. EAP-based Authentication Procedure. The EAP procedure will fol-

low the basic capability negotiation procedure. The ASN-GW acts as an authenticator of the EAP architecture, triggering the procedure by sending an EAP-request/identity, and the MS as a supplicant responds by providing the decorated NAI. After the AAA receives the access request from the ASN-GW carrying the EAP-response/identity, including the decorated NAI that the MS sent, the AAA starts the EAP method-specific procedure. An EAP conversation is performed among the MS, the ASN-GW, and the AAA. After the AAA successfully authenticates the MS, the ASN-GW downloads the policy and profile information for the given MS from the AAA.

- Phase IV. Security Association Negotiation with Key Agreement Handshake and Registration Procedure (Negotiations). Right after successfully performing the EAP conversation, the ASN-GW as the authenticator generates the AK (authorization key per the BS) and sends it to the BS to initiate the security association and key agreement three-way handshake, which called a SA-TEK three-way handshake. During this procedure, the MS and the BS negotiate the available security cryptographic suites and confirm the negotiated security parameters. After negotiating security parameters, the MS and the network negotiate MAC-level parameters by performing the registration procedure. During this procedure, the MS and the network negotiate various MAC parameters such as ARQ and convergence sublayer capabilities.

- Phase V. Preprovisioned Service Flow Establishment. After successfully completing the registration procedure, the MS and the ASN establish the preprovisioned service flows. During this procedure, the MS and BS set up the service flows that can carry the basic IP signaling; for example, DHCP and additional preprovisioned service flows that the AAA has configured for the MS. During this procedure, the BS and the ASN-GW also set up a GRE tunnel over the R6.

- Phase VI. Dynamic Host Configuration Protocol with Proxy Mobile IP Registration Procedure. Once the service flows have been established, the MS initiates a host configuration procedure by sending a DHCP DISCOVER message. In case the network has been configured to support Proxy Mobile IP, the ASN-GW sends a registration request to the HA. The HA allocates the home address for the MS and responds back to the ASN-GW by sending a registration reply. After the MS has been allowed to use the IP address by receiving the DHCP ACK, it now can access the Internet with the allocated IP address.

6.5 OVERVIEW OF MAJOR FEATURES IN RELEASE 1.5

WiMAX Network supports all the fundamental features in Release 1.0, but there are enhancements and additional features introduced and enabled in Release 1.5. In Release 1.5, three kinds of enhancements have been standardized (see Figure 6.10).

The first type of the enhancement is access network enhancement. In this category, the feature called "handover data integrity" has been designed in order for the remaining data buffer of the serving BS to the target the BS during active hard han-

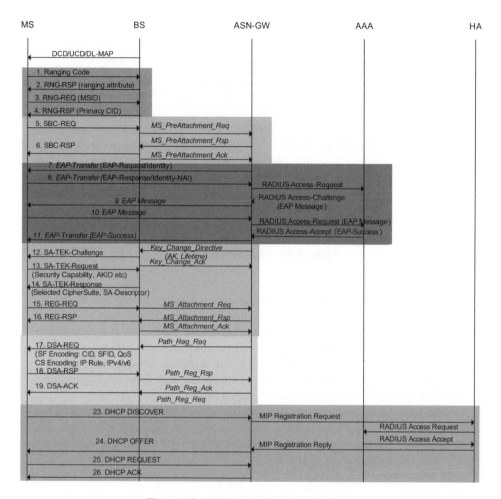

Figure 6.9. Initial network entry procedure.

dover. The second feature of this category is to enable robust header compression in the ASN-GW. The feature of robust header compression (RoHC) is to compress the IP and its upper layer header into one or two bytes so VoIP traffic can be utilized with this feature. The third feature of this category is to support dynamic establishment of QoS-enabled traffic connection for the mobile. It can be either network initiated or mobile initiated. The fourth feature of this category is to support the normative R8 reference point. The control signaling and bearer communication over this reference point was informative but in Release 1.5 the control signaling part became normative in the WiMAX network architecture so that the WiMAX network fully supports BS-to-BS communication in a standard way.

The second type of the enhancement is core network enhancement. The biggest change of the architecture is "simple IP" supports. In Release 1.0, the WiMAX net-

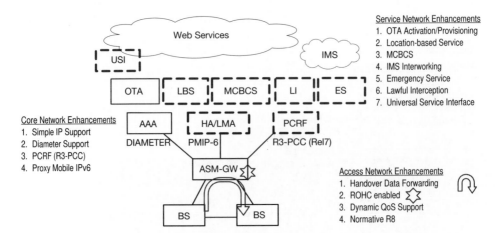

Figure 6.10. Features in Release 1.5.

work always has mobile IP-based communication between the ASN-GW and the HA. In Release 1.5, the traffic from/to the mobile can pass through the CSN without a mobile IP tunnel between ASN-GW and HA in the CSN. This architectural enhancement makes it network deployable without HA. The second enhancement is to support DIAMETER as the authentication, authorization, and accounting (AAA) protocol. In Release 1.0, only RADIUS was supported; in Release 1.5, the operators can choose one of the AAA protocols based on their needs.

The policy and charging control is also one of the biggest improvements of Release 1.5. As well as IP multimedia subsystem support, other application services that require over-the-air quality of service can utilize the PCC designed in Release 1.5. The PCC architecture of the WiMAX Network is based on 3GPP Release 7 with added mobility features as well as enhancements to support the WiMAX access network.

The IP-based mobility was based on Proxy Mobile IPv4 and Client Mobile IP in Release 1.0. In addition to the Release 1.0 IP-based mobility, Release 1.5 also has added the Proxy Mobile IPv6-based mobility feature based on the IETF-based PMIPv6 RFC.

The third type of the enhancement is service network enhancement, which usually requires a standalone server in the core network of the network service provider. Those features have been built on top of the underlying WiMAX access and connectivity services. In this category are over-the-air activation and provisioning, location-based service, muticast and broadcast service, emergency service, lawful interception, and IP multimedia services.

6.5.1 Over-the-Air (OTA) Activation and Provisioning

Over-the-air activation and provisioning service provides the online subscription activation for the subscriber and the provisioning service on the terminal that the subscriber is about to use. The OMA-DM protocol is chosen to manipulate the data objects stored in the terminal device so that the operators can provision and manage the configuration parameters and policy of the terminal device. During the over-the-air activation and provisioning procedure, the operators can configure and provision the user credential, device locking policy, channel plan, and CAPL/RAPL information to the terminal device. For more details on the OTA protocol in the WiMAX network, please see Chapter 7.

6.5.2 Location-Based Services (LBS)

Location-based service provides the location service for the terminal that belongs to the WiMAX network. Two kinds of mechanisms have been developed. One is to use GPS in order to get the exact location of the terminal. The other is to use TDOA. In order to get the location information for the three physical locations, the mechanism designed in the WiMAX forum and IEEE802.16 is used to support the TDOA feature. In the protocol perspective, for the location server to know the location of the terminal, OMA-SUP-L and IETF-based protocols have been developed. For more details on end-to-end support for LBS, please see Chapter 13.

6.5.3 Multicast and Broadcast Services (MCBCS)

A multicast broadcast service (MCBCS) involves a network's ability to provide common content to multiple users. MCBCS services can be broadly categorized into multicasting or broadcasting services. In multicast service, users may dynamically join and leave a multicast IP session and the system may monitor the number of users at each MCBCS zone to decide on data transmission and its mode. In particular, a subset of the multicast service is supported in Phase-1 MCBCS—static multicast service. In static multicast service, the content is always transmitted through one or more broadcast channels that have been preestablished prior to the user(s) joining and leaving a MCBCS session at one or more MBS zones that belong to the same MCBCS transmission zone. A subscription-based broadcast service is considered to be static multicast service. The broadcast service is a special type of MCBCS for which the content is always transmitted through broadcast channels by the access network without considering the number of MSs receiving the content from the BSs involved. For more details on end-to-end support for MCBCS, please see Chapter 12.

6.5.4 Emergency Services

Emergency services (ES) provide the framework for WiMAX networks to support WiMAX commercial VoIP services. Even though ES handling for specific VoIP technologies is not in the scope of the WiMAX specification, it can be supported in the way that IMS is supported. One important aspect to support emergency service is the location-determining feature, which is supported by the location-based service. It supports VoIP services provided over WiMAX by a VoIP service provider

(VSP) that can either be part of an NSP or can be a third-party commercial VSP with contractual relationship with the WiMAX NSP. The WiMAX emergency services framework supports communication between a VSPs VoIP server and the PSAP over both VoIP protocols or PSTN-based protocols. The specified scenarios and mechanisms clearly focus on the part that a WiMAX network, providing broadband packet data access needed to enable IP emergency calls. Also, the network reference model for the WiMAX ES architecture has been aligned with existing architectures to support IP emergency calls. JSTD036B specifies a network reference model that supports the interfaces between a circuit-switched cellular network and an ES network, and NENAi2 recommends a migratory solution architecture for supporting emergency calls originated in the IP domain (in this case the WiMAX network) and terminated using the PSTN ES infrastructure.

6.5.5   IP Multimedia Subsystem Support

The IP multimedia subsystem (IMS), mainly developed in 3GPP, is also supported in the WiMAX network. The WiMAX network can interwork with 3GPP Release 7 and the 3GPP2 IP multimedia subsystem (IMS). IMS is an open, standardized, multimedia architecture for mobile and fixed IP services originally defined by the Third-Generation Partnership Project (3GPP), and largely adopted by the Third-Generation Partnership Project 2 (3GPP2). The IMS is based on standardized IETF protocols (e.g., SIP, DIAMETER, and RTP). It is an access-independent platform providing services in a standardized way. One of the main purposes of the IMS is to provide different kinds of services from any location or access network in which the IMS can be reached. It also has the architecture designed to allow dynamic quality of service selection and flexible charging models (e.g., service-based and flow-based charging).

6.5.6   WiMAX Lawful Interception

The WiMAX lawful interception feature provides the WiMAX architecture for lawful intercept (LI) of WiMAX-based broadband access services and a conceptual overview of the LI service and LI model for WiMAX for voice, data, and IP-associated signaling. The current version of the LI feature in WiMAX covers only packet data services. Since laws of individual nations and regional institutions, and sometimes licensing and operating conditions, define a need to intercept telecommunications traffic and related information in modern telecommunications systems, it must be noted that lawful interception shall always be done in accordance with the applicable national or regional laws and regulations. Currently, WiMAX LI provides the detailed specifications for North American regulation.

6.5.7   Universal Services Interface (USI)

The universal services interface (USI) that WiMAX supports is an interface with trusted third-party iASPs and iASPs [SLA between the network service provider (NSP) and iASP may be required]. These network interfaces allow exposure of

WiMAX network capabilities and mobile user information between the NSP and iASP in a secure and controlled manner. The USI capability can be used as a differentiating feature in applications such as:

- Messenger with Wireless Paging Capability. When the sender finds that the receiver is in idle mode (turned off), he/she is able to click the paging icon of the receiver to wake up the receiver in order to participate in the current messenge service when the sender wants to send any message to the idle user. USI can provide the paging command from the Internet application service provider (messenger application) to the ASN to wake up the receiver.
- Video Streaming with Wireless QoS Capability. When the user is receiving video streaming in a best-effort manner and he/she wants to get higher speed and improved quality, he/she can click the QoS icon on the screen of the video source. Then the QoS with PCC can start being applied over the air and the video traffic can be transmitted with high speed and broadband bandwidth. USI can provide the QoS request from the iASP (video content provider) to the ASN to request QoS from the instant that the user wants to get. This mechanism can be easily applied to MCBCS with wireless QoS capability when the multicast and broadcast traffic needs to be transmitted in ertPS, not in best-effort mode.

Based on these and other examples, one can argue that the USI is intended:

- To create a solution for service providers to generate additional revenue
- To expose network-capability APIs to the iASP for dynamic service creation (in a controlled manneor for business purposes)
- To reuse network intelligence being built into WiMAX (e.g., LBS, MCBCS, and PCC)

The network reference model (nonroaming case) of USI is shown in Figure 6.11. A new CSN network element called the USI system has been added to the exist-

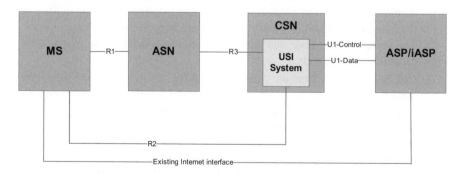

Figure 6.11. USI network reference model (nonroaming).

ing WiMAX NWG architecture. The USI system interfaces with the iASP through the U1 interface. The USI System can optionally interface with MCBCS servers, PCRF, AAA server/proxies, and location servers. The U1 interface carries control traffic (e.g., user information such as location/presence, QoS parameters, and accounting) between the USI system and the iASP. The R2 interface carries control messages between the MS and the USI system (e.g., USI registration messages). The U1 reference point is composed of U1-Control and U1-Data. The U1-Data interface on the downlink carries the data from the iASP, which is intended for the MS. Such data may then be delivered by the USI system to the MS by using the unicast or multicast data paths in the NAP and NSP. This is out of scope of the USI specification. On the uplink, the U1-Data interface may carry data traffic of the MS from the USI to the iASP. The data may be delivered from the MS to the USI system via the unicast or multicast data paths in the NAP and NSP. This is out of scope of the USI specification. The U1-Control interface is meant for carrying all the signaling and control-related exchanges between the iASP and the USI system.

For the roaming case and more detailed explanations of the USI, please refer to relevant WiMAX Forum specifications [4].

6.5.8   Miscellaneous Features Added in Release 1.5

Besides the basic and advanced features described previously, the following essential features are added in Release 1.5:

1. Keep Alive. With this feature, each network element can report its "aliveness" to the network node that it communicates with. For example, the BS may send the periodic report of its aliveness to the ASN-GW. Due to this feature, each network element can ensure that the communicating entity is alive and it may also inform the communicating entity the load information so that the information can be used for various purposes; for example, selection of the ASN-GW (see R6-Flex).

2. Version and Capability Negotiation. Capability negotiation signaling procedures can be used in order for the network entity to signal the version and capability support information to the network entity that it is intended to communicate with. The originator of the message lists (a) all the capabilities or (b) individual capabilities that it can support, together with release number. The corresponding node can answer the selected subset of the releases and capabilities of the node.

3. Certification and Version Signaling (CVS). Using this capability, the WiMAX Forum certified MS can inform the AAA that seves as operator's policy server of its certification status, by decorating the NAI with the MS's certification registry number. The AAA that is meant to communicate with the WiMAX Forum's certification database looks up the certification information about the device and delivers the list of certified and, therefore, acceptable features to the ASN so that the ASN-GW or the BS can enable or disable the feature for the MS as appropriate. This feature is essential to en-

able operation of retail distributed devices in WiMAX networks, without requiring the operators to have acceptance testing of such devices.

6.6 MAJOR FEATURES IN NETWORK RELEASE 1.6

Based on new service and deployment requirements in the WiMAX Forum, the work on Network Release 1.6 was begun in the first quarter of 2009. The five major work areas have been approved and are under development as follows.

6.6.1 Femtocell and Self-Organization Network (SON)

A femtocell is mainly intended to provide coverage over small areas such as the home and SOHO (small office, home office) environment. A femtocell system comprises of a femtocell access point and other additional network functions that may be necessary to provide service through that femtocell access point. Femtocells typically use a wired backhaul (e.g., DSL/cable connection at home) that is different from the wireless operator backhaul.

A key benefit to the end user is that he or she can have seamless mobility between the femtocells in the home, office, or a coffee shop and the base stations in wide area wireless network, since they use the same technology and share a common spectrum. A key benefit to the operator is that the operators can intelligently move the users from the base stations in the wide area to the femtocells as and when they see fit, so that they can reduce their load on their wide-area radio network and their backhauls, and, thereby, achieve better load balancing and capacity planning.

WiMAX Network R1.6 will support the femtocell system for, at a minimum, Release 1.5 and earlier WiMAX mobile, including following aspects:

- Femtocell system network reference mode and associated interface definitions
- Femtocell end-to-end quality of service
- Femtocell active mode mobility (i.e., hand in/out between Femtocell BS and macro BS)
- Femtocell idle mode mobility (i.e., rove in/out between Femtocell BS and macro BS)
- Femtocell BS certification, authentication, authorization and provisioning
- Femtocell subscriber and configuration management
- Femtocell configuration management

In addition to the Femtocell operation, self-organizing networks (SON) of femtocells are being designed to enable the radio and network components to interact among themselves, and to configure and tune the mobile system automatically in real time.

SON is a process that involves network elements (NEs) in radio access networks

(RAN) and core networks to enable automatic configuration, to measure/analyze KPI (key performance indicator) data, and to fine-tune network attributes in order to achieve optimal performance. It helps the network operators to reduce OPEX by minimizing human intervention in both deployment and operational phases. It is especially important to femtocells. Otherwise, operators will not be able to ramp up femtocell AP deployment to huge volumes quickly.

Self-organization network is includes the following technical aspects: automatic configuration, reconfiguration and update of PHY/RF parameters; MAC parameters and other parameters such as access control list, and time synchronization server address; automatic fault detection and recovery; self-test; and automatic data/information collection and reporting.

6.6.2 Architecture Evolution

The WiMAX network has to further evolve the baseline network specifications and the network architecture, including ASN, CSN, MS, and the NWG-specified reference points. The main enhancements of the architecture are as follows:

- R6-Flex. The BS can connect to one ASN-GW among multiple ASN-GWs within the same ASN. The selection of the ASN-GW for each MS can be decided either by the BS or by the default ASN-GW.
- R4 Optimization. In cases where the combined functions are collocated in the ASN-GW, the ASN-GW can relocate its authenticator, paging controller, and the anchor data path function in the combined way (Emergency Telecommunication Service Phase I).

6.6.3 WiFi–WiMAX Interworking

There is a need for WiFi and WiMAX systems to interwork with each other to improve the overall user experience. This can be realized in terms of better usability of the two networks and support for seamless mobility between the two systems. The WiFi system can use the authentication, authorization, and charging mechanisms provided by the WiMAX system, thus leading to better integration on the network side and giving a consistent experience to end users in terms of using the same set of credentials to access the two systems. The use of common or integrated charging mechanisms can result in a common bill to end users. The subscriber and operator may want to switch between WiMAX and WiFi to extend coverage, improve load balancing, and provide better cost structure, all targeting better user experience. Service continuity needs to be supported during such transitions, optimizing for time, performance, and power.

To achieve these goals, the following aspects will be developed in the WiMAX network:

- Common accounting. One bill from mobile operator for the users of both technologies.

- WiMAX-based access control. The IWK subscription should be anchored in the WiMAX NW using the same AAA and core network functionality.
- Access to WiMAX CSN-based Services. Allows an operator to extend WiMAX CSN-based services to the WiFi access.
- Service continuity. Maintain all active sessions during the transition without need to reestablish any service.
- Seamless handovers. No noticeable interruption of the active services during transition between access technologies.

6.6.4   IPv4–IPv6 Transition

Release 1 and Release 1.5 collectively support simple IP (for IPv4 and IPv6) and IP mobility for IPv4 and IPv6. However, the feature currently presumes end-to-end connectivity either for IPv4 or IPv6. To smooth the transition to IPv6 and make IPv6 virtually backward-compatible with IPv4, some additional tools are needed. WiMAX has a plan for the evolution of IPv4-only to dual-stacked nodes in the MS, ASN, and CSN, which will also support simultaneous IPv4 and IPv6 services to the same device.

6.6.5   OTA Evolution and Device-Reported Metrics and Diagnostics

The motivation for focusing on OTA in the WiMAX network was the need for a framework to activate generic and retail distributed devices to work with a WiMAX-compliant operator network. The result of the first stage of this effort was a powerful yet flexible device-management framework that uses generic, widely used protocols (e.g., OMA-DM and WEB) in a WiMAX-based, secure end-to-end system. Meanwhile, many other features are being standardized in the WiMAX forum and will in the future leverage the OTA framework to achieve their goal and/or require enhancements in the OTA functionalities. For example, the features in the WiMAX networks that are already known to be using the OTA framework are:

- DRMD (device remote monitoring and diagnostics). This uses the OTA framework to initiate a collection of parameters in the device, obtain them, analyze the device problems, and fix them remotely if possible.
- ANDSF (access network discovery support function). This uses OTA to allow the device to obtain all potential access networks (WiMAX, Wi-Fi, and Cellular) available in its surroundings that it can interwork with, a critical functionality for handover between access technologies.
- Device-initiated QoS. This provisions the device with the allowed QoS level it can request upon opening service, to ensure smooth operation.
- Fixed nomadic. This limits the device to act as a nomadic device and not a mobile, to ensure the expected behavior per type of account.

Practically any feature that requires configuration and parameter exchange between the device and core network can leverage and most probably will leverage the OTA framework to achieve its functional goals. Therefore, looking into the future, the OTA protocols and their enhancements are key parts of WiMAX network evolution.

6.7 COMPARISON OF MOBILE WiMAX AND 3GPP/SAE NETWORK ARCHITECTURE

In this section, we briefly compare the Mobile WiMAX network architecture and 3GPP SAE/LTE network architecture [5] because LTE (long-term evolution) is also one of the 4G technology candidates.

WiMAX and LTE are well positioned as 4G access technologies to drive the global evolution toward pervasive wireless broadband communications with market acceptance, rich ecosystems, and promising economies of scale.

4G networks are being specified and developed by vendors, even though operators are still deploying 1xEV-DO and HSPA networks. The data rates of 4G networks enable operators to offer broadband wireless access (BWA) to keep pace with wireline broadband networks. In addition, 4G networks will complete the process of evolution to all-IP architecture. As such, two major networks, WiMAX and LTE, have emerged for 4G deployment. Both networks are based on basic OFDMA technology despite differences in implementation.

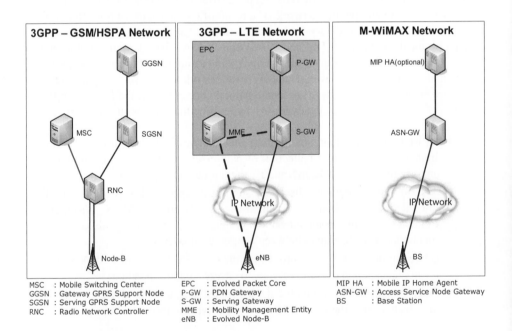

Figure 6.12. Similarity between the LTE network and the Mobile WiMAX network.

As seen in Figure 6.12, both M-WiMAX and LTE benefit from a simple peer-to-peer network architecture that features a flat, all-IP-based design. This enhanced, flat IP architecture significantly reduces the complexity of the network, offers the benefit of faster handovers, provides the foundation for versatile service delivery, and offers freedom in network scalability.

6.8 SUMMARY

Mobile WiMAX technology has been developed and promoted by a large system of operators and vendors aiming at low cost and scalable realization of fixed and mobile network convergence. The technology is utilizing advanced PHY and MAC techniques in radio to provide high-spectrum efficiency and QoS control, as well as an IP-based, flat network architecture supporting multivendor plug-and-play deployments. Mobile WiMAX has defined the technology evolution road map for the next few years, which includes, but goes beyond, further improvements in system efficiency and user experience. The next generation of mobile WiMAX is expected to provide flexible deployment solutions such as multihop, femtocells and multicarrier support, as well as optimized coexistence and interworking with other access technologies such as WiFI, Bluetooth, and 3G systems. While the combination of mobility and Internet will enable new killer applications and low-cost service models, making such services available to a wide range of embedded retail devices, including consumer electronics, is expected to bootstrap the global adoption of this technology in the next few years.

6.9 REFERENCES

[1] Air Interface for Fixed and Mobile Broadband Wireless Access Systems, IEEE 802.16e-2005

[2] WiMAX Forum™ Mobile System Profile Release 1.0. Approved Specification (Revision 1.4.0: 2007-05-02), WiMAX Forum.

[3] WiMAX Forum Network Architecture Stage 2 and 3 Documents—Release 1 (WMF-T33-001/002/003/004-R101v04_Network-Stage-3-Base and WMF-T32-001/002/003/004/005-R010v04_Network-Stage-2), http://www.wimaxforum.org.

[4] WiMAX Forum Network Architecture Stage 2 and 3—Release 1.0 (Revision 1.5), WMFT33-001/002/003/004/005-R015v01_Network_Stage2, www.wimaxforum.org.

[5] 3GPPTS23.401 v9.3.0, 2009-12, General Packet Radio Service (GPRS) Enhancements for Evolved Universal Terrestial Radio Access Network (E-UTRAN) Access, Release 9, 2009.

CHAPTER 7

OVER-THE-AIR (OTA) PROVISIONING AND ACTIVATION

AVISHAY SHRAGA

7.1 INTRODUCTION

One of the most important aspects of WiMAX operation is over-the-air (OTA) provisioning and activation, which enables a user to obtain WiMAX service and create business relationships with the WiMAX service provider of his choice in the immediate surroundings using his or her WiMAX device. OTA provisioning and activation (abbreviated as OTA in this chapter) also allows WiMAX service providers to acquire WiMAX subscribers and provision their devices automatically without manual tasks in service centers and shops.

OTA enables a device to establish a secure and authenticated WiMAX connection to this service provider, subscribe to this provider, and receive all relevant parameters and configurations for the WiMAX service using the WiMAX device. After the OTA process completes successfully, the device starts getting services from the service provider's network. An OTA-enabled device can be sold and distributed via the traditional subscription channels with a device subsidy from the service provider or via the open retail channel with no relationship to the service provider.

From the service provider's perspective, OTA is the ability to accept a connection request from any unknown WiMAX device or user, identify it, verify its certification, and associate the device with an account (paying customer) while provisioning the device with all relevant parameters to work effectively with its network and get all possible services such as specific authentication and roaming. At the end of the OTA flow, the service provider can start supplying services to the device.

The high-level OTA procedure creates the initial secured connection between the service provider and the device, fully activates flow under this connection, and enables normal connection afterward between both sides.

The OTA procedures touch on many aspects of network connection, such as network discovery and selection (ND&S) and authentication as well as recovering from failures and rolling back the process.

The WiMAX OTA standard defines two device models for service provisioning:

1. Model A for service-provider distributed devices, similar to the current cellular, cable modem, or DSL services-provisioning models. A dedicated stock-keeping unit (SKU) identifier is provided for each device at the point of manufacture to connect to one WiMAX network or a group of WiMAX networks. Model A may support self-subscription via OTA or a Web portal.

2. Model B for open retail devices with a generic SKU. OTA is needed to support over-the-air self-subscription, provisioning, and activation.

The main differences between the models from the OTA flows perspective are that Model A may support a subset (from null to almost full) of OTA flows, which must be supported by Model B, and that in model A the device is locked to one or more specific service providers during manufacturing such that it can work only with those service providers and can not be activated on any other service provider's network until it is unlocked by a proprietary mechanism.

In this chapter, we present the current status of the WiMAX OTA standard as defined in NWG Release 1.5, focusing on the OTA of Model B devices, as it represents the more complicated case. We will address the technical issues related to the OTA protocols using the Open Mobile Alliance—Device Management (OMA DM) protocol as the example, since OMA-DM is the mandatory OTA protocol for mobile devices. The WiMAX OTA standard also supports the TR69 protocol and the OMA DM protocol.

In the beginning of the OTA process, each device needs to connect to any service provider's network within reach in a secure manner, identify each other as valid (WiMAX certified), and establish an initial data link. A practical method to enable the above is to populate each device and each network with a certificate signed by the same certification authority (CA), which is authorized by the WiMAX Forum to supply these certificates. This method establishes the trust between the parties through both parties' trust of this specific CA using Extensible Authentication Protocol Transport Layer Security (EAP-TLS) authentication. The certificate includes all the relevant information for the parties to identify the certification level and capabilities of the other in order to be able to accept each other for connection. The two major challenges to this method are allocating the CA and distributing the certificate to network providers and device manufacturers in a cost-effective manner.

As the trusted CA signed certificate population was considered a crucial industry enabler, the WiMAX Forum selected a CA that was able to address the challenges. To keep the device certificate price low, an automatic Web-based distribution mechanism has been established.

Each certified WiMAX device that supports OTA is expected to have an embedded device certificate (with public key) as well as the root CA(s) certificate(s) of the WiMAX Forum licensed CA(s). Each certified WiMAX network is also expected to have a network (server) certificate and root CA(s) certificate(s).

When a certified device is connecting to a certified network for OTA activation, a mutual EAP-TLS authentication is used as part of the network entry, by which each peer sends its WiMAX certificate and the other peer verifies it using its WiMAX root CA certificates, resulting in mutual trust between the network and the device that both are WIMAX certified, and validating the identity of each peer (the name of the network and the MAC address of the device, which are part of the certificate). Using this trust and identification, the device and network create a business relationship that results in creating a device context in the network linked to a user account and a network context in the device. After both contexts are established, the device can receive services from the network as if it were directly purchased from the network provider.

Connecting to the network after the OTA process (to obtain services) can be done using the same certificates with mutual EAP-TLS or can be done using other authentication method, like EAP Tunnel Transport Layer Security (EAP-TTLS), and its associated credentials configured in the device during the provisioning process.

The three objectives of this chapter are:

1. Describe the key technical aspects of OTA and their coexistence with the normal connection
2. Describe the industry enablers of this feature
3. Provide examples of different uses of the infrastructure to supply different services (prepaid, postpaid, etc.)

The terms frequently used in this chapter are summarized below:

- Activation. The action of the network service provider (NSP) associated with a device signalling a user account (paying customer) that it is ready to provide service to this device.
- Activated. A state of the device in the network that means the device deserves service.
- Activated aware. An NSP state in the device that means the device is aware that it has an account with the NSP and can connect to this NSP as home NSP.
- Subscription. Creating a user account within an NSP network, choosing a service plan, and associating the account to a paying customer.
- Provisioning. populating the device and the network components (like the AAA) with data needed for the operation of the WiMAX device in the NSP's network.
- OMA DM. An Open Mobile Alliance standard for device management that is used as a protocol for the OTA activation. The OMA DM architecture includes a client in the device side and a server on the NSP side.

7.2   OTA HIGH-LEVEL OVERVIEW

7.2.1   User Experience

When person buys a notebook computer that includes a WiMAX device, this WiMAX device is generic and has no business relationship with any service provider. The use case includes the following steps:

1. The person powers up the notebook computer, opens the WiMAX connection manager user interface (abbreviated as UI in this chapter), and sees a list of all WiMAX service providers in the area—all of them marked as "other" (not a home NSP nor a visited NSP).

2. He or she chooses one of the service providers (say, the one that advertises an attractive offer) in the list and clicks the "connect" button.

3. The connection is established, the device gets an IP address, and the Web browser is opened (automatically or manually).

4. The person is shown the subscription page of the selected NSP where he or she learns about options and chooses a subscription plan such as $50 a month for unlimited connection with maximum 4 Mbps/1 Mbps (DL/UL) throughput. He or she puts in his or her credit card information and clicks the "submit" button on the Web page.

5. After the subscription step completes successfully, the person sees messages on the NSP's Web page that congratulates him on becoming a WiMAX subscriber and welcomes him to start enjoying the ubiquitous broadband experience.

6. Sometime later, the subscriber turns on his WiMAX device again and gets automatically connected to the NSP network he subscribed to (user policy may change it). When he opens the WiMAX UI, he will see this NSP marked as "home NSP" and may see some other NSPs' networks in the area, which he can connect to and can redo the activation flow as above.

7. When the subscriber opens his WiMAX UI in a different place where his home NSP has no coverage but has a roaming agreement with local visited NSP, the subscriber will see a list of all the available WiMAX networks in his area. Some of them are his home NSP's roaming partners. The roaming partner's information is available only after activation, and is based on the provisioned data. In the visited area, he can choose to connect to one of his NSP's roaming partners or to create another WiMAX subscription with any of the NSPs in the list.

7.2.2   OTA Logical Flow

The OTA logical flow is shown in Figure 7.1, where the steps involved in activating a device with the network are captured.

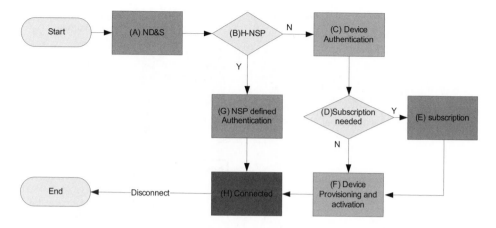

Figure 7.1. OTA logical flow.

The description of the steps is as follows:

1. An NSP is chosen to connect to through the network discovery and selection (ND&S). procedure.
2. The device checks whether it is the H-NSP (already activated). Depending on the check result, the proper authentication method is configured accordingly.
3. For an unactivated device, the device authentication method for the non-H-NSP is configured.
4. The NSP checks to see if a subscription is needed. A subscription may have been done earlier using another connection.
5. The user interacts with the NSP's subscription Web portal to create a subscription.
6. Based on the subscription plan, the network provisions the device with all relevant information and activates it.
7. For an activated device, the authentication configuration that was provisioned by the H-NSP during the activation phase is used.
8. The user is connected for broadband access via his device.

7.3 WiMAX NETWORK ARCHITECTURE FOR OTA

OTA activation and provisioning is an end-to-end feature that takes place between the NSP's CSN and the WiMAX device over R2 (see Figure 7.2). The specific protocol used for transport over R2 can be either OMA-DM or TR-69. The transport protocol is implemented in the provisioning server, which is one of the two OTA servers deployed in the CSN (see Figure 7.2).

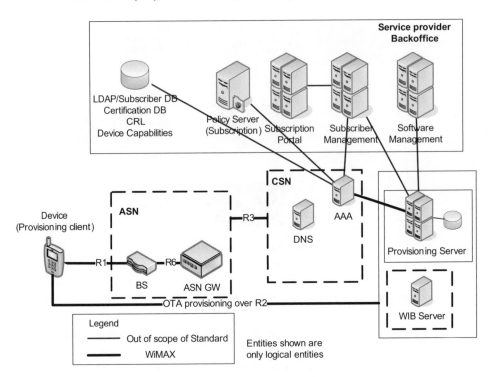

Figure 7.2. OTA end-to-end architecture.

The other OTA server deployed is the WiMAX initial bootstrap (WIB) server, which is a Hyper Text Transfer Protocol (HTTP) server used to bootstrap the provisioning client to connect to the provisioning servers. Other servers are required to support the OTA provisioning and activation processes, which are out of the scope of WiMAX and considered to be part of the NSP's back-office infrastructure. The reason that those servers are out of scope is because each NSP can integrate its WiMAX network with its existing back-office infrastructure.

The back-office servers include (see Figure 7.2):

- Subscription Portal—the server that presents the user with the Web-based subscription page and interacts with the user to create a subscription.
- Policy Server—the server that identifies the device and decides if a subscription can be offered and what kind it should be.
- Subscriber Management—the entity that manages the subscriber and accounts. It can be the home subscriber server (HSS) of the 3GPP.
- Software Management—the server that can download software as part of the provisioning process.

The transport of the provisioning protocol between the CSN and the device is HTTP/S (HTTP Secure) over IP, which uses the established data path of the device to carry this information.

The provisioning server has an IP address (which is most likely accessible only from the NSP network, and not from the Internet). The provisioning client in the device initiates a session toward the provisioning server IP address. Once it is established, there is a provisioning link between the CSN and the provisioning client. The bootstrap process delivers the server domain name to the provisioning client so that the provisioning client can store it internally and use it to issue the first provisioning session.

The protocol stack for OMA-DM-based provisioning is detailed in Figure 7.3.

The OMA-DM protocol runs over HTTP/S, which is over IP. The reason that the OMA-DM protocol runs on the HTTP/S data path over IP is that there are no management messages in IEEE 802.16e that can carry this protocol. Using the data path makes the OTA feature ASN agnostic and simplifies the supporting network requirements. The use of the data path forces the network to complete full network entry with the client and supply it with an IP address prior to establishing a business relationship with it. This requires the network to hotline the client data to be able to reach only the OTA servers and nothing else. Otherwise, the client will be able to get access to the Internet without a subscription for free. From the device perspective, this network entry for OTA is no different from the regular network entry.

7.4   OTA PROTOCOL

The OTA protocol consists of three major phases (Figure 7.4): bootstrapping, subscription, and provisioning (activation and activation awareness happen at the end of the provisioning phase). However, due to the protocol architecture that requires

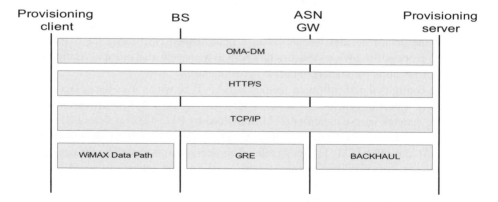

Figure 7.3. OMA-DM end-to-end protocol stack.

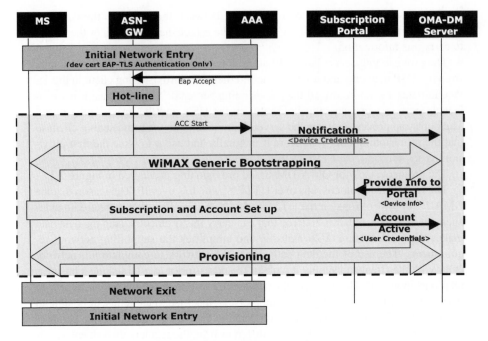

Figure 7.4. OTA protocol.

completing a secure connection to the network and establishing IP connectivity, there are two more phases prior to the OTA itself that must be done to enable OTA: authentication (network entry) and hotline. With that, the full OTA protocol flow from power-up contains five phases.

7.4.1 Authentication (Network Entry)

The first thing the provisioning client needs to do is to connect to the network. When the device decides to connect to a nonactivated NSP using the WiMAX ND&S process, the device identifies this NSP as a nonactivated one and configures the authentication parameters to enable network entry.

The configured authentication method is EAP-TLS, which uses the device certificate (a "generic" certificate that can be identified by all WiMAX-certified NSPs) and the network access identifier (NAI as defined by RFC 4282) is decorated with {SM=1} which signals the network that the device wishes to connect for activation. The NAI looks like {SM=1}User@realm. This decoration is treated by the network as a request from the client to enter for provisioning. The network uses this information to offer the correct authentication method and configures itself in the correct mode, that is, hotlining and triggering the activation flow.

The network may still configure itself to the same mode even if the device did not use the decoration in the case where the network does not identify the device as an activated device. Whenever the device uses the decoration, the NW must provision the device even if it thinks it is already activated. The reasoning behind this re-provisioning mechanism is that the device may lose provisioned information in such cases as operating system reinstalls.

This behavior introduces an attack opportunity on the network as the decoration is sent unprotected. The network should identify such attacks when they occur and ignore the decoration in this case.

7.4.2   Hotlining and Redirection

Hotlining means that the network allows the data from the device to reach only a certain part of the network and not all possible IP addresses and external addresses in particular. The network needs to hotline each device in the OTA process. Specifically, the network should let the device reach the OTA server but block it from getting free Internet access prior to activation. Hotlining is implemented with the hotline profile or rules that are downloaded from the AAA to the network access stratum (NAS) in ASN GW. With the hotline profile (or rules), only specific-destination IP addresses in the network can be routed from the device source IP address and all the other destination IP addresses are discarded.

Note that hotlining is different from redirection, which hijacks all the traffic of a specific type (like HTTP) and redirects it to a default IP address instead of the original destination. An example of redirection is when, in a WiFi hot-spot, all the Web traffic is redirected to the log-in page. Redirection is not required in WiMAX. However, it is a best practice for the network to implement it as a complimentary feature to hotlining.

With redirection, instead of discarding HTTP messages with IP addresses out of the hotline range, these messages will be routed to a default address of the NSP subscription portal. The reason to choose the subscription portal as the redirected point, instead of any other server supporting the OTA provisioning and activation process (all of them are reachable in hotlining), is that all the servers but the subscription portal are accessed by the applications integrated with the WiMAX stack. The messages directed to those servers can be assumed to have the correct address with no redirection.

In the provisioning and activation process, the Web browser is the only involved entity that the user may use to influence the destination address. So, most likely, a destination address outside the hotlining range can be dealt with only by the subscription server. Redirecting removes the burden to be aware of the need for performing the subscription service from the user and takes the user to the subscription Web page from any browser he uses. The OMA-DM protocol defines an automatic way for a Web browser to be opened with the initial address. Since the user can change the address or close the browser, redirecting helps recover from this.

7.4.3   Bootstrapping

Bootstrapping in the OTA protocol provides the provisioning client with the OTA server details (domain name/credentials) so that the client can initiate the session needed for the provisioning phase.

OMA DM defines a bootstrapping mechanism in which a bootstrap message is sent to a known port using the UDP protocol. When the message is received, the client initiates the session with the server using the received information. However, this mechanism alone is not enough for WiMAX OTA for two main reasons. The first reason is that WiMAX supports more than one provisioning protocol. So the server needs to know which bootstrap to send. The second (and more important) reason is that a lot of target platforms implementing WiMAX functional entities are open platforms that have a firewall that blocks any message originated outside the platform unless it is a response message to the internally initiated session.

To overcome these two issues, WiMAX defined the WiMAX Initial Bootstrap (WIB) protocol, which internally initiates the request for bootstrap and reports to the server the types of supported provisioning protocols. The WIB protocol is very simple and based on standard domain name system (DNS) queries and HTTP (see Figure 7.5). When an IP address is obtained by the device but the client cannot find provisioning server information for the connected NSP (or fails to create the provisioning session based on the information it has), the client simply requests the WIB server repeatedly for a bootstrap document and stops once it gets one (or the retry count exhausts). It is obvious that the client must not receive any bootstrap document and use it before verifying that it is a valid document from a valid source. This is why the bootstrap file is protected using an encryption key called the bootstrap encryption key (BEK). The BEK is derived from the extended master session key (EMSK) that is created during WiMAX authentication.

When the client is able to decrypt the bootstrap using the BEK, it can be sure that the bootstrap document comes from a trusted source and use it. This is referred to as

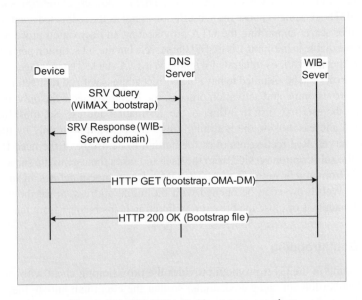

Figure 7.5. WiMAX initial bootstrap procedure.

a secure bootstrap. Clients and networks must support WIB for bootstrapping. The network should support UDP push as well. The client uses the first valid bootstrap it receives and initiates an OTA session.

7.4.4 Subscription and Account Setup

The subscription phase is the one in which the user creates a business relationship with the NSP, chooses the level of service he wants to get (throughput, QoS, or period), and commits to pay for the service by submitting his credit-card information or any other payment data.

Doing the online subscription is an essential OTA phase, which enables the user to order his services instantly. The subscription flow supports user interaction with the Web-based subscription portal right after the bootstrap phase and before the provisioning phase (which is done based on the choices in the subscription).

Online subscription gives the NSP a lot of freedom to implement its specific "mechanism" that uses the subscription portal as the way to interact with the user without contradicting the standard. An example of such a mechanism is to send a message to the user, let the user type in a code for prepaid service, or type his credential for the guest access the network. However, since Online subscription is not mandated, the NSP can choose to support other means of out-of-band (OOB) subscription based on its needs or regulatory requirements. Examples of other means to subscribe are other Internet connections, SMS, phone, and face to face in a store (in the case where the user must be identified in person). In these nononline-subscription cases, the network identifies the device when it enters the network (using the certificate, for example), retrieves the subscription information related to this device in the internal database, and goes directly to the provisioning phase, skipping with the Subscription phase. In some cases, the network may decide not to skip the subscription phase and, instead, show the user a subscription page that supports its OOB subscription, indicating that the user needs to enter a code it receives over the phone subscription.

7.4.5 Provisioning

Provisioning is the final phase of OTA, in which the server in the network downloads all the relevant parameters to the device for use in future connections to this NSP or its roaming partners. The downloaded parameters include, among other things, the lists of network access providers (NAPs) and channels the NSP uses, its roaming partners to enable a robust ND&S for this NSP, as well as subscription parameters, such as authentication method and credentials to use in regular connections. The last configured parameter is the "activated" flag which switches the device to an "activated aware" state, which means it is aware that this particular NSP is an H-NSP and the entire provisioned information is valid and should be used. The reason this parameter is the last one is to protect the process from deadlocks. As long as this "activated" flag is not set, the device ignores the provisioned parameters and act as not activated. In contrast, when it is set, the device acts as activated and uses the provisioned parameters. With that, if the session breaks in the middle of the provisioning phase, the device is still in not-activated state and may start over with a clean slate.

Once the provisioning phase ends, hotlining is removed and the device can get full service. If this is not possible to do while the device is connected (due to NSP's network architecture), the network will disconnect the device after provisioning has ended; when the device connects again, it will use all provisioned parameters and get full service.

There are two more issues of provisioning that are supported by both the client and the server. The first one is ongoing provisioning, meaning that some of the provisioned parameters are changed and the server wants to update the client with those changes. This may happen in parallel with any normal connection (which has no need for special network entry or hotline). In ongoing provisioning, the device or the server simply initiates a provisioning session and updates the required parameters in parallel with any other user activity such as video or e-mail. The second one is called reprovisioning, which is needed when the client in the device has lost all or some of its provisioned information (reset to factory defaults by the user is an example). In this case, the device asks for full reprovisioning by using the {SM=1} decoration in network entry and the network fully provisions the device in response. Since the subscription in the network is not erased, it is most probably not necessary to go through the subscription portal and the network can treat it as out-of-band subscription. In any case, it is up to the network to decide whether to include the subscription page in reprovisioning, based on its usage models.

7.5 OMA DM

OMA DM is the device management protocol designed by OMA orientated toward mobile devices and accessing in a single-parameter level in order to optimize the usage of the air link for updates. The protocol defines the device database in an extensible markup language (XML) tree structure (see Figure 7.6) and five commands between the server and the client to perform actions on a node or a group of nodes: add, replace, get, delete, and execute.

The protocol also includes four aspects of security for each node to ensure that only the allowed server can reach the node and the node executes only the allowed actions. The first aspect is authentication between the client and the server to ensure that the server is allowed to access the client. The second aspect is encryption of the content between the server and the client. The third aspect is the specification of which server is allowed to access each node, since not all servers are allowed to access all nodes (an example is that a server of one NSP will not be able to access the nodes of another NSP). The forth aspect is the definition of which actions are allowed on each node (an example is that if a node is not allowed for a "get" command, the server that is allowed to access this node cannot read the content).

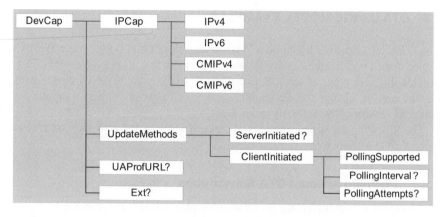

Figure 7.6. XML tree for device capabilities.

The OMA-DM protocol deals with transferring information between the server and the client, but not the meaning of the content and the associated action (see Figure 7.7). In the WiMAX Forum, the OMA-DM protocol is adopted for exchanging relevant OTA parameters between the OTA client on the device and the OTA server in the network. The OTA specification also defines the management objects for OTA in the XML structure that includes all the parameters relevant for individual NSPs, called WiMAX_supp MO, and the structure that describes the device, called WiMAX_MO. In each client, WiMAX_MO appears only once, whereas WiMAX_Supp MO may have several NSP instances, one for each NSP the device is activated on.

The specification also includes the definition of the expected behavior of WiMAX entities based on the content of the node. However, such behavior definition is only applied to some of the nodes, which must be interpreted in the same way by both sides for interoperability. The behavior definition for the rest of the nodes (most of them) is implementation specific.

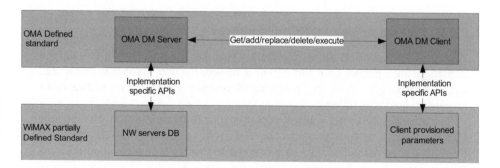

Figure 7.7. OMA-DM logical layers.

7.6   OTA USAGE MODEL EXAMPLES

The OTA infrastructure based on a Web-based subscription portal and a WiMAX signed device certificate is a very flexible framework that allows different NSPs to provide different types of services and accounts to the same WiMAX device without the need to change the device architecture or behavior. This is a unique and powerful feature of the WiMAX technology.

The following are some examples of types of services and how they are created using the OTA framework.

7.6.1   Postpaid Account OTA Subscription

The postpaid account offers the user a monthly plan that is paid for month by month until the user asks to terminate the business relationship. The OTA flow to establish a postpaid account is as follows.

The user sees the list of available networks (that he has not subscribed to) on the UI and chooses a network to connect to (say, based on an advertisement he saw on TV). The device connects to this network using the WIMAX certificates. A connection is successfully set up because the WiMAX certificate is accepted by all devices and networks. Then the user is presented with a subscription portal in his or her Web browser.

The user sees a list of possible account types offered by the NSP and chooses the postpaid plan with monthly payment at the Silver service level (for example). The user is asked for payment method details. After the payment method details are entered by the user, the network creates a context that links the device, user, and payment method. The network provisions the device with the NSP and subscription information and creates the network context in the device. The device may start getting service or may be requested to reconnect using the provisioned information.

The next time the user scans for available networks, he will see the activated network marked as H-NSP (in the nonroaming case). When the user chooses to connect to the H-NSP (or when the device automatically connects to the H-NSP using its defined policy), the device is connected for service as the network identifies the user and treats the device as a valid device by the H-NSP .

7.6.2   Prepaid Account OTA Subscription

A prepaid account offers the user a monthly plan that is paid month by month for a limited, defined period; once this period is passed, the user account is invalidated and user cannot get service until renewing the account. The OTA flow to establish a prepaid account is as follows.

The user sees the list of available networks (that he has not subscribed to) on the UI and chooses a network to connect to. The device connects to this network using the WIMAX certificate and the user is presented with a subscription portal in its Web browser. In the subscription portal, the user sees a list of possible account types offered by the NSP and chooses a plan, for example, three consecutive months of Gold-level service. The user is then asked for payment method details. Once the payment method details are entered and validated, the network creates a context that links the device, user, and payment method.

The network provisions the device with the NSP and subscription information, including user credentials to be used with EAP-TTLS, and creates a network context in the device. The device may start getting service or may be requested to reconnect using the provisioned information.

The next time the user scans for available networks, he will see the activated network marked as H-NSP. When the user chooses to connect to the H-NSP (or the device automatically connects to H-NSP using its defined policy), the device will connect using the provisioned credentials. When authentication is completed, within the three-month subscription period, the device can be directly connected for service. When the device attempts to connect to the NW after the subscription period has ended (more than three months), the network will identify that the payment details in the device context are invalid and will block the user from service and present it with the subscription portal offering the user to renew the subscription.

7.6.3 Usage-Based Account—Prepaid Card Purchased in a Shop

This account offers the user a prepaid card that contains a fixed amount of money, time, or data volume, and a card number for account validation; the card is purchased in a shop. The OTA flow to establish a usage-based account is as follows.

The user purchases a prepaid card issued by a specific NSP. He sees the list of available networks (that he has not subscribed to) on the UI and chooses the network that he purchased the card from. The device connects to this network using the WIMAX certificates and the user is presented with a subscription portal in its Web browser.

In the subscription portal, the user sees a list of possible account types offered by the NSP, including a button for the prepaid card account. The user clicks on prepaid card and is asked to enter the card number for validation. Once the card number is identified, the user starts receiving service and the network deducts the amount of money, time, or data volume used during the connection period from this card.

The next time the user wants to connect using the prepaid card, he will repeat exactly the same process as above. The device may have a policy to automatically connect to this network without user intervention, but the rest of the flow is the same as in the case of manual connection

7.6.4 Postpaid Account—Out-of-Band Subscription (Office/Phone)

This account offers the user a monthly plan that is paid month by month until the user asks to terminate the business relationship. The subscription is done prior to connecting the WiMAX device for the first time. The OTA flow to establish a postpaid account with out-of-band subscription is as follows.

The user creates an account with a NSP by an out-of-band subscription method, such as calling a phone number, going to a NSP office, or browsing the Web using any Internet connection via a computer or device that is not the mobile WiMAX device to be provisioned.

When the user attempts to connect using his WiMAX device, he sees the list of available networks (that he has not subscribed to) on the UI and chooses to connect to the network it created the subscription with. The device connects to this network using the WIMAX certificates. The network identifies the device as a subscribed device (based on the MAC address that was acquired by the NSP during the subscription, for example) and the user is presented with a subscription portal in his Web browser. In the page presented to the user, all he is asked to do is to enter a code that was given to him when he did the out-of-band subscription. Note that the entire subscription interaction can be skipped if the NSP does not require this code verification.

When the correct code is entered, the network provisions the device with the NSP and the subscription information and creates the network context in the device. The device may start getting service or may be requested to reconnect using the provisioned information.

The next time the user scans for available networks, he will see the activated network marked as H-NSP. When the user chooses to connect to the H-NSP (or the device automatically connects to H-NSP using its defined policy), the device is directly connected for service as the network identifies him and treats his device as a valid device in the H-NSP.

7.6.5   Postpaid Account—Guest Subscription

A guest account means there is no linkage between user and device. The user can access the account using any device and pays for the usage. This practically requires the network to identify both the device and the user since there is no one-to-one relationship between the user and the device. The OTA flow to establish a postpaid account with guest subscription is as follows.

The user creates a "guest" account with a specific NSP. When he wants to connect to the network, he gets access to a WiMAX device. On the device, he sees the list of available networks on the UI and chooses the network he wishes to connect to as a guest. The device connects to this network using the WIMAX certificate and the user is presented with a subscription portal in his or her Web browser. The user sees a list of possible account types offered by the NSP and a button for guest connection. The user clicks on the guest connection button and is asked to enter his user name and password. When the user is identified and authenticated based his user name and password, the user start receiving service and the network charges the user subscription (no charges on the device). The next time the user wants to connect using the guest account, he will repeat exactly the same process.

7.7   SUMMARY

The WiMAX Forum has defined a unique framework to achieve over-the-air (OTA) provisioning and activation of generic devices. OTA provisioning enables the generic devices in the retail distribution model to progress into a state that has the identical behavior and user experience as the devices that are customized, sold, and subsidized by the NSP itself (the traditional device distribution model), without the need for NSP stores, service centers, or installing hardware or software components in the device.

The driving force for the OTA framework is a real business need of 4G access technology in order to have an offering with great operational value to the market beyond the improvement in data rates. The impact of the OTA specification is a revolutionary mode of operation in the wireless communication world, from both technical and business perspectives. We believe that this novel approach will be adopted by more and more wireless technologies in the 4G and beyond.

7.8 REFERENCES

[1] WMF-T33-001-R010v04_network-stage3-base, July, 2009.

[2] WMF-T33-103-R015v02_OTA-general, November 2009.

[3] WMF-T33-104-R015v02-OTA-OMA-DM, November 2009.

CHAPTER 8

MOBILITY IN WiMAX NETWORKS

SHAHAB SAYEEDI AND JOSEPH R. SCHUMACHER

8.1 INTRODUCTION

IEEE 802.16 was conceived of as a wireless metropolitan area network (WMAN) access technology [1] that was originally focused on point-to-multipoint applications for frequencies greater than 10 GHz. Subsequent amendments and revisions to the original base standard added profiles for operation in different frequency bands and new features. However, the original standard had no provisions for handover, the ability of a subscriber station to cease operation with one base station and commence operation with a different base station without dropping network services.

The addition of handover presented a challenge in terms of backward compatibility. The IEEE 802.16 standard already had numerous mature features and practices (e.g., RF link maintenance and security) that were standardized without consideration for support of subscriber mobility. For example, the basic practice of requiring a subscriber station to request a bandwidth grant for sending a MAC message can be viewed as ill-suited to handover implementation. Handovers are (almost by definition) initiated in areas with low RF coverage and should, therefore, be completed as quickly as possible; the latency of the bandwidth grant mechanism is contrary to that goal. Thus, adding handover without comprehensively reworking the MAC protocol was a major problem in developing handover support.

IEEE 802.16 Amendment "e" [2,3] (hereafter, simply 802.16e) was initiated to provide support for subscriber stations moving at vehicular speeds. In addition to providing physical layer enhancements for operation in mobile wireless channels, 802.16e added the signaling and features to support handover. A number of corrections and improvements to the handover process were introduced in 802.16-2009 [4].

WiMAX Technology and Network Evolution. Edited by Kamran Etemad and Ming-Yee Lai
Copyright © 2010 the Institute of Electrical and Electronics Engineers, Inc.

This chapter provides an overview of handover support based on 802.16e/.16-2009 as well as network support for end-to-end mobility.

8.2 NETWORK TOPOLOGY

The base air interface standard supports both mesh and point-to-multipoint modes of operation. This section will only consider the point-to-multipoint mode, as support for mesh topology has not been pursued in mobile WiMAX and it has been removed from the IEEE802.16-2009 standard.

The point-to-multipoint mode is based on a single BS providing service to multiple MSs. Although not specifically stated in the standard, it is assumed that each BS within the set of BSs covering a geographic area is connected with the other BSs in the set via a backbone network to form a continuous coverage zone.

Each BS periodically broadcasts downlink channel descriptor (DCD) and uplink channel descriptor (UCD) messages that include parameters to be used by the MS for access and operation within the coverage area of that BS. These parameters include thresholds, triggers, and action types that the MS uses when evaluating handover decisions. In addition, each BS transmits (either broadcasts to all MSs or unicasts to a particular MS) a MAC management message, called MOB_NBR-ADV, which contains DCD and UCD information for nearby BSs ("neighbor" BS), including the thresholds, triggers, and action types advertised by the neighbors in their own DCD/UCD messages. By associating these entries in the MOB_NBR-ADV message with neighbor BSs, the MS is able to determine the network topology and is subsequently able to quickly synchronize with those BSs when a handover is required or when the MS needs to scan the BS to determine its suitability as a handover candidate.

8.3 HANDOVER MODES

The 802.16 standard supports three modes of handover: hard handover (HHO), macrodiversity handover (MDHO), and fast base-site switching (FBSS) (Figure 8.1). The WiMAX System Profile [5] mandates support for HHO, whereas MDHO and FBSS are left as optional handover modes.

HHO is a break-before-make technique in which traffic is interrupted while the handover is in progress. HHO can be initiated by either the MS using the MOB_MSHO-REQ MAC management message or by the BS using the MOB_BSHO-REQ message. Generally, the handover is considered complete when the BS receives the MS_HO-IND message from the MS.

Both MDHO and FBSS are macrodiversity techniques in which multiple BS are designated as HO candidates by being included in a diversity set. The diversity set is built from the set of neighbors advertised in the MOB_NBR-ADV message, along with neighbor BSs either added or subtracted through the MOB_BSHO-RSP and MOB_MSHO-REQ messages. In MDHO, the MS may communicate with one

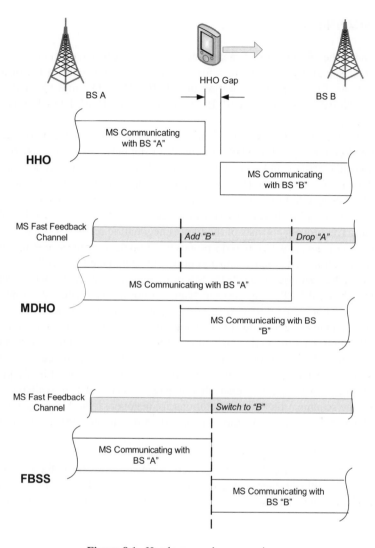

Figure 8.1. Handover modes comparison.

or more BS in the diversity set simultaneously, whereas in FBSS, the MS communicates with a single BS drawn from the diversity set. In either case, the MS indicates its BS selection through a fast-feedback channel, as opposed to sending a MAC management message. This allows lower latency in completing a handover choice. However, each mechanism imposes much greater resource demands on each BS in the diversity set and on the backhaul connecting the members of the diversity set.

In WiMAX Mobile System Profile Release 1.0 [5], MDHO and FBSS are optional features. As such, they have not experienced the same level of development as HHO. For that reason, the remainder of this chapter will focus on HHO.

8.4 SCANNING

Scanning is the MS process of measuring carriers of nearby base stations to determine their suitability as candidates for handover. The MS builds a list of nearby base stations, first through monitoring the MOB_NBR-ADV message and, subsequently, through monitoring MOB_BSHO-RSP and MOB_MSHO-REQ messages (Figure 8.2).

It is assumed that data reception and scanning are mutually exclusive activities performed by the MS. This relaxes the need for duplicate receiver resources dedicated to data reception and scanning activities (although the standard does not limit the MS to a single receiver). However, it does require the BS to have knowledge of when the MS is able to receive data and when it is performing scanning. For this reason, the protocol includes provisions for negotiating scanning intervals and scan iterations. The process consists of the typical negotiation scheme using MAC management messages, in which the MS proposes parameters for the scanning intervals and scan iterations, and the BS either approves the parameters by sending those parameters in the response or rejects the MS proposal and supplies its own values. In a similar fashion, the MS may perform scanning during negotiated sleep intervals, but the MS is not required to perform scanning while in sleep mode. Also, the standard only mandates support for the negotiation of the scanning intervals and iterations (although scanning in this fashion is not required); it does not impose requirements on the MS for the order in which neighbor BSs are scanned or any other details of MS scanning performance. The MS is also free to scan other frequencies to determine if an alternate access technology (e.g., 3G cellular) is available.

Once the scanning interval and iterations are negotiated, the MS can begin scanning. It may do so in an autonomous manner, that is, without further direction from the BS.

The MS sends scan reports in the MOB_SCN-REP message. The message is typically sent autonomously when the received signal from a BS in the neighbor list crosses a threshold, triggering the MS to send the report. However, the MS may also be instructed to send scan reports periodically. The BS may also send an unso-

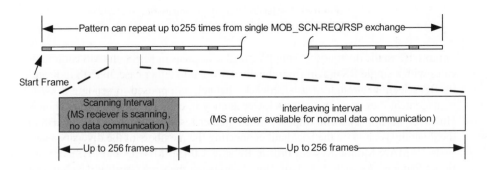

Figure 8.2. Scanning pattern.

licited scan response (MOB_SCN-RSP) message at any time to instruct the MS to send a scan report for a particular BS or set of BSs, irrespective of a prearranged scan period or the status of any report triggers. An unsolicited scan response could be considered as a prelude to a BS-initiated handover.

The actual measurement that a MS performs during scanning depends on the type of scan report directed by the BS. Report types include received signal strength indication (RSSI), carrier-to-interference plus noise ratio (CINR), relative delay, and round-trip delay (RTD).

8.5 BASIC HANDOVER MECHANICS

A handover begins when either the BS or the MS sends a MAC management message announcing its intention to move connections to a new BS. In the standard nomenclature, the BS that is providing service to the MS before the handover is defined as the *source BS* and the BS that is identified as the candidate for completion of the handover is defined as the *target BS.*

If the source BS initiates the HO, it sends the MOB_BSHO-REQ message to the MS. This message may contain zero, one, or more candidate target BSs for the MS along with parameters indicating the handover capabilities of each target BS (e.g., indication of the degree to which the network can support the handover). The message includes an action time, which indicates when the MS can expect an allocation for its first ranging message with the target BS. If a source BS sets the action time to zero, it signals that the MS must perform contention for a ranging slot on the target BS, just as the MS would be forced to do if it were performing network reentry. The MS responds by sending the MOB_HO-IND message.

If the source BS initiates the HO without including any candidate target BS in the MOB_BSHO-REQ message, the source BS expects the MS to provide its preferred candidate target BS for HO. In this case, the MS responds with either a MOB_MSHO-REQ or MOB_HO-IND messaging containing the MS preferred target BS for handover.

Figure 8.3 illustrates a network-initiated handover in which the MS has sent its scanning results to the source BS. The source BS initiated a handover for the MS and included the reported candidate target BS in the MOB_BSHO-REQ message.

If the MS initiates the HO, it sends the MOB_MSHO-REQ message. In the MOB_MSHO-REQ message, the MS presents a list of HO candidate target BSs to the serving BS. The candidates are presented along with the relevant scanning metric collected by the MS for each candidate. For instance, if the MS was instructed to measure CINR during its scanning intervals, then the MS would send the measured CINR for each of the candidates listed in the MOB_MSHO-REQ. The serving BS responds by sending the MOB_BSHO-RSP message, which includes a list of one or more candidate target BSs. A typical case would be one in which the MS presents several candidate target BSs in the MOB_MSHO-REQ, and the serving BS responds with a single candidate in the MOB_BSHO-RSP. Regardless, the MS responds to the MOB_BSHO-RSP with the MOB_HO-IND message.

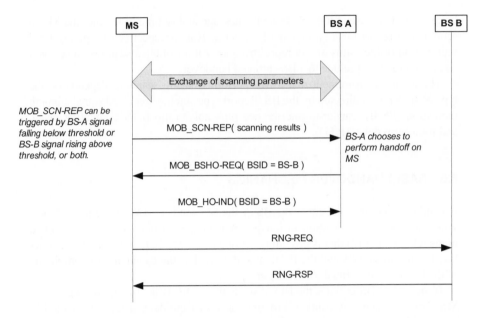

Figure 8.3. Typical handover transaction (BS-initiated).

For both MS- and BS-initiated HOs, the MS typically ends the HO transaction with the serving BS by sending the MOB_HO-IND message. The MS can choose one of three response types to send in the MOB_HO-IND message: *serving BS release, HO cancel,* or *HO reject.* Note that only one of these response types signals a successful HO transaction that immediately results in ending service with the serving BS. The HO cancel and HO reject response types both indicate that the HO transaction did not succeed, which implies that the MS will continue to seek some level of service from the serving BS for the short term. The HO cancel response type indicates that the MS wishes to stop the HO process; although not specifically standardized, the MS may exercise this option in case it decides that an HO is no longer necessary or in case it decides that it cannot tolerate the service discontinuity resulting from a handover at that time. The HO reject response type is an indication from the MS that it has evaluated the list of HO target BSs presented by the serving BS in the MOB_BSHO-RSP and has failed to find any suitable HO candidates, possibly based on very recent scan results for those BSs. In this case, the MS may include its preferred target BS in the MOB_HO-IND, which the network may propose as a new candidate target BS, to the MS for completing the handover.

The availability of both MS-initiated and BS-initiated HO techniques reflects different philosophies of how mobile networks should be managed. The MS-centric view holds that the MS is better situated to evaluate the RF link and should, therefore, be in control of HO decisions. The BS-centric view is based on maximizing total throughput and fairness in distributing resources to all MSs in the coverage area, which suggests more centralized control. The standard supports implementations that lean toward either view.

After successfully completing a HO transaction with the serving BS, the MS begins its connection with the selected target BS, exchanging the RNG-REQ/RSP messages with the target BS. Depending on the level of synchronization between neighbor BSs for a given deployment, this exchange can either consist of a simple handshake between the MS and target BS, or, for deployments with a lower level of synchronization, the MS and target BS may begin the *ranging process*. Ranging is a feedback process in which the MS and BS exchange RNG-REQ and RNG-RSP messages, the RNG-REQ messages being pilot messages sent by the MS and used by the BS to judge MS time and frequency synchronization, and the RNG-RSP messages provide incremental time- and frequency-correction commands to the MS. Ranging is also part of network entry. In network entry, there is no trust relationship between the BS and MS, so the MS performs ranging anonymously, that is, it must contend with other MSs in the contention ranging slot. For HO, since the MS has already been admitted to the network and since network resources exist whereby the source BS can notify the target BS of the impending arrival of the MS, the target BS can provide an exclusive bandwidth allocation for the RNG-REQ from the MS. In the most optimized case, the BS can include a *fast-ranging IE* in the UL MAP that signals the slot allocation to be used by the MS for sending the RNG-REQ. This avoids the latency involved in contending for the attention of the target BS in performing the handover.

After handover signaling is complete, the MS and target BS must synchronize their respective contexts. In this usage, context is used to describe the state of the MS/BS connection in terms of negotiated parameters and states derived from the state of the session as it existed between the MS and source BS before the Handover. Context consists of information such as the state of ARQ variables and operational parameters negotiated during network entry. Each of the service flows established prior to the handover also has its own context. Part of that context is the mapping between service flows and their respective connection identifiers (CIDs) and security association identifiers (SAIDs).

For a given BS, CIDs are drawn from a common number space for all MSs. A conflict can arise where a MS can enter a target BS in which the CIDs used by the MS are already in use for service flows connected with a different MS. In anticipation of this problem, 802.16e provided signaling (in several forms) to be used during handover to update (reassign) CID numbers. This process was streamlined in [4], and will be discussed later in this chapter.

8.6 WiMAX NETWORK SUPPORT FOR HANDOVERS (ASN ANCHORED MOBILITY)

WiMAX networks include support for mobile and network-initiated handovers based on 802.16 handover specifications. The WiMAX network currently only supports hard handover. Network support for MDHO and FBSS handovers is not considered at this time. Handovers in WiMAX networks may be characterized as "ASN-anchored" or "CSN-anchored" mobility.

CSN-anchored mobility refers to moving the point of attachment within the ASN or, specifically, the foreign agent (FA). The WiMAX network layer is based on IETF standards for mobile IP, hence, CSN-anchored mobility results in the network moving the MIP tunnel from the old FA to a new FA. Both client mobile IP (CMIP) and proxy mobile IP (PMIP) based mobility procedures are supported.

ASN anchored mobility involves moving the network point of attachment at the ASN. ASN-anchored mobility may require an intra-ASN handover or a handover in which the MS moves from a serving BS to a target BS controlled by an ASN-GW logically located within the same ASN or inter-ASN mobility, where the MS moves to a target BS controlled by an ASN-GW logically located in a new or target ASN. The following sections discuss WiMAX handovers based on ASN-anchored mobility in more detail.

WiMAX handovers in the network are characterized as either controlled or uncontrolled handovers and are collectively referred to as ASN-anchored mobility. The type of handover, whether controlled or uncontrolled, is determined by whether the WiMAX network is able to notify and prepare prospective target BSs of an impending handover from a mobile station prior to its arrival at the target BS.

If an MS signals its intent to handover to a target BS offered by the serving BS (source BS) by sending an 802.16 MOB_HO-IND message in response to either a MOB_BSHO-RSP or MOB_BSHO-REQ message (for MS or network-initiated handover, respectively), the handover is considered a *controlled handover*. In this scenario, network signaling is used to notify potential target BSs, and potentially target ASNs in the event of an inter-ASN handover, of a potential handover from the MS.

In contrast, when an MS initiates a handover at a target BS either autonomously without requesting a handover from the serving BS, or after sending a handover request to the serving BS but then initiating handover at a target BS not offered to it by the serving BS, the target BS selected by the MS has not been prenotified of an impending handover from the MS and is unprepared to accept the MS. This scenario is known as an *uncontrolled handover* in WiMAX networks. If the target BS decides to accept an uncontrolled handover from the MS, the handover will be completed with greater latency as the network context for the MS must be sent by the serving BS to the target BS during handover after the MS has already moved to the target BS.

8.6.1 Controlled Handovers

The notification of candidate handover target BSs and preparing them for a potential handover from an MS is known as the *handover preparation* phase in WiMAX. The handover preparation phase is triggered by a handover request from either the MS or network and allows the WiMAX network to notify candidate target BSs of a potential handover from the MS, and receive acknowledgements from candidate target BSs that the MS will be accepted for handover before the target BSs are offered as handover candidates to the MS. The WiMAX handover preparation phase includes transfer of authenticator context for the MS to the target BS and may also include preestablishment of bearer paths between the network DP (data path) function and potential target BSs in anticipation of a handover from the MS.

If the MS finds one or more of the candidate target BSs offered to it by the serving BS for handover acceptable, it sends a MOB_HO-IND message to the serving BS indicating its selected target BS. This notification triggers the WiMAX network handover *action phase*. During this phase of the network handover procedure, the serving BS notifies the target BS selected for handover by the MS that the MS has selected it. The serving BS also notifies the other candidate target BSs that may have been notified of a potential handover during the handover preparation phase that they were not selected for handover by the MS so that they can release any network resources that may have been allocated to the MS during the prior handover preparation phase. Meanwhile, the MS begins exchange of the RNG-REQ/RSP messages at the selected target BS.

Handovers in the WiMAX network may be intra-ASN or inter-ASN depending on whether the selected target BS is controlled by the same ASN-GW controlling the serving BS, or an ASN-GW logically located with the same ASN, or an ASN-GW logically located at another ASN.

Initially, when a WiMAX service flow is initiated, the data path function associated with the foreign agent supporting the MS's session and authenticator function are collocated at the serving ASN-GW controlling the serving BS. If the target BS selected by the MS for the first handover is also controlled by the same ASN-GW, no R4 signaling is required and an intra-ASN handover will occur via R6 signaling procedures.

Figure 8.4 illustrates an example in which the serving and target BS are controlled by the same ASN-GWs logically located in the same ASN as the serving and target BS.

Such a network configuration may occur during handover if the MS has not performed an inter-ASN handover after the current session was created.

As an MS experiences mobility and successive handovers occur, the authenticator and data path functions may eventually become anchored at other ASNs. In fact,

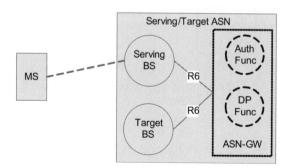

Figure 8.4. Intra-ASN handover.

up to four ASNs may be involved in a handover. This includes a serving ASN that controls the serving BS, an Authenticator ASN in which the authenticator function is anchored, an anchor ASN in which the FA is located and data path function is anchored, and a target ASN that controls the target BS. The anchor ASN and authenticator ASN may be collocated in the same ASN, or the authenticator ASN and serving ASN may be collocated, or any other combination of collocation may occur. In Figure 8.4, such a network configuration could also occur if previous inter-ASN handover did occur but the anchor authenticator and/or data path functions were relocated after handover completed.

Figure 8.5 illustrates an example of an inter-ASN handover in which the serving and target BS are controlled by different ASN-GWs logically located within different ASNs, but the network functions are anchored at the serving ASN. Such a network configuration may occur during handover when the MS is in the process of completing its first inter-ASN handover, or, if previous inter-ASN handovers have occurred, the anchored authenticator and or DP functions were relocated over to the target ASN upon completion of the inter-ASN handover.

Figure 8.6 illustrates an example of an inter-ASN handover in which the serving and target BS are controlled by two different ASN-GWs logically located within separate ASNs, and the authenticator and DP network functions are anchored and logically located in yet two more ASNs. Such a network configuration may occur during handover when the MS has completed multiple inter-ASN handovers and the authenticator and/or DP functions were not relocated upon completion of the handover.

The WiMAX standard refers to these inter-ASN handovers as "ASN-anchored mobility" and supports the many potential configurations that may occur. R4 sig-

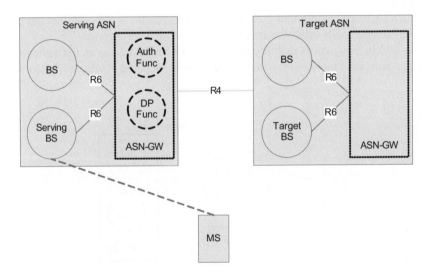

Figure 8.5. Inter-ASN handover—network functions anchored at the serving ASN.

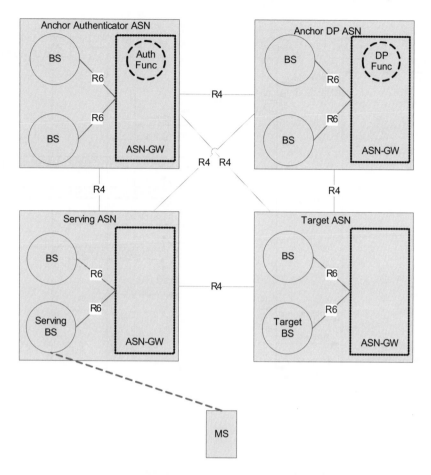

Figure 8.6. Inter-ASN handover—network functions anchored in other ASNs.

naling between the respective ASNs in conjunction with R6 signaling between the ASN-GW controllers and the serving and target BSs they control are used to support anchored ASN mobility.

Figure 8.7 provides an illustration of the WiMAX network handover preparation procedures that are triggered when an MS initiates a WiMAX handover. Four ASNs are shown in this figure to illustrate ASN anchored mobility but, as previously explained, the number of ASNs involved can include only one for a basic intra-ASN handover to as many as four as shown in this inter-ASN handover example. In the event of intra-ASN handover, only R6 signaling from the serving ASN-GW would be required to complete the handover and R4 signaling would not occur.

In an MS-initiated handover, the MS triggers a handover in the WiMAX network by sending a MOB_MSHO-REQ message to the serving BS in step 1. The message may or may not include any target BSs. If it does not, the network must

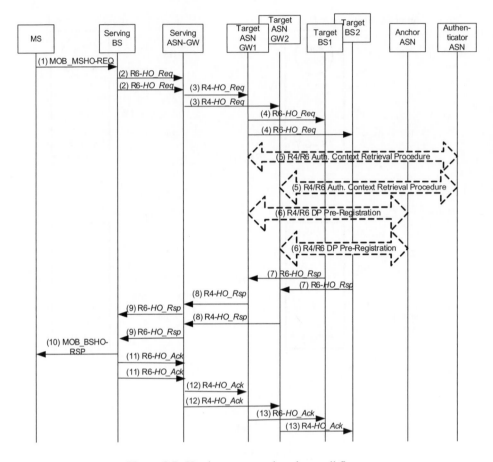

Figure 8.7. Handover preparation phase call flow.

first determine suitable handover candidates based on scanning reports from the MS. The serving BS sends a handover request message over the R6 RP (reference point) to its ASN-GW controller in step 2. The ASN-GW selects candidate handover targets BS1 and BS2, which are controlled by ASN-GW controllers in other ASNs, and sends a handover request message over the R4 RP to each of the ASNs in step 3. The ASN-GW controllers generate handover requests for each of their respective BSs in step 4. The target BSs request authenticator context for the MS from their respective ASN-GW controllers via R6, which then forward the request on to the authenticator function that is anchored in an Authenticator ASN over R4 in step 5. The authenticator function returns the authenticator context for the MS to each of the target BSs. In step 6, the candidate target BSs initiate preestablishment of a data path for the MS at each of the candidate target BS with the data path function at the anchor ASN. The request and response traverse R6 between the target BSs and their respective ASN-GW controllers, and R4 between each target ASN and the anchor ASN, respectively. Both target BSs and their respective controllers agree to accept the MS for handover in this example and indicate this by responding with R6/R4 handover response messages in step 7, which are then forwarded to the Serving ASN in step 8 and then down to the serving BS over R6 in step 9. Upon reception of the handover response messages confirming the acceptance of the MS for handover by the respective target ASNs and BS, the serving BS/ASN selects one or more candidate target BSs and sends this to the MS in the MOB_BSHO-RSP message in step 10. At the same time, the serving BS and ASN acknowledge the target BS/ASN's acceptance of the handover via a three-way handshake by sending the R4 handover acknowledgement message over R6, R4, and R6 to each of the candidate target BSs in steps 11–13 and the handover preparation phase is completed. Although this example illustrates the case in which each target BS/ASN agrees to accept the handover, it is possible that one or more of the selected targets, either BS or ASN-GWs, may reject the request due to load or other reasons. In the event this occurs, the serving ASN may initiate a request to additional potential candidate target BS and ASNs.

The WiMAX network handover action phase is normally triggered by the sending the MOB_HO-IND message by the MS and its reception at the serving BS following completion of the handover preparation phase. However the handover action phase can also be triggered if the MS initiates handover ranging at the target BS in the event the MOB_HO-IND message was not received at the serving BS, for example, due to poor radio conditions. Figure 8.8 provides an illustration of the WiMAX network handover action phase procedure. In this example, target BS1 was offered by the serving BS as a handover candidate during the handover preparation phase procedure and subsequently selected by the MS to handover to.

The MS sends a MOB_HO-IND message to the serving BS confirming the selection of a target BS negotiated during the handover preparation phase. The serving BS sends a handover confirm message to its ASN-GW controller over R6, which is then relayed to the selected target BSs ASN-GW controller over R4 and then to the target BS over R6 in steps 2, 3, and 4. The target BS reconfirms its willingness to accept a handover from the MS and acknowledges the received handover confirm message with a handover acknowledgement message over R6 to its ASN-GW controller, which relays it serving BS ASN-GW controller over R4, which relays it on to the serving BS over R6 in steps 5, 6, and 7. If authentication context was not received or retrieved during the handover preparation phase, the selected target BS initiates the authentication context retrieval procedure for the MS with its ASN-GW controller over R6, which relays authentication context retrieval from the anchor authenticator over R4 in step 8. Additionally, if the target BS did not perform data path preregistration during the handover preparation phase, the target BS also initiates data path preregistration with the data path function in the anchor ASN over R6 and R4 in step 9, and awaits the arrival of the mobile via handover.

The serving ASN-GW controller for the MS also sends a handover confirm message with a cancellation indication to handover candidate target BS2, ASN-GW2 controller over R4, which is then relayed to target BS2 in steps 10, 11, and 12 to notify it that was not selected for handover by the MS. This message is sent soon after

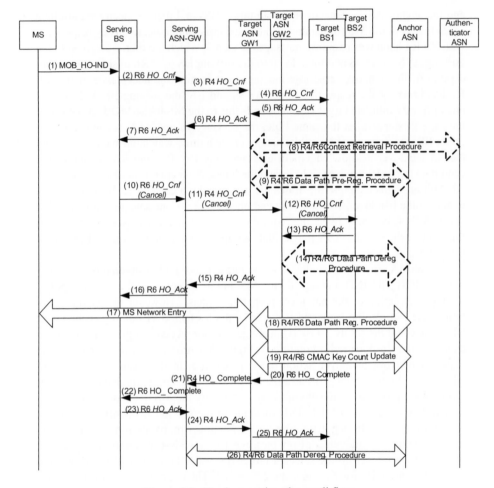

Figure 8.8. Handover action phase call flow.

step 2 so that network resources allocated by unselected handover target BS2 and its ASN-GW controller, and other unselected target BSs and their controllers, can be freed up and made available for handover from other mobiles. If data path pre-registration was requested by target BS2 during the handover preparation phase, the anchor data path is requested to be released to data path in step 14. This step may actually occur as soon as target BS2 is notified that it was not selected by the MS. The target BS also responds with a handover cancellation acknowledgement message to its ASN-GW2 controller in step 13, which is then relayed to the serving BS via the target and serving ASN-GW controllers in steps 15 and 16.

The MS initiates handover network entry with selected target BS2 via the 802.16 air interface procedure in step 17. The target BS2 then proceeds to initiate the data path registration procedure in step 18, updates the anchor authenticator with the new CMAC key count in step 19 (see Chapter 9 for details on WiMAX security), then notifies the serving BS that the MS was successfully acquired with a handover complete message that is relayed by the respective ASN-GW controllers in steps 20, 21, and 22. The serving BS relays acknowledgement of the handover complete message by responding with its own handover acknowledgement message, which is relayed to target BS1 in steps 23, 24, and 25. It further proceeds to initiate release of its data path connection for the mobile in step 26. At this point, the MS has successfully completed a controlled handover to BS2. Target BS2 may proceed to optionally relocate the authenticator and DP functions to its own ASN-GW1 controller upon completion of the handover procedure to free up resources in the other ASNs.

8.6.2   Uncontrolled Handovers

WiMAX uncontrolled handovers refer to ASN-anchored mobility scenarios in which the MS selects a target BS after moving from an area served by its serving BSs, moving into an area served by the selected target BS, and providing little or no notification to its previously serving BS. The handover is uncontrolled from the network perspective since the MS initiates the 802.16e handover ranging procedure at its autonomously selected target BS, which received no prior notification via network backbone signaling of the impending arrival of the MS from the serving BS and, hence, is unprepared to complete a seamless handover. From a WiMAX network perspective, the target BS did not participate in the handover preparation phase for the MS, if one occurred. Uncontrolled handovers may be intra-ASN or inter-ASN. Some examples of uncontrolled handover scenarios in WiMAX include the following:

- The radio signal deteriorates so quickly on the serving side (for example, if the user enters an elevator) that the MS does not have time to initiate a handover request at the serving BS and simply appears at the target BS and starts handover ranging. In this scenario, WiMAX network signaling supports the "pulling" of network context by the target BS from the serving BS.
- The radio signal deteriorates so quickly on the serving side that the MS only has time to send the 802.16e MOB_HO-IND message, which includes its autonomously selected target BS, to the serving side, then departs for the indicated target BS. In this scenario, WiMAX network signaling supports the "pushing" of the MS-related context by the serving BS to the target BS.
- MS initiates a handover request at the serving BS, then either leaves prior to completing the handover signaling protocol or ignores the target BS recommended by the serving BS at the conclusion of the procedure and departs for a target BS that had not been offered by the serving BS. In this scenario, the serving BS may "push" or the target BS may "pull" network context from the serving BS depending on whether the serving BS knows which target BS the MS is departing for via the MOB_HO-IND message.

Though controlled handovers provide a more seamless and "controlled" handover in WiMAX, uncontrolled handovers can still be successfully completed without requiring a new network entry at the target BS. The MS triggers an uncontrolled handover at its autonomously selected target BS according to one of the scenarios described above by sending a RNG-REQ message indicating that it desires a handover to the target BS. The MS includes its previous serving BS identity in the RNG-REQ message.

If the serving BS has not already pushed the MS context to the MS's selected target BS after previously being notified by the MS of the target BS, the target BS, which was previously unaware of the impending handover from the MS, uses the serving BS identity provided by the BS in the RNG-REQ message to contact the serving BS and request the MS context for the MS. The MS context includes the 802.16 MAC context; authenticator context if available or the authenticator network identifier, otherwise and the data-path function identifier.

Regardless of whether MS context was pushed by the serving BS or pulled by the target BS, the MS next contacts the authenticator, if an authenticator identifier was sent instead of authentication context, to retrieve the authenticator context for the MS, then establishes a data path for the MS, with the data path function identified by the data-path identifier. The serving BS is then informed of the successful completion of the handover via a handover-complete message, which proceeds to release its data path for the MS. The serving BS may also notify other candidate target BSs that may have been alerted of a potential handover from the MS if a handover preparation phase occurred using the network signaling procedures described in the handover action phase.

8.7 SECURITY CONSIDERATIONS

Handover has two security problems: authentication (MS to the target BS and target BS to MS) and reestablishment of preexisting security arrangements.

Authentication is a matter of demonstrating that an entity really is the entity it claims to be. In 802.16 terms, a device demonstrates authenticity by producing the result of a computation that is equivalent to what is expected when the expected security keys are applied to the specified security algorithms.

Authentication is initiated between an MS and the BS during network entry using the process defined in Chapter 9. Although the BS and MS can be configured to repeat this process at handover, this process consumes too much bandwidth and incurs too much delay to be practical for a system that supports any kind of real-time or near-real-time services. For this reason, WiMAX systems can accomplish handover authentication simply by signing the RNG-REQ and RNG-RSP messages with CMAC tuples.

The standard supports the option of Layer-2 encryption on traffic service flows. Whereas the presence or absence of encryption for a given service flow persists through handovers, the key for encrypting the service flows does not. The target BS typically sends new traffic encryption keys (TEKs) to the MS after handover. The TEKs are encrypted by the key-exchange key (KEK) before being sent over the air.

There are no requirements on the target BS to choose the TEKs in any particular fashion, although there is an algorithm for choosing TEKs based on security information to support seamless handover (see the following section).

8.8   SEAMLESS HANDOVER

Seamless handover (Figure 8.9) is a group of new concepts introduced in 802.16 standards [4] to improve handover performance. Specifically, the goal was to streamline handover signaling and minimize disruptions to data flow. Three aspects of handover are addressed:

1. CID update
2. TEK update
3. Backward compatibility

The CID update problem has already been described; CID values used by an MS in a source BS might be in use by another session with another MS in the target BS. The mobility additions to [3] provided a solution in the form of messages and information elements (IEs) that were used to explicitly map CIDs in the source BS to new CIDs in the target BS. Data cannot be transferred on the target BS until the CID remapping is complete.

Reference [4] provides a preallocation mechanism for assigning traffic CIDs based on the basic CID assigned to the MS. Prior to a handover, the MS can be provided with the new basic CID that it will use in the target BS in the MOB_BSHO-REQ message. The MS can then calculate the traffic CIDs to be used in the target BS by applying in an equation its new basic CID along with parameters broadcast in the DCD.

Seamless handover performs a TEK update in a similar fashion. Instead of waiting for the handover to complete before assigning new TEKs, seamless handover defines an algorithm for TEK derivation that is based on the target BS BSID [from which the access key (AK) is derived], the CMAC key, and the CMAC key counter (which essentially counts the number of handovers since the CMAC was first derived for the current session). The MS can easily derive the new TEKs prior to the handover, and the target BS can also derive the new TEKs when it receives the MS context through the backhaul.

The main goal of seamless handover is minimal interruption of data flow during handover. A third concept that was introduced is to allow substitution of a data exchange in either direction as the initial handshake instead of requiring a RNG-REQ/RSP exchange. This allows data transfer to commence immediately upon entry to the target BS. Note that immediate data transfer implies that the BS has preallocated bandwidth for the new BS; this allocation is just a simple reuse of the fast-ranging IE. To preserve the authentication requirement, the data exchanges are signed with a CMAC, just as the RNG-REQ/RSP would be. If L2 encryption is en-

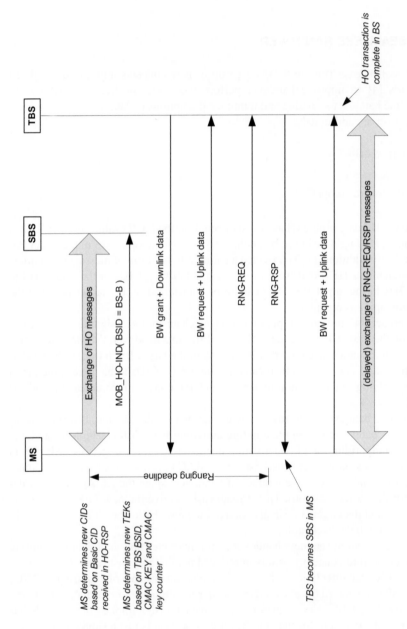

Figure 8.9. Example of seamless handover.

abled, the data exchanges are encrypted using the TEKs belonging to the target BS. However, in the interest of preserving as much legacy development as possible, it was recognized that the RNG-REQ/RSP exchange is an important event in existing designs and that the absence of that exchange could present major design problems for moving forward. For that reason, the RNG-REQ/RSP exchange is still required at some point following the handover. If seamless handover is enabled, a deadline for the completion of this exchange is carried in the MOB_BSHO-REQ and MOB_BSHO-RSP messages.

8.9 HANDOVER OPTIMIZATIONS

The 802.16 standard provides significant flexibility in HO processing. The flexibility allows wide variation in product capabilities and complexity.

As noted previously, in the most basic case, a handover can be processed as a completely new initial network entry. Although that might accomplish a basic need for mobility, it would likely cause an unacceptable interruption for most service types. On the other hand, a completely seamless handover would require significant distribution of information regarding the state of each MS and a high degree of synchronization between BSs, which may be impractical or overly complex for most applications. In order to allow real deployments to be tailored to successfully balance implementation trade-offs, a list of HO optimizations has been standardized. An HO optimization bitmap has been defined in the standard. It identifies, by either MS or BS, the specific optimizations supported.

Specific optimizations include:

- Omission of capabilities exchange (SBC-REQ/RSP messages). Instead of requiring an MS to explicitly state its capabilities to a target BS after handover, it is likely faster and more efficient for the network to cache MS capability information and distribute that information to a new target BS as the MS moves through the network. Implicit in this assumption is that the MS also stores BS capabilities.

- Omission of registration (REG-REQ/RSP messages). Similar to the omission of capabilities exchange, registration information exchanged in network entry can be cached in the network so that it does not need to be repeatedly transferred in the air protocol for each handover.

- Omission of need for unsolicited SBC-RSP or REG-RSP messages. It is possible that the capabilities of the target BS do not match the capabilities of the source BS. If this is the case, the target BS can send an unsolicited SBC-RSP or REG-RSP to update the BS capabilities cached in the MS. However, the MS needs to know if this is a possibility and must, therefore, wait for these messages. This waiting period can be eliminated if the MS knows that all BSs in the coverage area support the same capabilities.

- Control of reauthentication on the target BS. As noted previously, authentication may or may not be necessary upon MS entry to a new BS based on the network architecture and the trust relationship between BS attached to the network. Reauthentication can typically be avoided if all BSs in a coverage area are part of the same network. On the other hand, if reauthentication is desired, the HO optimization bitmap can be used to indicate the desired procedure for reauthentication.

- ARQ context transfer. If an MS has a service flow in which ARQ is enabled, it is possible that the serving BS (i.e., the serving BS before the handover) was not able to satisfy all of the retransmission requests made by the MS. If the ARQ context is retained through the handover, then the target BS is able to satisfy the MS retransmission request without restarting the ARQ state machine. This saves time and potentially avoids data loss due to handover.

- Elimination of need for IP address assignment.

- Elimination of network acquisition. This includes acquisition of network parameters such as network address and time of day.

8.10   INTERACTION WITH OTHER FEATURES

In general, any features used for a connection between an MS and BS must be negotiated between those two entities. When an MS hands off to a new BS, there is no guarantee that the features negotiated in a previous connection will be present in the new connection. The optimizations discussed above are examples where support for features is guaranteed through the handover (due to specifics of the system deployment) and, thus, support does not need to be renegotiated.

8.10.1   Sleep Mode

Sleep mode is a mode of operation in which the MS may physically remove or reduce power to parts of its radio transceiver for the purpose of extending battery life. Sleep mode also benefits system performance by allowing an MS to temporarily release uplink bandwidth without requiring subsequent negotiation to regain the use of the bandwidth.

In terms of the protocol, sleep mode is a defined period during which no communication is possible between the BS and the MS on a given connection. The parameters of the period (start time, duration, repetition interval) are negotiated for each connection between the MS and BS before the MS enters sleep mode. Note that sleep mode is negotiated for connections, not on a per-MS basis; an MS with multiple connections may need to manage a different sleep configuration for each connection. Once sleep mode is successfully negotiated for a connection, that connection becomes a member of a *power savings class* or PSC. Connections sharing similar characteristics (e.g., VoIP connections) may be joined to the same PSC since similar characteristics may imply similar sleep mode operation and, thus, similar performance improvement opportunities.

The aggregation of connections into a PSC is problematic for handover. An MS needing a particular kind of sleep mode characteristic may hand off into a coverage area where the PSC for that characteristic does not exist, or it may hand off into a coverage area where that PSC is oversubscribed. In either case, it is not possible to continue the sleep mode characteristics for that connection in the new coverage area and, thus, no prenegotiation is possible. In addition, the CID of a connection may be renegotiated as a result of handover. Since sleep mode is defined per connection, the CID belonging to a connection in a particular PSC may not be valid after the handover. For these reasons, sleep mode must always be renegotiated following a handover.

8.10.2   Idle Mode

Idle mode is another mode in which the MS is unavailable for communication with the BS for a negotiated period of time in the interest of saving power and system resources. Whereas sleep mode was intended for short periods of inactivity between traffic bursts, idle mode is more suited for situations in which the MS is essentially not in use but not turned off.

In terms of idle mode protocol, the MS transitions between the *MS paging unavailable interval* and the *MS paging listening interval* in a periodic manner. The MS is unavailable for communication during the MS paging unavailable interval and may use that interval to deactivate its transceiver for power saving or it may perform scanning, including scanning other access technologies in other frequency bands. During the MS paging listening interval, the MS must decode the DCD and UCD of its preferred BS (established on entry to idle mode) and attempt to decode a MOB_PAG-ADV message if one is present.

Upon entry into idle mode, the MS chooses a preferred BS. The MS signals its preferred BS by performing ranging with that BS. The preferred BS is, in turn, part of a *paging group*. A paging group is a group of BSs that forms a continuous coverage area and is served by a common *paging controller*. When data arrives in the network for a particular MS, the paging controller instructs all BSs that are members of the paging group to page the MS. The size of a paging group should be large enough to minimize the overhead associated with tracking each MS, but not so large that the number of MSs moving within a paging group is unmanageable.

While in idle mode, mobility consists mainly of *location updates* sent by the MS. Location updates are RNG-REQ messages that indicate nothing more than the MS location; a RNG-REQ sent for location update does not result in network entry. A location update is necessary when an MS leaves a paging group and enters a new paging group. This updates the MS location in the system location register. Location updates are also required when an MS powers down and are required periodically to verify that an idle-mode MS is still within its last paging group. An MS is also free to send a location update at any time for any reason.

In terms of network impact, once the RNG-REQ message is received from the MS, the BS sends a location update request message to the paging controller in the ASN-GW. If the paging controller is located in another ASN, the location update request will also pass over R4 to reach the "anchor PC ASN." When the message is received by the PC, it first retrieves the authenticator context associated with the MS from the MS's authenticator, which may be colocated at the same ASN-GW as the PC function, or must send signaling over R4 if the authenticator is anchored at another ASN. Once the MS is authenticated, the PC updates its location register for the MS so it knows where to page the MS when needed, and then confirms the location update with a location update confirmation message to the BS.

Idle mode mobility can also trigger paging controller relocation in the network. Either the serving ASN, that is, the ASN-GW controlling the current serving BS, or anchor PC may request paging controller relocation, which, if performed, results in moving the PC function to the serving ASN-GW.

8.11 SUMMARY

The addition of mobility to 802.16 was a major undertaking. It required the addition of signaling and procedures for handover and required revision and rethinking of preexisting features such as ranging, sleep mode, and idle mode. It also required the development of an entire network infrastructure capable of tracking MS devices, making handover decisions, and implementing handovers when necessary. Although these techniques may resemble use in cellular technologies, the final techniques are unique to 802.16.

An important area of future work is handover to and from other network technologies. This is important for operators that need to manage a smooth transition between their legacy technologies and WiMAX. It also helps enable the vision of the always-on Internet, anywhere.

8.12 REFERENCES

[1] IEEE Std 802.16-2004, Air Interface for Fixed Broadband Wireless Access Systems.

[2] IEEE 802.16-02/48r3, PAR 802.16e: Amendment for Physical and Medium Access Control Layers for Combined Fixed and Mobile Operation in Licensed Bands (2002-12-16) [approved by IEEE-SA Standards Board, December 11, 2004].

[3] IEEE Std 802.16e-2005 Air Interface for Fixed and Mobile Broadband Wireless Access Systems.

[4] IEEE Std 802.16-2009, Part 16: Air Interface for Broadband Wireless Access Systems.

[5] WiMAX Forum™ Mobile System Profile Release 1.0. Approved Specification (Revision 1.4.0: 2007-05-02), WiMAX Forum, 2007.

[6] WiMAX Forum Network Architecture Stage 2 and 3—Release 1.0 (Revision 1.2), WiMAX Forum, 2009.

CHAPTER 9

WiMAX END-TO-END
SECURITY FRAMEWORK

SEMYON B. MIZIKOVSKY

9.1 INTRODUCTION

The end-to-end (E2E) security framework of the WiMAX system is the focus of this chapter. As will be shown, expected security is preserved by applying a combination of protection methods: some mandatory to establish a minimum baseline level of expected security and resource preservation, and some optional, which can be used depending on a particular business and deployment strategy of the WiMAX operator. The E2E security framework covers the entire link from the client (user) to the server and all the way back from the server to the client.

Additional virtual private network (VPN) tunneling mechanisms between the client and the application server are not a part of this discussion, as they can be applied above and beyond the WiMAX system to ensure complete privacy and security expected by the application. Likewise, the end-to-end security of client–server-based multimedia applications, such as in IMS (IP multimedia subsystem, VoIP), are also outside of this discussion.

9.2 WiMAX SECURITY REQUIREMENTS

Due to unusual initial requirements placed on the WiMAX system from its infancy, security solutions made to protect its owners and users vary in many respects from those defined for other commercial mobile technologies, like GSM, UMTS, Cdma 1xRTT, or 1xEV-DO.

Specifically, in the open access model, the mobile terminal user may choose any available network access provider (NAP) to access the network service provider of their choice without prior contractual preprovisioning. In the gradual evolution of the client-centric Wi-Fi operational model toward the mobile-centric WiMAX model, attempts were made to add mobility to the core IEEE protocols (IEEE 802.16) and requirements to maintain a focus on data-centric services while maintaining close alignment with the IP-based architecture.

Hence, requirements were defined for the user's access terminal—the mobile station (MS). The MS must:

- Be dynamically provisioned or reprovisioned with identity and security credentials allowing access authorization to the chosen networks and services
- Inform the chosen home core serving network (CSN) of the network service provider (NSP) of its identity, capabilities, credentials, and genre, with guarantees of nonrepudiation
- Be able to verify that the accessed chosen CSN is really the one which the user chose to access
- Be able to establish a secure wireless link with the access serving network (ASN) in order to protect the transferred information, including control and data traffic, from eavesdropping and manipulation

Similarly, requirements were defined for the chosen home serving network, which must:

- Be able to dynamically provision, in a secure fashion, the access security credentials in a terminal after securely validating its presented claimed identity and capabilities
- Be able to verify and validate already established security credentials presented by the provisioned terminals at the time of service access
- Be able to provide the ASN with necessary parameters to establish and maintain the secure session with the mobile terminal

Requirements were also defined for the access serving network:

- Establish secure communications links, including ciphering and integrity protection of the transferred information, with the access terminals, using session-related security parameters provided by the home CSN
- Allow access to the IP-based service only to the properly authenticated terminals and only through the secure connection wireless connection
- Provide secure and efficient mobility, including fast access, handovers, and macromobility for bearer services within and between the ASNs, while protecting from channel and service hijacking by unauthorized terminals

Based on the above requirements, the IEEE 802.16 and WiMAX Forum standardization bodies made unique development choices for the WiMAX security
framework described in the following sections.

9.3 END-TO-END SECURITY ARCHITECTURE

The WiMAX Forum specifies the architecture, protocols, and procedures that all
WiMAX systems must adhere to in order to facilitate interoperability. The general
architecture of the WiMAX system is shown in Figure 9.1.

In this architecture, secure relations are established between the MS and the
AAA, the authentication, authorization, and accounting server in the CSN that belongs to the home network service provider. In fact, once these relations are established, that is, once the subscription record is set up in the AAA database, the NSP
owner of this database becomes the home NSP. Chapter 6 provides an overview of
WiMAX network Architecture.

The MS establishes a subscription with the Home NSP (H-NSP) that owns the
home CSN (H-CSN) shown in Figure 9.1, but accessed from the bottom-left ASN.
The link for validating authenticity and authorization of the MS will be provided by
the bottom-left ASN through its own default visited CSN in V-NSP (R3 interface)
and via the "visited" AAA server through the R5 interface to the home AAA server
in the H-CSN. Further breakdown of the ASN is shown in Figure 9.2.

In the above end-to-end architecture, there are multiple points of vulnerability:
the air link (R1), the backhaul links (R6, R4), the links between ASN and CSN (R3,

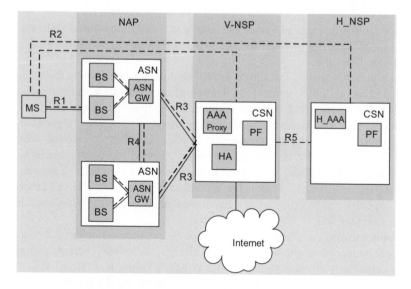

Figure 9.1. End-to-end network architecture.

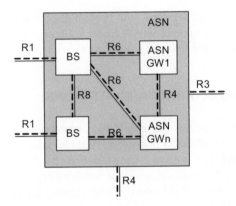

Figure 9.2. General ASN architecture.

R5), and the bearer traffic. Potential weaknesses of these interfaces are discussed in Section 9.4.3.

9.3.1 Development Choices for WiMAX Security

9.3.1.1 Digital Certificates for Device and Network Authentication

But how would the accessed chosen NSP establish a secure connectivity with a previously unknown and possibly unprovisioned terminal in order to establish secure relations? How could such connectivity even be established through the IP network if the terminal is not allowed to receive IP services over the unsecure wireless and backhaul connection?

This conundrum was addressed by turning to a certificate-based authentication and key establishment protocol well known in the industry and widely used for Internet-based services.

Each WiMAX terminal would be provisioned manufacture with a digital certificate based on the X.509 protocol [1]. This certificate contains terminal digital identity (MAC address); genre; and manufacturer's specifics, including model, type, basic capabilities, and validity period. The certificate is securely "signed" by the manufacturer; such digital signature is further certified—"cosigned"—by the WiMAX Forum, whose certification is further attested to by an industrially reputable and trusted "Root" Certification Authority (CA), such as Verisign. This triple-signed certificate is presented by the MS to the home AAA for validation and, once validated, can be accepted as a guarantee that all information contained in it is true and unmodified.

Symmetrically, the certificate would also be provisioned at the AAA servers with information that attests to the NSP identity, its validity period, and so on. These certificates are also signed by the WiMAX Forum and "Root" CA, so the mobile terminal receiving it could validate it to ensure that the accessed NSP is indeed the one chosen by the user.

9.3.1.1.1 *What is an X.509 Certificate?* A certificate is a digital "document" that contains both an identity and a public key, binding them together by a digital signature. This digital signature is created by an organization called a certification authority (CA). This organization guarantees that upon creating the digital signature it has checked the identity of the public key owner (e.g., by verifying credentials of a mobile device's vendor, or a service provider–owner of the AAA server) and that it has checked that this public key owner is in possession of the corresponding private key. Anybody in possession of the CA's public key can verify the CA's signature on the presented certificate. In this way, the CA guarantees that the public key in the certificate belongs to the entity whose identity is in the same certificate.

The worldwide accepted standard for digital certificates is called X.509 [1]. An X.509 certificate contains the information that is signed and, therefore, is guaranteed to be true and unmodified. This information may contain serial number; validity period; issuer name (this is the identity of the issuing CA); subject name (this is the identity of the party being certified; this can be a device, operator's AAA server, or even the issuing CA itself); subject public key (the public key of the party being certified); and a path length field, indicating the maximum number of intermediate CAs that are allowed between the "root CA" and the end user.

The digital signature is produced using the issuing CA's private key.

By the digital signature in a certificate, the CA guarantees that it has checked (according to its policies) that the subject identity mentioned in the certificate is indeed the owner of the public key that is in the same certificate, and that this public key owner also is in possession of the corresponding private key kept in secret by an owner. The signature in the certificate can be verified by anybody, by using the issuing CA's public key.

9.3.1.2 *Authentication Protocols—EAP*

Several existing protocols were considered for exchanging certificates, and the preference was given to those that, in addition to providing mutual certificate validation, thus authenticating the end points, would also generate the mutually authenticated session keys that can be used for establishing session security.

Several years ago the IETF (Internet Experts Technical Forum, an industry acclaimed forum that governs the protocols used in the Internet), defined the efficient and flexible framework called the Extensible Authentication Protocol (EAP) [2]. The EAP allows various authentication methods adopted for it to be transparently transported over the communications network, between the end points, without the need for this network to be aware of specifics of those transported methods. That is, the MS and the AAA as the end points would agree on the authentication method, which would be transparently transported between them over the ASN and CSN without any reprocessing, or even the knowledge of what is communicated.

As the result of executing the method within the EAP, the end points could mutually verify the authenticity of each other, agree on the security context for the upcoming session, define credentials for the future sessions, and so on. Based on the success of executing the EAP method, the Authorization Server (AAA) provides session authorization that is delivered to the serving network. If necessary, the session-specific security context can also be delivered to the access system (ASN) for protecting radio links, IP traffic, mobility signaling, and so on.

The WiMAX Forum specifically requires that the EAP methods used for the EAP access autentication be capable of generating the master session key (MSK) and the extended master session key (EMSK). The MSK is exported by the AAA to the authenticator in the ASN for further use for access security, and the EMSK is retained by the AAA for further use for mobility security.

9.3.1.3 Device Authentication—EAP-TLS For Initial Provisioning

One EAP-based method that allows mutual validation of certificates as well as generates mutually authenticated session keys, namely EAP-TLS [3], was chosen as a default authentication protocol for WiMAX terminals that need to connect to the network for initial provisioning, without prior or valid subscription. The process proceeds as follows.

The MS accesses the ASN in an insecure mode. As long as any WiMAX device is expected by default to support the EAP protocol, the special function in the ASN called the authenticator invokes the EAP protocol by requesting the MS Identity. This identity is expected to be in the form of a NAI (Network Access Identifier) specified by IETF [4]. The NAI format username@realm clearly identifies the NSP the MS is attempting to receive service from. Through the DNS discovery, the ASN finds the CSN that belongs to the selected NSP and presents the NAI reported by the MS. The NAI is communicated to the target AAA in the EAP payload of one of the supported AAA signaling protocols, such as RADIUS or DIAMETER, also defined by the IETF [5].

The AAA searches its subscription database and recognizes that the MS does not have an associated subscription record and, thus, is a target for provisioning. The EAP server at the AAA initiates the EAP process and returns back to the MS, transparently to the ASN, the command to invoke the EAP-TLS method. The MS responds back to the AAA with the acceptance of the EAP-TLS.

The multistep EAP-TLS procedure as it is executed over the IEEE 802.16 air interface is shown in Figure 9.3. The WiMAX transport is not shown on this figure for simplicity. It is enough to assume that the EAP-TLS signaling of messages 1 and 2 is executed between the MS and the ASN-GW (authenticator), and signaling of messages 3 to 9 is executed between the MS and the HAAA. The EAP success included in message 9 is also observed by the authenticator, which is informed of the success of the EAP process and allows access.

The figure also shows that EAP messages are using the PKMv2 transport [6].

During execution of the multistep EAP-TLS method (see Figure 9.3), the server certificate is delivered to the MS, the device certificate is requested from the MS and is delivered to the AAA, these Certificates are validated to ensure the authenticity of claimed identities, an ephemeral protection tunnel is established between the MS and the AAA by running the Diffie–Hellman public key exchange [7], random nonces are exchanged between the MS and the AAA under cover of this tunnel to ensure the re-

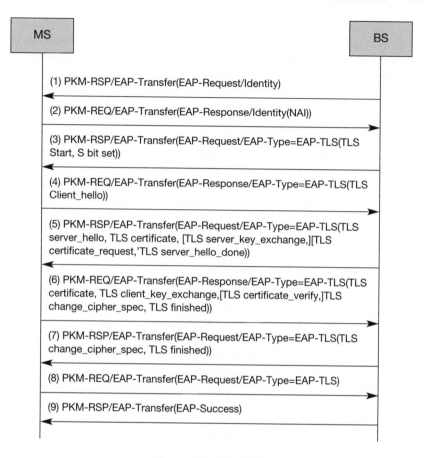

Figure 9.3. EAP-TLS.

play protection, and then, once the success of the EAP-Method execution is verified by checking the expected results of cryptographic computations, the symmetrical master session key (MSK) is generated independently by the MS and the AAA.

This success of the EAP method signifies success of device authentication, which is reported to the authenticator by the AAA via the AAA signaling (e.g., RADIUS access accept). Along with this signaling, the EAP payload carrying the success indication is included as well. The MSK is sent to the authenticator by the AAA with the security guarantees of the AAA signaling protocol (RADIUS or DIAMETER).

Once the authenticator at the ASN receives the EAP success as well as the MSK, it further derives the whole range of security keys for the access protection, delivers them to the serving base station in the ASN, establishes a secure communications link over the air, and sets up secure mobility service to the over-the-air provisioning (OTA) server.

As described in Chapter 7, the OTA server now provisions the subscription parameters into the MS. These parameters, among others, include the permanent subscription user identity in the form of the NAI with the USERID established for the subscription record, and the realm of the NSP that the user chose to act as its "home" NSP. As a result, the reported NAI will henceforth uniquely identify the user record in the specific subscription database. For example, if the user reported mickey.mouse@orlando.disney.com, the ASN would resolve "disney.com" to the specific AAA server that serves the Orlando region of the NSP identified as disney.com, and the accessed AAA would easily find the subscription record of one mickey.mouse that has established a valid subscription.

Along with identity, other unique and secret subscription parameters can be provisioned into the MS during the OTA process. Among them are the subscription password and symmetric secret keys for subscription authentication. The utility of these parameters will be seen from the following sections.

9.3.1.4 Subscription Authentication

The WiMAX Forum mandates support and execution of subscription authentication with the home CSN, which differs from the device authentication described in the previous section by validation of provisioned subscription credentials. Detailed description of some preferred subscription authentication methods will be presented in the following sections. It should be noted, however, that EAP presents a flexible extensible framework that supports many other methods, and a wide range of other methods exists that satisfy WiMAX requirements of efficient mutual authentication and secure exchange of mutually authenticated session keys.

9.3.1.4.1 EAP-AKA for Subscription Authentication. The EAP-AKA was defined by the IETF [8] to allow reuse of the AKA protocol that is used by the Third-Generation Partnership Project (3GPP) standardization body for third-generation commercial wireless systems, such as WCDMA/UMTS. It is also accepted by the 3GPP2 organizational partners for the Cdma2000 1xRTT system. Having being deployed in millions of mobiles and multiple serving systems, it became an easy choice of protocol for many WiMAX operators, especially in Asia and Europe. The EAP-AKA protocol proceeds as shown on Figure 9.4.

When the HAAA receives the MS identity, it analyzes the subscription record and recognizes that the MS supports the EAP-AKA. The AAA generates the AKA authentication vector (AV). The format of the AV is shown in Figure 9.5.

To generate the AV, the AAA uses the symmetric secret key Ki provisioned in the subscription record associated with the user identity, and the randomly generated nonce RAND that guarantees uniqueness of each vector. The RAND is the challenge element of the AV.

The same symmetric secret key Ki is also provisioned in the MS at the time of OTA provisioning, or can be provided to the User preprovisioned on the removable WiMAX subscriber identity module (WSIM), or user identity credentials card (UICC).

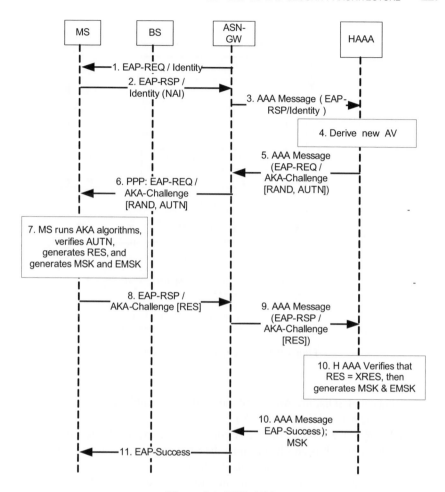

Figure 9.4. EAP-AKA.

To ensure that the vector cannot be repeated by any unauthorized network entity, the AAA also utilizes the special sequence counter (SQN) that is incremented with every successful completion of AKA protocol. A similar counter is independently maintained at the MS or inside the WSIM or UICC. As seen in the AV format, the SQN can further we concealed in transmission by covering it with the anonymity key (AK) generated as the part of the AV creation.

The AV also contains an authenticated message field (AMF), which can carry limited, but assured, control information between the AAA and security module at the MS.

The integrity of the SQN and AMF, and thus validity of the AAA that created the vector, is attested to by generated message authentication code (MAC) that is computed on critical elements of the vector delivered to the MS.

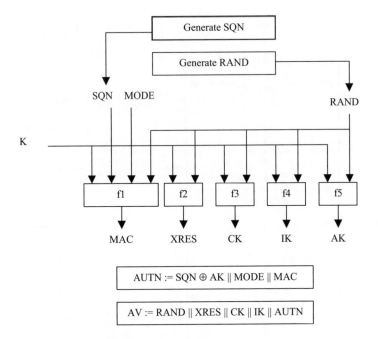

Figure 9.5. AKA authentication vector.

The concealed SQN, the AMF, and the MAC are combined into the vector element called the network token (AUTN).

In addition to the AUTN, the AAA computes the expected response from the MS, the XRES, which is checked against the actual response (RES) computed and returned by the mobile in response to the challenge.

The RAND and AUTN are sent to the mobile inside the EAP capsule by the AAA, transparently, through the network. The mobile validates the received MAC included in the AUTN, decrypts the SQN and checks that the SQN is in the expected range. If both checks are positive, the mobile computes the RES and returns it to the AAA inside the EAP capsule. The AAA validates the RES and, if it checks, computes the 512-bit MSK (Master Session Key), which is sent to the authenticator along with the success indication of the EAP process. A copy of the MSK is not retained at the AAA.

Once the mobile receives the EAP success, it also computes the MSK.

In addition to the MSK, both the AAA and the mobile also compute the extended master session key (EMSK), which is retained by the AAA and the mobile. This key, as will be shown later, is used for the mobility security.

9.3.1.4.2 EAP-TTLS for Combined Device and Subscription Authentication. EAP-TTLS [9] is an EAP method that utilizes TLS to establish an initial secure connection between a client (mobile) and server (AAA) first, through which the subscriber authentication can be then executed. A Typical EAP-TTLS procedure is shown on Figure 9.6.

The initial TLS handshake may mutually authenticate client and server by validating presented X.509 certificates, or it may perform a one-way authentication of a server only, in which only the server is authenticated to the client. Whether or not the device certificate needs to be requested by the AAA is left to the operator's discretion.

The secure connection established by the initial handshake may then be used to allow the server to authenticate the client using one of well-known user authentication protocols, such as MS-CHAPv2 [10]. If authentication of subscription auser credentials is a requirement, than these credentials, such as username, and secret password, have to be provisioned into the mobile as well as into the AAA subscription database. When these credentials are transported over the established TLS tunnel, they are concealed from an unauthorized eavesdropper, thus providing additional user anonymity. The identity sent for setting up the "outer" TLS tunnel can simply have a random number in the USERID part of the "outer-NAI."

The widely deployed authentication infrastructures such as RADIUS are used to

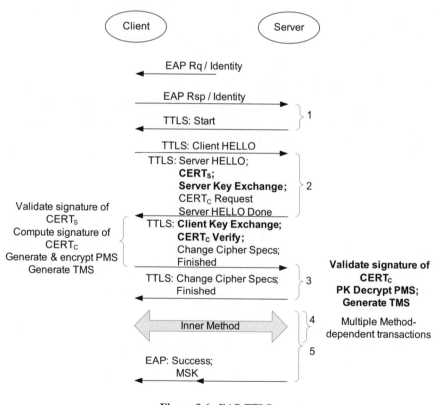

Figure 9.6. EAP-TTLS.

transport the protocol over the R3, and the EAP Authentication Relay Protocol defined for R4 and R6 is used to transport it over the WiMAX network.

Once the inner authentication method succeeds, the AAA server and the mobile compute the common MSK and EMSK using the outer TLS protocol credentials, similar to the manner described in Section 9.3.1.3.

It is notable that the use of EAP-TTLS allows combined device and subscription authentication in one completed execution of the EAP procedure if the device certificate is requested by the server.

9.3.1.4.3 EAP-TLS for Subscription Authentication. Some deployments do not require concealment of the user identity, nor do they require setting up any user-specific subscription. In such cases, the subscription is tightly linked to the device that is used for the WiMAX access and, therefore, the mandatory subscription authentication requirement can be satisfied by executing authentication of the WiMAX device. The EAP-TLS protocol is suited for these deployments.

The AAA server authenticates the Device X.509 Certificate and verifies that the device identity (MAC address) presented in the certificate it that expected to be registered in the subscription record of the database. Similarly, the mobile verifies that the X.509 Certificate presented by the AAA is valid and belongs to the expected home NSP.

Both MSK and EMSK are computed as described in Section 9.3.1.3.

9.3.2 Security Protocol Stack

The Security protocol stack is shown in Figure 9.7. As may be seen, the authentication method is transported within the EAP framework that is extended between the MS and the authentication server (AAA) in the home CSN.

This EAP payload is transported specifically by the AAA signaling protocol

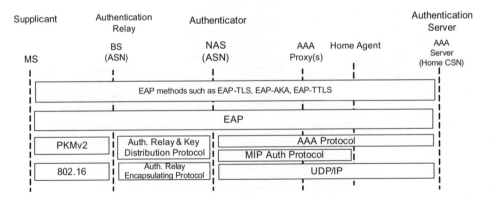

Figure 9.7. Security protocol stack.

(RADIUS or DIAMETER) on the R3/R5 interface between the AAA in the CSN and the authenticator in the ASN.

The authenticator relays this protocol over the WiMAX backhaul (interfaces R4 and R6) using the Authentication Relay Protocol. This protocol transparently transports the EAP payload between the ASN gateways (ASN-GW) in different ASNs and between the ASN-GW and the serving base station (S-BS) in the same ASN.

The PKMv2 Protocol (Protection Key Management) is defined between the BS and the MS on the air link. This protocol is run over the 802.16 air interface defined by the IEEE.

The Key Distribution Protocol is defined by WiMAX to create and distribute the access security keys for air interface ciphering and integrity protection. This protocol distributes the keys generated by the authenticator to the serving and target base stations that serve the MS. Associated keys are computed by the MS with the assistance of the PKMv2 Protocol used for key validation and cryptographic synchronization.

The MIP Authentication Protocol is used for preserving security of the mobile IP application. It is extended between the mobile IP client in the MS or the proxy MIP client at the ASN-GW and the home agent in the CSN. The Figure 9.7 does not show that for the Client MIP case, that is, when the MIP client is a part of the MS, the MIP Auth Protocol is carried in 802.16 signaling and is transported to the ASN-GW using R6 and R4 signaling.

The use of specific security protocols is described in subsequent sections.

9.3.3 Security Keys

Figure 9.8 shows the security key hierarchy of the WiMAX system. Use of the keys shown in this figure is further described in detail in Section 9.4.

9.3.4 Security Algorithms

IEEE 802.16 and the WiMAX Forum define the wide range of cryptographic algorithms to support required security. Specific algorithms are invoked based on bilateral negotiations between the mobile and the serving ASN, or based on preconfigured subscriptions between the mobile and the home CSN.

9.3.4.1 Algorithms Used for Authentication

For example, there is no need to standardize the cryptographic algorithms used by the end points of the subscription authentication, that is, the mobile and the home AAA, as there is no need to limit support for just a few authentication methods. These methods and algorithms can be left to the provisioning functions or agents involved in setting the subscription. However, to allow for the "open access" business model preferred by WiMAX operators, the minimum set of cryptographic algorithms was defined for the mandatory EAP-TLS protocol. These are: TLS-PRF-64 and TLS-PRF-128 hash functions, and TLS_RSA_WITH_AES_128_CBC_SHA

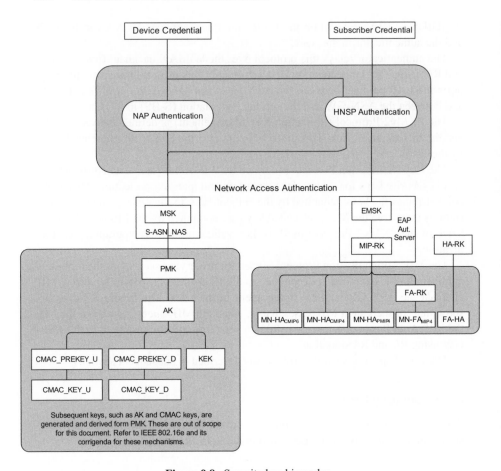

Figure 9.8. Security key hierarchy.

and TLS_RSA_WITH_3DES_EDE_CBC_SHA for the encrypted TLS tunnel, specified in the [3].

Functions and algorithms for the EAP-TTLS are not limited by the WiMAX Forum, but are expected to follow the cryptographic suits defined in the EAP-TTLS Internet Draft and the [9] for MS-CHAPv2.

Functions and algorithms for the EAP-AKA are not specified by the WiMAX Forum, and are expected to follow the [8]. Specifically, the MILLENAGE algorithm is expected to be used for the AKA methods functions, and HMAC-SHA-256 for EAP protection and MSK/EMSK generation.

9.3.4.2 Algorithms Used for Access Security

Security algorithms used between mobiles and ASNs must be standard to ensure interoperability. These algorithms are specified by the WiMAX Forum for security key hierarchy and by IEEE 802.16 for airlink protection.

Specifically, the WiMAX Forum specifies the HMAC-SHA1 based on the IETF [11] secure hash function to generate all keys in the key hierarchy from the EMSK (see Section 9.3.4), that is, the mobility keys. This function generates the 160-bit output given specific defined inputs. This 160-bit result is then used as the secret key for the specific security function.

The IEEE 802.16 specifies

- The 3xDES in CBC mode and AES in CCM mode algorithms for encrypting the TEK when transmitted from the BS to the MS (see Section 9.4.1.4)
- The SHA-1 based Dot16KDF algorithm construction for creating the AK, HMAC_KEY, and CMAC_KEY keys to protect management signaling messages, and the KEK (see Section 9.4.1)
- The CMAC-AES construction for generating the CMAC signatures of the management signaling messages
- DES and AES algorithms in CBC [12] mode for encryption of air interface traffic

Negotiation of the specific cryptographic suite is done during the initial phases of accessing the network by the mobile, through exchange of capabilities. These SBC request–response negotiations lead to selecting specific algorithms for integrity protection (HMAC or CMAC) and encryption (DES, AES, or none).

9.4 SECURITY ZONES

9.4.1 Airlink Security

WiMAX provides for embedded security capabilities through 802.16 security architecture standards. Due to the mobility and roaming capability in 802.16e and the fact that the medium of signal transmission is accessible to everyone, there are a few extra security considerations applied to 802.16e. These extra features include PKMv2, EAP authentication method, and AES encryption wrapping.

The overall airlink access security process is shown on Figure 9.9. Use of the specific steps in this process is described in the following subsections.

9.4.1.1 *Master Session Key (MSK)*

Airlink security fully relies on the access authentication conducted using the EAP protocol. As shown before, the successful EAP authentication (Step 2) results in generation of the MSK that is delivered to the authenticator in the ASN-GW (Step 3). This 512-bit key is further truncated and split into the 160-bit pairwise master key (PMK) and the 160-bit EAP integrity key (EIK) that can be used for the subsequent EAP reauthentications.

9.4.1.2 *Authentication Key (AK)*

The PMK serves as the main session-related root key for the access security. By cryptographically hashing the PMK with the identities of the MS and the BS, the

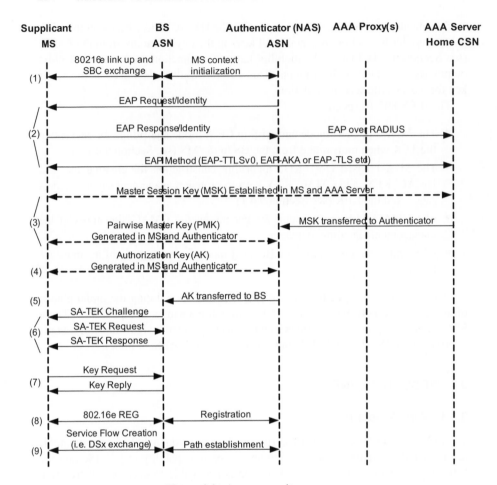

Figure 9.9. Access security process.

authenticator derives the pair-specific authentication key (AK) (Step 4), which is sent to the BS expected to serve the MS (Step 5). This AK is a part of the Security Context that also contains the AK identity and AK lifetime. Please note that the AK sent to different base stations will be different for serving the same mobile. This is done to ensure cryptographic isolation between different base stations.

Symmetrically, the MS also generates the MSK, derives the PSK from it, and computes the AK that belongs to the BS the MS is communicating with.

To ensure that both MS and the BS have the same AK, the mutually secure PKMv2 procedure is defined, called three-way handshake (3WHS) (Step 6). During this procedure, both MS and BS exchange random nonces and, using the locally derived AK, compute the cryptographic signatures of these nonces. The signatures are then exchanged between BS and MS, who verify them. Successful verification assures validity and correctness of the AK.

The AK is subsequently used to generate the whole range of security keys. But two main security services that need to be highlighted here are the security of management messages and airlink ciphering.

9.4.1.3 Management Message Security—CMAC and HMAC

There are multiple messages exchanged between the MS and chosen BS in order to establish, maintain, restore, and terminate the communication links over the air interface. These messages are defined by the IEEE 802.16 standard. According to the 802.16 access model, the MS initially selects the BS for access by ranging it, that is, sending the RNG-REQ message.

Obviously, before the MS has been properly authenticated, there is no established security association between the MS and the BS. Therefore, initial ranging cannot be protected in any way. However, once the MS has been authenticated at least once, that is, the security context at the authenticator in the ASN has been established, it can use this security association to cryptographically sign its transmitted messages. It is expected that the accessed BS will request the necessary security keys from the authenticator and validate the access.

The cryptographic signature of the management messages is created by using one of two negotiated cryptographic hash algorithms: HMAC or CMAC. As the use of the HMAC algorithm is not a preferred mode of operation in WiMAX, we will describe the use of CMAC.

The CMAC is the hash algorithm based on the advanced encryption standard crypto function (AES) currently standardized by multiple industries worldwide for encryption. This function in certain configurations can also be used for creating the secure hash or secure representation of the message. The properties of this hash guarantee that any manipulation or modification of the message will result in a different hash output, that it is infeasible to find another message that results in the same hash output, that analysis of the hash cannot lead to recovery of the message itself, and that use of the secure secret in it guarantees that an attacker cannot generate the correct hash without knowing the secret key.

The secret key for the CMAC, CMAC_KEY, is generated from the AK by the MS and BS. To ensure that the message with secure hash signature cannot be simply replayed by an attacker, the replay protection using the combination of packet number (PN) and an access counter CMAC_KEY_COUNT is defined. The access counter CMAC_KEY_COUNT is used for protecting against replayed messages that lead to the system access. The scheme is shown on Figure 9.10.

In this scheme, the CMAC_KEY_COUNT is maintained by the MS and the authenticator. It is reset to 0 at every successful EAP authentication or reauthentication, and is monotonically incremented by the MS for every access to the BS, either new or old. This counter is sent by the MS to the accessed BS included in the RNG-REQ message. The BS verifies with the authenticator that the value of the received counter is either equal to or larger that that expected by the authenticator, and, if so, generates the CMAC_KEY and validates the CMAC of the message. Only if this is correct is the message accepted. Successful access also triggers the

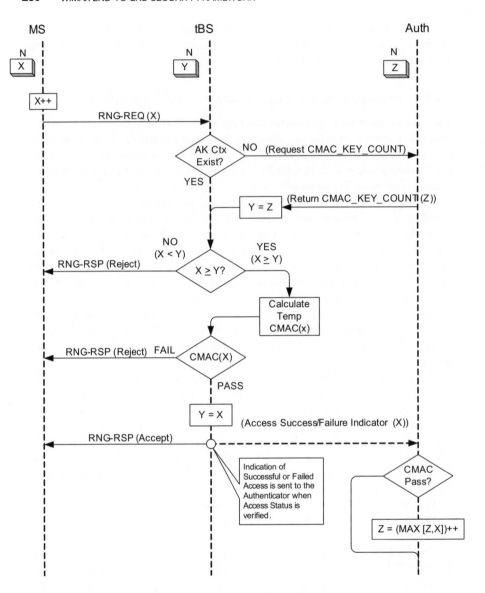

Figure 9.10. Replay protection of signaling messages.

counter increment at the authenticator to the value that will be expected of the MS at the next access.

Once the access is granted, each subsequent message transmitted between the MS and this BS will be protected from repetition attacks by the incremented packet number counter (PN). This counter is monotonically incremented for each transmitted message, independently of the direction of this message, and is reset to 0 with every access to another BS.

In summary, the combination of CMAC_KEY_COUNT and PN is always unique and is not repeatable. When this combination approaches a limit of a counter close to its rollover, the full EAP reauthentication is expected, thus restoring the full replay protection.

The BS also signs the messages sent to the MS, thus establishing the bidirectional authenticated signaling path for management messages on the airlink. Actually, to be accurate, both MS and BS generate the pair of CMAC keys: one for the uplink (in the direction from the MS to the BS), and one for the downlink (in the direction from the BS to MS). The CMAC_KEY_U and CMAC_KEY_D are used for encryption and decryption, respectively.

Integrity protection of management signaling messages ensures that no signaling from the unauthenticated or unauthorized mobiles or base stations can affect legitimate communications and cause the loss of NAP resources.

9.4.1.4 Airlink Ciphering

The airlink transport is ciphered with one of negotiated encryption algorithms, such as AES, 3xDES. The key for this ciphering is called the TEK (Traffic Encryption Key). The TEK is randomly selected by the BS and is simply sent to the MS over the air. To ensure its secrecy in transport, it is encrypted by the special key generated only for this purpose, the KEK (key-encryption-key), which is derived from the AK by the BS and MS, and is used by the BS to encrypt the TEK with the special mode of the AES algorithm.

The TEK is usually different for each BS when it communicates with a specific mobile. Transmission of the TEK to the MS is shown in Step 7 of Figure 9.9. However, to accommodate the fast handoff between neighboring base stations, specifically the FBSS mode (fast BS selection), the same TEK can be shared by these base stations, so the MS would not need to maintain two different sets of encryption keys. This is an exception mode of operation and will not be discussed here any further.

As with integrity protection, data ciphering is protected from replays by using the frame counters of transmitted data frames as cryptographic synchronization parameters, so no two messages, even with the same content, will be encrypted producing the same result. Also, the TEK in forward (BS-to-MS) and reverse (MS-to-BS) direction are different.

9.4.2 Mobility Security

Support for mobility in WiMAX relies on the Mobile IP protocol specified in IETF [13] for Mobile IPv4 and IETF Mobile IPv6 in IETF [14]. WiMAX mobile IP security relies on the authentication protocol specified for both MIPv4 and MIPv6.

9.4.2.1 MIPv4 Security

9.4.2.1.1 Client MIPv4 (CMIPv4) Security. MIPv4 security is achieved by attaching specific authentication extensions to the MIPv4 signaling. Such extensions are generated using secure hash algorithms specified in [13] and established security associations (symmetric keys) between MIPv4 entities.

WiMAX mandates that there is a symmetric security association between the MIPv4 client and the home agent. This MN-HA security association, or MN-HA key, is generated (bootstrapped) as the result of successful EAP access authentication of the mobile. Specifically, it is generated from the EMSK (Section 9.3.1.2) by first running the cryptographic isolation procedure that produces the MIP-RK and then generating the MN-HA-MIP4 key specific for the pair of MIPv4 Client–HA.

At the time of MIP-RK generation, the security parameter index (SPI) is also generated to uniquely identify the security association. This SPI (MN-HA-MIP4-SPI) is computed from the MIP-RK and is unique for the MIP session associated with the mobile. Computation of the SPI at the mobile and the AAA by using the same computational rules ensures that resulting value of SPI are the same on both sides, and do not have to be communicated between them prior to establishing the MIP binding.

To establish the MIPv4 binding and to request the IP address (HoA), the CMIPv4 in the mobile assembles the MIP registration request (RRQ) toward the foreign agent (FA) as follows:

- The CMIPv4 mobile node (MN) includes its identity (NAI), using the first NAI ever sent to the ASN for this access, which could be an "outer" NAI included in the EAP-TTLS request at the initial access to the network.

- The MN includes the indication that it requests the dynamic assignment of the IP address, so the HoA attribute in the RRQ is set to the "all-zero-one" address as specified in the [13].

- The MN includes the known address of the HA. If MN does not get preconfigured with the HA IP address, it includes the "all-zero-one" address, indicating the request for dynamic assignment of the HA-IP address.

- The MN then computes the MN-HA authentication extension (MN-HA AE) using the hash-based procedure specified in [13], and includes this extension in the RRQ. Note that if the MN is not aware of the HA-IP address, it will compute the MN-HA AE based on the "all-zero-one" HA-IP address (HA-IP-RRQ).

Because IP services for mobile WiMAX are provided using the Mobile IP Protocol, the assignment of the HA IP address for the session becomes the essential mandatory step of the WiMAX access sequence, and it is accomplished during the initial EAP Access Authentication. During the initial EAP access authentication, the authenticator in the ASN sends the RADIUS access request to the HAAA. If this request traverses the AAA in the visited CSN (V-CSN), the visited AAA (VAAA) may include the suggestion for a local HA in the V-CSN. This vHA-IP address (suggestion) is sent to the HAAA. The assignment of the HA-IP address is sent back to the ASN in the EAP access accept that carries the indication of successful completion of EAP access authentication.

Specifically, there are three different ways of getting the HA-IP assigned:

1. If HAAA policy does not allow assignment of the HA in the V-CSN, the HAAA simply returns the assignment of the HA in the H-CSN (hHA-IP), ignoring the suggestion of the VAAA. This assignment in the RADIUS access accept is forwarded by the VAAA to the authenticator in the ASN.
2. If HAAA policy allows assignment of the HA in the V-CSN, the HAAA loops back the HA-IP address suggested by the VAAA (vHA-IP). This assignment in the RADIUS access accept is forwarded by the VAAA to the authenticator in the ASN.
3. If the HAAA policy allows assignment of the HA in either the V-CSN or H-CSN, the HAAA returns the HA-IP address in the H-CSN, and also loops back the suggested HA-IP address in the V-CSN. Both HA-IP addresses are included in the RADIUS access accept and are forwarded by the VAAA to the authenticator in the ASN. The ASN then makes a selection of one HA-IP address, either in the V-CSN or the H-CSN.

Once the FA in the ASN receives the RRQ from the MN, it forwards this RRQ to the HA based on assigned HA-IP address. The MN-HA AE, according to the [RFC3344], also contains the SPI identifying the MN-HA security association used to generate the authentication extension.

Since this is the initial RRQ that is used for creating the mobility binding on the HA, the HA does not yet have the MN-HA key needed to validate the MN-HA AE. To obtain this key, the HA sends the RADIUS access request to the HAAA, including the MS identity (NAI) and the SPI received in the MN-HA AE as the part of RRQ. The HA also includes the HA-IP address that was a part of the RRQ, and also its own IP address.

The HAAA verifies that the SPI received from the HA is indeed associated with the active session for this mobile, that is, there is a current valid EMSK and MIP-RK associated with the SPI for this NAI. If this verification checks, the HAAA uses the MIP-RK to compute two MN-HA keys: (1) the MN-HA-RRQ based on HA-IP-RRQ, or the HA-IP included by the MN in the RRQ, and (2) the MN-HA based on the real assigned HA-IP.

Both keys are returned to the HA. The HA uses the MN-HA-RRQ to validate the MN-HA AE in the received RRQ from the MN; if it checks, the HA assembles the MIPv4 registration reply (RRP). The RRP contains the assigned home IP address for the MN (HoA), the associated care-of-address of the FA (CoA), and the HA IP address (HA-IP). The HA then uses the MN-HA key based on the real assigned HA-IP to generate the MN-HA AE for the RRP. This "signed" RRP is returned to the MN.

The MN first computes the real MN-HA key based on the HA-IP received in the RRP, and then validates the MN-HA AE of the RRP. If MN-HA AE checks, the MN accepts the binding, including the HoA, HA-IP, and CoA for the session.

For subsequent reregistration, whether the MN moves from one FA to another or simply conducts the timer-based periodic reregistration, the MN-HA AE is generated and validated for both RRQ and RRP using the already available MN-HA key, so the HA does not need to request it from the HAAA.

9.4.2.1.2 Proxy MIPv4 (PMIPv4) Security. If the mobile does not support the CMIPv4, the functionality of the MIP client is assumed by the ASN on behalf of the mobile. This mode of mobility is called the proxy mobile IP, because the ASN-GW actually becomes the proxy MIP client to compensate for the mobile's deficiency.

Because the ASN-GW does not have the EMSK, it cannot compute the MIP-RK and associated SPI, MN-HA keys, and so on. This is why the proxy MN-HA key is simply sent to the authenticator by the HAAA with the last RADIUS access accept at the end of successful EAP access authentication. The PMIPv4 MN-HA SPI is returned as well, in addition to the HA-IP address.

Having the proxy MN-HA key, the PMIPv4 client at the ASN generates the MIP4 RRQ, including the MS's NAI, assigned HA-IP, and request for dynamic assignment of home IP address for the mobile session ("all-zero-one" value of the HoA). The RRQ includes the MN-HA AE generated using the received proxy MN-HA key, and contains the MN-HA-SPI-PMIPv4.

As in the CMIPv4 case, the HA initially requests the MN-HA key from the HAAA to validate the MN-HA AE in the RRQ and create the MN-HA AE in the RRP, as described in the previous subsection. For subsequent reregistration, whether the PMIPv4 client moves from one ASN-GW to another or simply conducts the timer-based periodic reregistration, the MN-HA AE is generated and validated for both RRQ and RRP using the already available MN-HA key, so the HA does not need to request it from the HAAA.

9.4.2.1.3 MIPv4 Security between MN and FA. Several optional security enhancements are also specified by WiMAX Forum. The secret key for association between the MN and FA (MN-FA key) is generated from the EMSK branch through two cryptographic separations: first through the MIP-RK, and, second, through the FA-RK computed from it using the identities of MN and FA. This MN-FA key is always generated by the the MN in the mobile, but may or may not be sent to the FA in the ASN-GW depending on the local policy. If policy requires security between MN and FA, the MN-FA AE is included in the RRQ by the MN and in the RRP by the FA, and is validated by the recipient.

9.4.2.1.4 Internode MIPv4 Security between FA and HA. The secret key for association between the FA and HA (FA-HA key) is established on a per-node basis, and is not associated with any specific mobile session. In fact, protection of all signaling between a given pair of FA and HA is done using the same FA-HA tunnel key. This additional FA-HA tunnel security is used only if there is no other protection on this link, that is, no network-based backhaul security is applied on R3 interface (Section 1.4.3).

To establish this key, the HAAA generates the random HA-RK that is associated with the authenticator in the ASN, which is used to establish the tunnel to a specific HA.

This HA-RK is delivered to the ASN with the last RADIUS access accept at the end of successful EAP access authentication. The associated HA-RK-SPI is generated and sent together with the HR-RK. Before sending the RRQ to the HA, the FA requests the FA-HA key from the authenticator. The authenticator generates the FA-HA key using the HA-RK associated with the assigned HA and the FA IP address. Therefore, the FA-HA key returned to the FA is unique for the tunnel between a given FA and given HA.

FA creates an additional FA-HA AE for the RRQ using the FA-HA key. The FA-HA AE also contains the FA-HA-SPI associated with the HA-RK key. If the HA does not already have this HA-RK key, it requests it from the HAAA and, once received, it validates the FA-HA AE. The HA also uses the FA-HA AE to authenticate the RRP and MIP revocation messages sent to the FA.

To avoid possible race conditions, the lifetime of HA-RK is usually chosen to be much longer than any currently active MIP session on the given HA. It is also recommended that the same fresh HA-RK is sent to the authenticator with every RADIUS access accept carrying the success of the EAP access authentication. This will ensure that any given authenticator and any assigned HA will always have an active HA-RK for the per-node tunneling.

9.4.2.1.5 MIPv4 Key Distribution. Generalized case of MIPv4 key distribution is shown on Fig. 9.11. Once the EAP authentication is successfully completed (Steps 1 and 2), the MN initiates the MIPv4 RRQ (Step 3), which is forwarded to the HA (Step 4). The HA requests the MN-HA key from the HAAA by sending it the NAI and SPI used by the MN (Step 5), and receives in return the MN-HA and HA-RK keys (Step 6). The HA returns the RRP signed by MN-HA AE and FA-HA AE (Step 7), and FA returns the RRP to the MN signed with MN-HA AE and MN-FA AE.

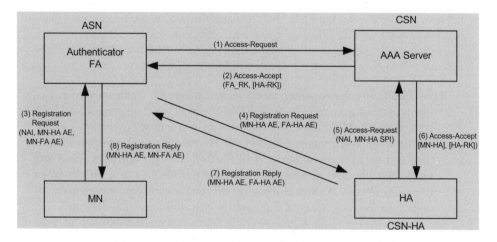

Figure 9.11. MIPv4 key distribution.

9.4.2.2 MIPv6 Security

In addition to fully supporting the protection of MIPv6 signaling between the client MIPv6 MN and HA with the IPSec, as specified in the IETF [14], the WiMAX also defines support for the MIPv6 Authentication Protocol that use MN-HA authentication option (MN-HA AO) in binding updates (BU) and binding acknowledgement (BA). For this, WiMAX defines generation (bootstrapping) of the MN-HA-MIP6 key from the MIP-RK, as well as mutual computation of the MN-HA-MIP6-SPI at the MN and the HAAA.

Similarly to the CMIPv4 procedure, the MIPv6 MN generates the MN-HA-AO for the MIPv6 BU using the MN-HA-MIP6 key and sends it to the assigned HA. The HA requests the MN-HA-MIP6 key from the HAAA based on MN-HA-MIP6-SPI, and validates the BU. The BA is then sent back to the MN containing the MN-HA-MIP6 AO.

A similar scheme is now being defined for the proxy MIPv6, which at the time of this writing was not yet fully standardized by the WiMAX Forum.

9.4.3 Backhaul Security

In order to ensure end-to-end security of the NWG architecture, security of each reference point must be considered. Privacy, authentication, integrity, and replay protection must be ensured either at the lower layers (PHY, MAC, or network layer) or at the higher layers. Security at the lower layers comes in the form of a secure channel that can be utilized by any one of the signaling protocols and data traffic running above it.

It should be noted that the lower layer security and the higher layer security are complementary. Absence of one should be compensated for by the presence of the other. At times, both may be present. Lower layer security between two end points can be a substitute for the higher layers that terminate on the same end points. If the end points are different, the substitution may not apply. For example, a secure channel between the BS and the ASN GW alleviates the need to secure any R6 signaling, but pass-through R2 signaling cannot rely on this security.

Deployments must be aware of the necessity and availability of layered security for each reference point using the following defined guidelines:

R1—The 802.16 primary management connection over R1 is authenticated and integrity and replay protected at the IEEE 802.16 MAC layer upon successful EAP authentication. All the subsequent R1 messaging over these connections can rely on this lower layer cryptographic security. On the other hand, transport connections may not be crypto-protected at all. For that, any signaling protocol and data traffic that run above these connections must provide their own security when necessary. (Note: Although enabling security on the transport connections is optional, it is recommended that deployments take advantage of this feature).

R2—This reference point may not have an end-to-end secure channel. It is assumed that the lower layers are insecure and the signaling protocols and data traffic will provide their own security when necessary.

R3—This reference point may not have an end-to-end secure channel. It is assumed that the lower layers are insecure and the signaling protocols and data traffic will provide their own security when needed. Examples are Mobile IPv4 using authentication extensions comprising the MIP Authentication Protocol, RADIUS using authentication attributes and encrypted critical attributes, and DIAMETER relying on the underlying IPSec.

R4—This reference point is defined between two trusted ASNs and, therefore, expected to have a secure protected link between them. This end-to-end secure channel is expected to include privacy and integrity protection. The channel security may be implemented using physical security, IPsec or SSL VPNs, and so on. The VPN end points may be collocated with the R4 end points, or be on-path between the two to ensure end-to-end security.

R5—This reference point may not have an end-to-end secure channel. It is assumed that the lower layers are insecure and the signaling protocols and data traffic will provide their own security when needed. Expectations for this reference point are the same as for the R3. Examples include the RADIUS authentication attribute.

R6—This intra-ASN reference point is expected to be defined between individual base stations and their respective trusted ASN-GWs. Transport links in the intra-ASN domain are usually well controlled by the owner–operator of the NAP, and so it is expected that R6 has an end-to-end secure channel, including privacy. The channel security may be implemented using physical security, IPsec or SSL VPNs, and so on. The VPN end points may be collocated with the R6 end points, or be on-path between the two to ensure end-to-end security.

R8—This reference point between neighboring base stations that belong to the same NAP is expected to be tightly controlled by the operator–owner of the NAP. Therefore, it is expected that R8 has an end-to-end secure channel, including privacy. The channel security may be implemented using physical security, IPsec or SSL VPNs, and so on. The VPN end points may be collocated with the R8 end points, or be on-path between the two to ensure end-to-end security.

9.5 SUMMARY

IEEE 802.16 and the WiMAX Forum defined a comprehensive set of security measures to protect owners and users of the licensed WiMAX-802.16 networks from realistic attacks. Even though complete end-to-end security is not specified, relying in part on the specific deployment choices of the network owners and operators, the critical reference points of the network—air interface, system and service access, IP assignment—can fully rely on strong, modern security protocols and cryptographic suites mandated for the operation. In addition, strong security measures are recommended for specific deployments for the intra- and interdomain security to allow realistic protection of valuable resources.

9.6 REFERENCES

[1] IETF RFC-5280, Internet X.509 Public Key Infrastructure Certificate and Certificate Revocation List (CRL) Profile, May 2008.

[2] IETF RFC-5247, Extensible Authentication Protocol (EAP) Key Management Framework, August 2008.

[3] IETF RFC-5216, The EAP TLS Authentication Protocol, March 2008.

[4] IETF RFC-4282, The Network Access Identifier, December 2005.

[5] IETF RFC-3579, RADIUS (Remote Authentication Dial In User Service) Support For Extensible Authentication Protocol (EAP), September 2003.

[6] IEEE Std. 802.16e-2006, Standard for Local and Metropolitan Area Networks Part 16: Air Interface for Fixed and Mobile Broadband Wireless Access Systems, P802.16Rev2/D2, December 2007.

[7] IETF RFC-2631, Diffie-Hellman Key Agreement Method, June 1999.

[8] IETF RFC-4187, Extensible Authentication Protocol Method for 3rd Generation Authentication and Key Agreement (EAP-AKA), January 2006.

[9] IETF RFC-5281, Extensible Authentication Protocol Tunneled Transport Layer Security Authentication Protocol Version 0 (EAP-TTLSv0), August 2008.

[10] IETF RFC-2759, Microsoft PPP CHAP Extensions, Version 2, January 2000.

[11] IETF RFC-2104, HMAC: Keyed-Hashing for Message Authentication, February 1997.

[12] FIPS Publication 197, Advanced Encryption Standard (AES), November 2001.

[13] IETF RFC-3344, IP Mobility Support for IPv4, August 2002.

[14] IETF RFC-4285, Authentication Protocol for Mobile IPv6, January 2006.

CHAPTER 10

QUALITY OF SERVICE (QoS) IN WiMAX NETWORKS

MEHDI ALASTI AND BEHNAM NEEKZAD

10.1 INTRODUCTION

IP networks were originally designed for supporting best-effort data traffic, but now Internet users expect to run various services with distinct requirements, for example, voice, video, interactive gamming, and data services, over a single IP network and expect to receive an appropriate level of quality of service (QoS). Basically, QoS is a set of capabilities and functionalities in a network to provide appropriate levels of services to applications and support a satisfactory level of end-user experience. Although QoS is one of the major concerns in an IP network, unlike the circuit-switched networks, supporting end-to-end QoS for real-time and nonreal-time traffic over a converged packet-switched network is a challenge that requires "authorizing" the services requested by a user depending on the subscriber's contract, "delivering" the service to the subscriber, and also "accounting" for subscriber services, according to the subscriber's contract.

Supporting QoS provides significant advantages for service providers:

- QoS enables the service providers to protect their infrastructure investments by efficiently utilizing costly network resources for accommodating the promised service portfolio with minimum need for bandwidth overprovisioning.
- Supporting QoS permits a service provider to attract more customers by broadening their service portfolio and differentiating themselves by support-

ing specific applications. A good example is support of Push to Talk (PTT) service by Nextel.

• Offering QoS allows a service provider to increase its revenue by providing different service plans for different subscribers (e.g., Gold, Silver, Bronze, etc service plans) and billing them accordingly. A student might only need inexpensive, simple best-effort connectivity, but an executive is willing to pay more in order to receive high-quality services.

Different applications have distinct QoS requirements. For instance, VoIP and interactive gaming are low-bit-rate traffic with short packets that are latency and jitter sensitive, whereas VoIP packets are issued at periodic intervals (especially during a talk spurt period), and interactive gaming sends one packet after a thinking period. Video is high-bit-rate real-time traffic of variable-sized data packets issued at periodic intervals. On the other hand, data applications are nonreal-time traffic of possibly long data packets with no strict latency constraint. In order to meet requirements of different applications, network QoS should manage the network resources appropriately to provide controlled latency and jitter, dedicated bandwidth, improved loss characteristics, fairness, and network availability.

It is fairly complicated to support broadband services over a wireless link. Throughput, latency (delay), jitter (delay variation), and transmission errors vary much more widely over a wireless connection than over a wired connection because of constantly changing radio signal conditions (even if the user is not moving) and extensive digital radio processing. Standard Internet protocols, designed for use over a more stable wire-based connection, are not well suited to handle these variations.

Any broadband wireless service is further complicated by the shared nature of the service; at some point in the network, traffic from many different applications is sharing the same network path. For example, each user session, consisting of Web browsing, e-mail, sending and receiving files, and real-time communications via instant messenger or VoIP, involves many different types of data and protocols. Some applications require short, time-sensitive, bursty data transmissions, whereas other applications require large, error-sensitive, continuous data transmissions. Additionally, network capacity is often overprovisioned on wired networks to compensate for less-than-perfect traffic management. Since wireless broadband technologies must make efficient use of limited radio spectrum, the wireless network needs to rely on more sophisticated QoS techniques that also account for the unique characteristics of the radio link.

Mobile WiMAX technology offers wireless Internet with comprehensive network control over mobile device, network, and service performance to enable the desired throughput, latency, and fairness among the services and subscribers, thereby providing full QoS support and enhanced end-user experience. Very-high-capacity air link, fine resource granularity, and flexible resource allocation mechanisms supported by Mobile WiMAX enable the operators to offer a broad range of applications and services with different throughput and latency/jitter requirements. The Mobile WiMAX QoS framework intelligently creates the means to offer strong flow-based as well as subscriber-based QoS capabilities. Indeed, WiMAX technology enables mobile service providers to deliver high-quality integrated services such as VoIP and other mission-critical applications, gaming, multimedia, and Internet data services to their subscribers.

This chapter demonstrates the current preprovisioned and the future dynamic QoS frameworks and their associated architecture and interfaces, features, and capabilities supported by Mobile WiMAX technology. In addition, the chapter presents the common concepts and means used for QoS support in packet-switched networks.

10.2 AN OVERVIEW OF QoS IN PACKET-SWITCHED NETWORKS

10.2.1 The Basic QoS Models

Quality of Service (QoS) refers to the ability (or probability) of the network to provide a desired level of service for selected traffic on the network. Service levels are specified in terms of throughput, latency (delay), jitter (delay variation), and packet errors or losses. Fundamentally, a network element (or a network segment) in a packet-switched network follows one or a combination of the following three types of QoS models:

1. Best effort, which provides basic connectivity
2. Differentiated services (soft QoS) which is connectionless and is usually priority-based
3. Guaranteed services (hard QoS) which is connection oriented and is usually reservation-based

10.2.1.1 The Best-Effort Model

In the best-effort (BE) model, all the packets of different applications reside in a single FIFO buffer and are treated identically, which simplifies this model dramatically. In order to support the applications QoS requirements within BE model, very high bandwidth overprovisioning (on the order of ten to twenty times) is required to guarantee virtually no packet blocking in the buffers and to support the promised services. Figure 10.1 demonstrates the BE model in which a number of applications with entirely different requirements are mapped to the same BE FIFO buffer but, since the link capacity is large (overprovisioned), the packets experience very small latencies.

In network elements or segments with abundant bandwidth, the BE model is the preferred solution due to its simplicity.

10.2.1.2 Model-Differentiated Services or Soft QoS Model

In the soft QoS framework, the network offers a number of classes of service with different priority levels. This QoS model uses a class-based queuing mechanism

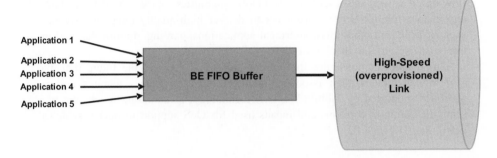

Figure 10.1. An overprovisioned best-effort model to support QoS.

with a number of queues, each associated with a class and assigned a certain priority level. With the soft QoS model, the traffic entering a particular network element is classified and mapped into appropriate classes and packet treatment depends on the class of service: VoIP and other mission critical services are mapped to the highest priority class, video services to the high-priority class, and data applications to the lowest priority class. The classification is performed based on few classification rules to identify the application type and the subscriber profile. Usually, these classification rules rely on a combination of source and destination IP addresses, source and destination port numbers, and a ToS (type of service) field in the IP header. The ToS fields of the IP packets mapped to a certain class are usually marked or remarked to indicate the class of service they belong to.

Figure 10.2 demonstrates a simple example for illustration of soft QoS with three classes in a network element. In this example, Class 1 is assigned to low-latency VoIP (App1) and the other mission-critical applications (App2), which can potentially receive up to 20% of the link capacity; Class 2 is allocated to other real-time applications [e.g., video telephony (App3), video streaming (App4), and audio streaming (App5)], which can potentially be allocated up to 40% of the leftover link capacity; and Class 3 is for BE data and all other applications, which will receive the leftover capacity of the link. Roughly speaking, the soft QoS framework of Figure 10.2 controls the relative packet latency of Class 2 to be "three" times shorter than packet latency of Class 3; however, it does not guarantee the absolute packet latencies.

The soft QoS model is connectionless, since it does not require reservations, and the granularity of its operation is based on classes and not flows. Since the soft QoS architecture does not maintain the state of the flows, it is scalable. However, with the soft QoS model, packet blocking may still occur across different packets of each class. Therefore, in order to offer the necessary QoS guarantees, the soft QoS framework still requires some level of network bandwidth overprovisioning. Due to simplicity and scalability, this QoS model is widely deployed in the packet networks both at Layer 2 in Ethernet using IEEE 802.1p and at Layer 3 in IP networks using IETF DiffServ.

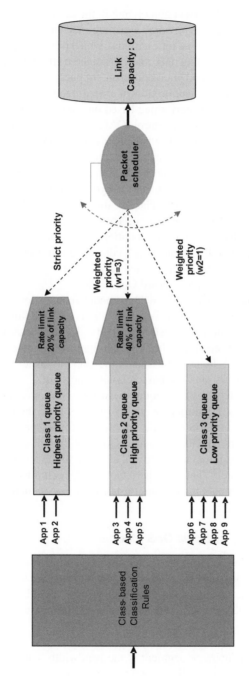

Figure 10.2. An example of soft QoS.

10.2.1.3 Guaranteed Services or Hard QoS Model

With the hard QoS model (connection oriented), the network makes resource reservations on a per-flow basis using some signaling mechanisms. In other words, the network establishes a virtual pipe for each flow, which provides significant granularity of the QoS control, including maximum sustained traffic rate, minimum guaranteed rate, maximum latency and jitter, and so on, for packets belonging to the flow. A flow can be defined as an application flow or an aggregation of a number of applications in which aggregation is performed based on some classification rules. In the hard QoS model, a call admission control (CAC) module in the network checks the reservation requirements against the available resources, investigating whether there are sufficient resources available to support the requested service. After admission by CAC, the network establishes an appropriate connection with right set of QoS attributes, based on the reservation request. The network identifies the incoming packets belonging to a flow based on some classification rules, usually a combination of source and destination IP addresses, source and destination port numbers, and so on, and maps the packets based on the classification rules to the appropriate connection. The resource allocation of the connection is controlled using a number of tools to be explained later in this chapter. IP integrated service (IntServ) architecture and ATM virtual connections are examples of hard QoS models.

Figure 10.3 demonstrates the hard QoS model (connection-based) in which a network serves different flows. It makes reservations for each flow separately based on each flow's QoS requirements, establishes virtual connections, and creates separate buffers (queues) for each specific flow, configures the packet scheduler to support the appropriate QoS treatment based on the QoS parameter set associated with each flow, and configures the flow-based classification rules.

The hard QoS model does not require network overprovisioning and provides fine QoS granularity as it controls the quality of every flow and the fairness among them. Therefore, the hard QoS model is appropriate for segments in which the network resources are limited. However, hard QoS has limited scalability because of the need for the reservation mechanism and also due to the fact that every flow has to be controlled separately. Thus, the hard QoS model is limited to the scenarios in which the number of flows is not large, usually the access networks and the last mile in a network.

10.2.1.4 Summary of Basic QoS Models

Wireless networks are comprised of three different segments: air interface, backhaul transport, and core transport. Generally, the air-interface segment has the lowest bandwidth and the core transport has the highest bandwidth.

The appropriate QoS model for the air interface is the connection-oriented (hard) QoS model since:

- Usually the air-interface segment is the bandwidth bottleneck of a wireless network and the connection-oriented QoS model enables efficient use of spectrum.

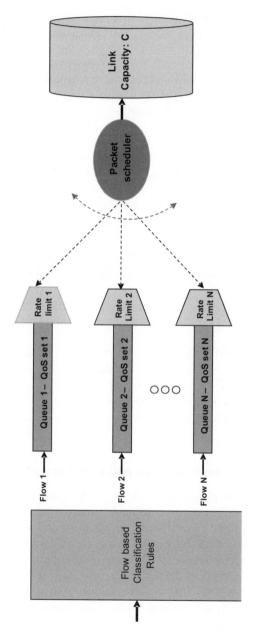

Figure 10.3. Hard QoS model to support *N* flows: Flow 1, Flow 2, . . . , Flow *N*.

- The air interface is the segment dealing with the fewest number of flows and, thus, lack of scalability of the connection-oriented QoS model is not a big issue over the air interface.
- Delivering fairness and the performance requirements to users with poor RF conditions at the cell edge without the connection-oriented QoS model is practically impossible.

Backhaul transport is one of the most expensive segments of a wireless network. Usually, the appropriate QoS model for backhaul transport is the connectionless class-based (soft) QoS model. This is because:

- The emergence of new backhaul technologies such as "microwave links" and "optical links" enables the mobile operators to provide more bandwidth and some level of overprovisioning on backhaul transport. However, the existing backhaul links in many locations in North America are still T1/T3 links and, in general, the bandwidth overprovisioning on the backhaul is insufficient to support QoS with the BE model.
- The major trend of backhaul transport is migrating to packet technologies (Layer 2 Metro Ethernet or Layer 3 IP backhauls) and these technologies inherently support class-based (soft) QoS through use of IEEE 802.1p or DiffServ.
- Backhaul transport deals with aggregation of traffic of a number of sectors, and supporting hard QoS over backhaul has scalability issues.

Since the core transport is constituted of high-capacity links, the BE model is the most appropriate model for this segment. The diagram in Figure 10.4 illustrates the three important segments of a wireless network, their relative associated bandwidth, the recommended QoS model, and the QoS granularity for each segment.

The QoS model in different network segments will be discussed in greater detail in later sections.

10.2.2 QoS Management, Control, and Enforcement

All of the three important planes of a wireless network—management plane, control and signaling plane, and bearer plane—are involved in the end-to-end QoS. The management plane defines the service policies and the accounting profile associated with a subscriber. The control and signaling plane allows the network to perform the subscriber's service authorization, CAC, QoS negotiation, resource reservation, policy enforcement, and QoS mapping within the network. The bearer plane provides the mechanisms to control network resources for appropriate QoS treatments within the network, queuing, packet marking, packet scheduling, traffic shaping, active queue management (AQM) such as random early detection (RED), and so on.

Figure 10.5 illustrates a high-level description of QoS management in a network. The subscriber profile repository (SPR) maintains the subscribers' service profiles in the management plane. The policy decision point (PDP) in the control plane per-

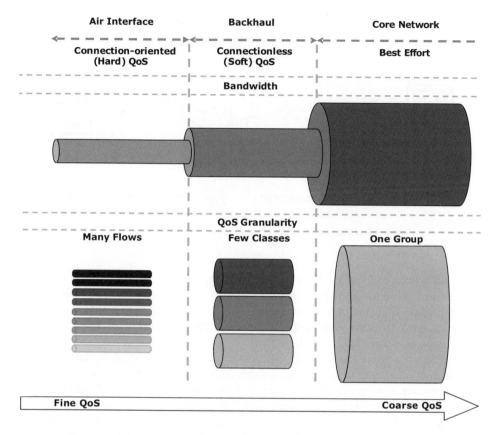

Figure 10.4. The three major segments of a wireless network, their relative bandwidth, their QoS granularity, and appropriate QoS model for them.

forms service authorization and maps the subscriber policies in the SPR to the appropriate QoS parameters. The policy enforcement point (PEP) performs the packet treatment based on the QoS model and the QoS policies received from the PDP.

10.2.3 Service and Accounting Profiles

In order for the mobile operator to attract different subscribers with different expectations and affordances, the operator offers a number of service profiles, each with different accounting profiles. Therefore, people can subscribe to the appropriate service profile based on the services they need and how much they are willing to pay for. Let us clarify this with the following example.

A student and an executive have different expectations from the network. The student might only require an inexpensive network connection for his/her searches or e-mails, but the executive expects to receive a high-performance service from the network not only for searches and e-mails (possibly with large attachments), but

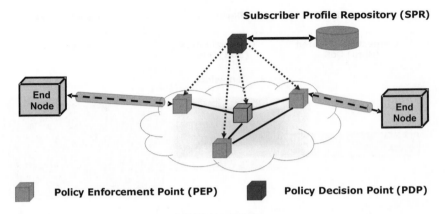

Figure 10.5. High-level description of QoS management in a network.

also for voice and video calls, video conferencing, and so on. Also, the executive expects to have a higher speed link for fast searches and e-mails and is willing to pay for that. On top of that, the executive expects the network to be highly available for him/her regardless of network load, but the student can wait to try connecting to the network when it is less highly loaded.

Different service profiles with different accounting profiles can be defined by operators in order to offer a variety of service plans to customers. A service plan may include a group of authorized service flows and associated classification rules and accounting profile (e.g., session-based). For example a service plan can include two service flows for real-time traffic (e.g., VoIP) and two service flows for other data (e.g., Web browsing and e-mail) in DL and UL directions with a specified session-based accounting.

10.3 WiMAX QoS ARCHITECTURE OVERVIEW

This section analyzes the WiMAX end-to-end QoS framework while the overall network architecture and the layer protocols are discussed. This includes the required functions and network elements for provisioning and management, control and signaling mechanisms, configuration, and delivering of end-to-end QoS in order to offer a broad range of high-performance applications and services to the end user as well as the charging mechanisms supported by WiMAX.

A subscriber can be in his or her home network (his/her operator network) or in a visited network (the network of a roaming partner); regardless, the subscriber expects to receive the performance he/she has subscribed to and pays for. Figure 10.6 shows WiMAX end-to-end QoS picture in the home network.

The WiMAX QoS framework is three-dimensional, with different planes, different network segments, and different layers in the protocol stack. WiMAX QoS functionalities exist in the management plane, signaling and control plane, and

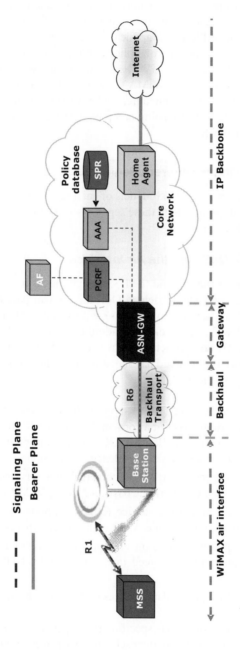

Figure 10.6. WiMAX end-to-end QoS picture.

255

bearer plane. The subscriber profile repository (SPR) in the provisioning and management plane (Figure 10.6) is in charge of maintaining subscribers' service and accounting profiles based on the contract between the subscriber and the operator. In the control and signaling plane, authentication authorization and accounting (AAA) and the policy control and charging rules function (PCRF) read the service and subscriber profile from the SPR and/or application function (AF) and then perform service authorization, admission control, policy enforcement, QoS parameter mapping, and configuring PEPs (i.e., ASN-GW, base station, and MS) through QoS signaling. On the bearer plane of the WiMAX end-to-end QoS picture, the QoS treatments (including classification, queuing, packet marking/remarking, rate capping, packet scheduling, and traffic shaping and policing) are provided within the ASN-GW, base station, and MS.

10.3.1 WiMAX QoS in Different Segments

This section demonstrates the QoS models of each segment in end-to-end WiMAX in greater detail, and explains how these segments work to deliver the required performance to the subscribers.

10.3.1.1 WiMAX QoS over the Air Interface

This segment connects the MS and the BS through the air link (R1 interface). The air interface uses a connection-oriented QoS mechanism based on unidirectional virtual pipes known as "service flows" (SFs), with specific set of QoS attributes described by the IEEE 802.16e standard. An admitted SF is identified by a 32-bit "service flow ID" (SFID) and an active SF is identified by a 16-bit connection ID (CID). The IEEE 802.16e standard has specified five types of SFs, each appropriate for a certain class of services:

1. Unsolicited grant service (UGS) supports flows transporting fixed-size data packets on a periodic basis at a constant bit rate (CBR), for example, wireless T1 or wireless fractional T1. Also, UGS SF is appropriate for voice services using fixed-rate vocoders with no silence suppression.
2. Real-time polling service (rtPS) supports real-time services with variable-size data packets issued on a periodic basis, for example, MPEG video.
3. Enhanced real-time polling service (ertPS) supports latency-sensitive real-time applications, for example, VoIP with silence suppression.
4. Nonreal-time polling service (nrtPS) supports delay-tolerant data traffic that requires a minimum guaranteed rate, for example, FTP or movie upload/download.
5. Best effort (BE) supports data services such as Web browsing and e-mail.

The IEEE 802.16e standard has specified five types of data delivery services for downlink flows and five corresponding scheduling services for uplink flows. In other words, uplink is differentiated from downlink because all uplink service flows

except UGS involve some form of request/grant mechanism for resource allocations. Each data delivery service flow consists of a specific set of QoS parameters
and is designed to support different types of traffic streams. This means that each
SF is characterized by a set of QoS parameters to be selected appropriately to deliver the desired QoS for each SF. A number of these QoS parameters will be described later in this section. Table 10.1 illustrates DL delivery and UL scheduling
services and some targeted traffic types.

Each SF is assigned a buffer before packets are forwarded over the air interface.
The QoS of the SFs are maintained through an appropriate bandwidth-allocation
mechanism among the SFs. On the DL, the bandwidth allocation is performed by
the air-link scheduler, and on the UL it is done by bandwidth request from the MS,
and bandwidth assignment is done by the air-link scheduler at the base station and
the scheduler at the MS. The scheduling decision is determined based on the SFs
QoS needs, the SFs buffer lengths, and the RF conditions of different MSs.

The SF framework provides an excellent QoS granularity on the air interface and
also inter-SF isolation. In such a SF-based framework, traffic is classified to service
flows based on classification rules for DL and UL directions by the ASN-GW and
BS, respectively. Downlink traffic is scheduled by the base station according to service flow type, and uplink traffic is scheduled by base station according to re-
quest/grant procedure for service flow. This means that real-time service flows receive unsolicited grants of air interface resources, whereas nonreal-time service
flows must request an air-interface resource.

WiMAX allows the network to establish virtually unlimited numbers of SFs with
flexible combinations of different types and different QoS attributes for each subscriber. For instance, it allows a mobile operator to be able to establish two BE SFs
for a subscriber. If the operator would like to partner with a list of Internet services,
such as specific search engines, news blog, and so on, in order to encourage its subscribers to use these services, the operator establishes two types of BE SFs: basic
BE SF and preferred BE SF, in which the packets of a preferred BE SF receive relatively better treatment compared to the packets of a basic BE SF and are then for-

Table 10.1. IEEE 802.16e service flows for DL and UL directions and some targeted
applications

DL data delivery service	UL scheduling service	Targeted traffic type
Unsolicited grant service (UGS)	Unsolicited grant service (UGS)	Constant-bit-rate (CBR) services, TDM services
Extended real-time variable rate (ERT-VR)	Extended real-time polled service (ertPS)	VoIP with silence suppression/ activity detection
Real-time variable rate (RT-VR)	Real-time polled service (rtPS)	Streaming audio and video
Nonreal-time variable rate (NRT-VR)	Nonreal-time polled service (nrtPS)	File transfers
Best Effort (BE)	Best effort (BE)	Web browsing, e-mail

warded to the preferred BE SF. A mobile operator can differentiate between different subscribers. For example, let us assume that the operator intends to offer several service profiles, such as Gold, Silver, and Bronze. The operator can establish different numbers of SFs of appropriate types and QoS parameters to differentiate these service profiles. WiMAX offers tremendous flexibility for the mobile operator to allocate almost any number of SFs, of any combination of SF types, with any QoS parameters, to any subscriber. This enables an operator to differentiate between its subscribers based on their profiles and also differentiate between different services based on its partnerships.

Figure 10.7 illustrates an example in which two subscribers, MS1 and MS2, receive service from the BS. Figure 10.7 only shows the SFs on the DL, but the same concept works for the UL in the reverse direction. In this example, the network establishes four SFs to MS1: UGS1, ertPS1, Preferred BE1, and Basic BE1. The network also establishes two SFs to MS2: UGS2, and Basic BE2. The rate cap of each SF is captured by the width of the pipe representing the SF. As shown in Figure 10.7, the number of SFs and the rate cap and the other QoS parameters of SFs allocated to MS1 and MS2 are different. The packets associated with each SF reside in the buffers assigned to the SF at the base station before getting forwarded over the air interface by the scheduler. The implementation of the air-interface scheduler is not specified by the standard and WiMAX vendors can differentiate themselves through their proprietary techniques.

10.3.1.1.1 WiMAX QoS Parameters. WiMAX defines QoS parameters for each service flow type:

Maximum Sustained Traffic Rate (MSTR). MSTR caps the traffic of an SF to a certain level but does not provide any guarantees on minimum rate. In other

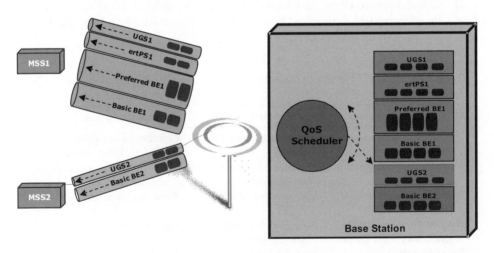

Figure 10.7. An example of two subscribers and their associated SFs.

words, even if the network is lightly loaded, the rate of SF cannot exceed its assigned MSTR. Mobile operators are interested in controling their subscribers' traffic.

Maximum Reserved Traffic Rate (MRTR). MRTR specifies the minimum rate that is guaranteed to an SF if the SF has traffic to send, even under heavy network traffic load.

Maximum Latency. Controls the maximum delay of the SF packets.

Maximum Jitter. Controls the maximum jitter of the SF packets.

Traffic Priority. This very interesting QoS parameter allows the mobile operator to assign a priority to SFs based on a combination of the subscriber's profile and the services mapped to the SFs; for example, basic BE and preferred BE SFs can be assigned different traffic priorities. The air-interface scheduler uses this parameter to adjust the priority of packets of different SFs.

10.3.1.1.2 Service Flow QoS and Application QoS. This section explains the relationship between SF QoS and application QoS. It should be noted that not all of these QoS parameters are defined for each service flow type. Also, each specific application has its own QoS requirements but all the applications can be classified into few classes such as CBR, real-time voice and video, and gaming. Each one of these classes can be appropriately treated by an SF type. Table 10.2 summarizes the SF types supported by IEEE 802.16e and a number of their important QoS parameters. For instance, both VoIP and gaming can be served by ertPS, although they have different QoS requirements. In this example, the network can establish two ertPS SFs with appropriate QoS parameters, one for VoIP and one for gaming. Clearly, establishing an SF per application might cause SF scalability issues. As will be explained later in this chapter, this issue is more significant within the pre-

Table 10.2. IEEE 802.16e service flow types and their QoS parameters

Class	Description	Minimum reserved traffic rate	Maximum sustained traffic rate	Max latency	Jitter	Traffic priority
UGS: unsolicited grant service	VOIP, T1, CBR		X	X	X	
rtPS: real-time polling service	Streaming audio/ video	X	X	X		X
ertPS: enhanced real-time polling service	VoIP with activity detection	X	X	X	X	X
nrtPS: nonreal-time polling service	FTP, movie upload, download	X	X			X
BE: best effort	Data transfer, Web browsing, e-mail, etc.		X			X

provisioned QoS framework and is less significant within the dynamic QoS framework that will be in future WiMAX releases. In order to resolve the SF scalability issue, the network can establish a limited number of SFs and several applications can share the same SF. Sharing of an SF by a number of applications might cause some challenges. For example, let us assume that VoIP requires a 100 Kbps ertPS SF and gaming requires a 50 Kbps ertPS SF. In order to support both VoIP and gaming, the following cases are possible.

1. Establish two ertPS SFs: a 100 Kbps ertPS SF for VoIP and a 50 Kbps ertPS SF for gaming. This guarantees QoS for both VoIP and gaming as these two applications are isolated with two distinct SFs. However, this solution has scalability issues and in cases where only one application is active or none of them is active, two SFs are unused.
2. Establish one ertPS SF, an ertPS SF, and map both VoIP and gaming to same SF. Since VoIP and gaming with different QoS requirements are sharing the same SF, in order to guarantee QoS for both of the applications, either the SF should be overprovisioned or a QoS mechanism must appropriately provide relative priority to the packets of two applications. To avoid the inefficiencies of overprovisioning on the air interface, WiMAX allows the scheduler to prioritize the packets of different applications mapped to the same SF over the air interface. This requires the BS to assign a buffer to each application. Figure 10.8 illustrates an example in which the VoIP and gaming share one ertPS SF and the BS gives a higher priority to VoIP packets.

As a result, WiMAX provides a scalable solution with minimal overprovisioning to support application QoS over the air interface using the SF QoS framework.

10.3.1.1.3 Role of Scheduler in Delivering Air-Interface QoS. The QoS of each SF is maintained by the BS scheduler. Basically, the network resources over the air interface are subchannels. Each subchannel is a two-dimensional collection of subcarriers on the frequency domain and a number of OFDMA symbols on the time domain. The number of bits carried by a subchannel is a function of modulation and coding rate, and the repetition of these depends on the RF condition. In WiMAX, each MS reports its RF condition through a CQICH channel over the fast-

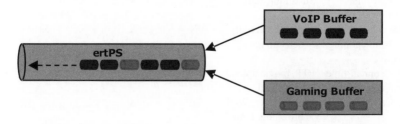

Figure 10.8. Example of two applications sharing the same ertPS SF.

feedback channel on the UL-to-BS scheduler. The scheduler uses the knowledge of RF channel conditions, the buffer lengths, and the SF's QoS requirements to make its scheduling decisions. Fundamentally, in a wireless network, the poorer the RF condition, the more air-interface resources needed to transmit the same number of bits. In WiMAX, an MS in a poor RF condition requires more subchannels to meet its QoS requirements compared to an MS in a strong RF condition. Since RF condition is a random process and randomly varies with time, the air-interface capacity in bits per second varies randomly with time.

The scheduler controls the QoS requirements of each SF as well as the fairness among all the SFs by assigning a utility function to each SF, which translates to a priority level for the scheduling decisions. The utility function depends on the SF type; for example, for ertPS the utility function could depend on the packets' elapsed time, the short-term assigned rate, the CQICH, and so on, whereas for the BE, the utility function could depend on the assigned short-term rate and the CQICH. For instance, the utility function of BE SF could be defined as follows:

$$U_{BE_i}(t) = TP_i \frac{CQICH_i(t)}{R_i(t)} \tag{1}$$

In this equation, $CQICHi(t)$ represents the channel capacity of BE SF i at time t, $R_i(t)$ represents the short-term average rate allocated to BE SF i, and TP_i is the traffic priority weight assigned to SF i. This utility function, which is also known as weighted proportionally fair, provides weighted fairness among the SFs while maximizing the sector throughput. Based on this utility function, the scheduler gives higher priority to SFs of MSs with good RF conditions to improve sector throughput. Also, it gives higher priority to SFs with low short-term average rate to avoid SF starvation. Finally, the traffic priority modifies the priority level. The utility functions defined in the scheduling algorithm at the BS are outside the scope of the WiMAX standard and different vendors use their proprietary algorithms. An ideal scheduling algorithm provides the QoS requirements of each SF and fairness among the SFs while maximizing the sector capacity.

Briefly, the resource scheduler in the base station allocates air-interface resources differently for each service flow type. Traffic within the same service flow type may be further differentiated by traffic priority values assigned to users. The resource scheduler also dynamically accounts for users' signal conditions when allocating nonreal-time resources. In general, allocations to users with better signal conditions are weighted so they receive more resources (to maximize overall sector throughput) but an operator may adjust fairness to ensure that users with poor signal conditions receive reasonable throughput.

10.3.1.2 *WiMAX QoS Over the Backhaul Transport*

The backhaul transport connects the BS and the ASN-GW through the R6 interface. Backhaul transport is one of the most expensive segments of a mobile network and, depending on its type, might have limited bandwidth. The backhaul transport model could be LEC (local exchange carrier), AAV (alternative access vendor), or microwave. In the LEC model, the mobile operator leases T1 links, DS3 links, or fiber links from the local carriers. The operator converts its traffic from the base station on the UL and from the ASN-GW on the DL to TDM for transmission over the TDM links. With AAV, the backhaul provider connects the base station located at the cell site to the ASN-GW located at the mobile switching center (MSC) through virtual circuits (VCs) in a Metro-Ethernet. In this model, the mobile operator converts its traffic to Ethernet before transmission over the backhaul transport. In the microwave backhaul model, the mobile operator owns the microwave links connecting the cell sites to the MSCs and to each other. Usually, the microwave link is a TDM pipe and the traffic has to be converted to TDM before transmission. In all the cases, the traffic coming from the ASN-GW on the DL and from the base station on the UL has to be converted before getting forwarded over the backhaul transport.

Fundamentally, the QoS over the backhaul transport is connectionless based on packet marking, classification, and prioritization. The backhaul QoS framework defines a number of classes, each with specific priority level. The packets of different SFs (and, therefore, different applications) receive different priority levels mapped to appropriate class and so receive appropriate treatments. Due to the connectionless nature of the backhaul QoS framework, backhaul requires some level of over-provisioning and so requires a higher bandwidth compared to the air interface. Since WiMAX average sector throughput can be as high as 15–20 Mbps, the backhaul to handle traffic of a cell site (three sectors) is expected to be as high as 20–50 Mbps. Note that due to the oversubscription and the statistical multiplexing gains, the capacity to handle the traffic of three sectors is less than three times the capacity of one sector. Basically, the backhaul QoS framework is as follows.

In the DL direction, the ASN-GW classifies and queues the packets into their SFs. The ASN-GW rate-caps traffic of each SF based on its MSTR; then it marks the packets of the SF appropriately and forwards them over the backhaul transport. The backhaul transport maps the packets into appropriate classes specified by their markings and treats the packets of different classes according to a priority mechanism.

In the UL direction, the base station classifies and queues the packets into their SFs. The SF level rate-capping over the UL was already performed by the base station over the air interface based on SF MSTR. Then the packets sitting in the queues inside the base station are marked and mapped to appropriate classes. The packets are forwarded over the backhaul transport and receive appropriate treatment based on the class they are mapped to.

The total traffic of a class as well as the total traffic of the R6 interface can be rate-capped over the backhaul transport within the routers/switches in both DL and UL directions.

The QoS functions of the backhaul transport can be performed in Layer 2 (L2-Ethernet) or Layer 3 (L3-IP). The L2 QoS framework over the VALNs is specified through IEEE 802.1p by marking the three bits of the COS field appropriately, which assigns a priority level to each packet. The L3 QoS is based on the differentiated services (DiffServ) model, which is a policy-based approach and is specified by IETF. DiffServ marks the 6-bit DSCP (DiffServ code point) field in the IP packet header, which determines the class the packet belongs to and service type by specifying a per-hop behavior (PHB) implemented in a network element. DiffServ defines three types of classes:

1. The EF (expedited forwarding) class is appropriate for high-priority traffic such as VoIP, mission-critical traffic, management and control traffic, and so on.

2. The AF (assured forwarding) class is appropriate for better than BE traffic such as video applications, streaming, and so on. There might be a number of AF classes with different priority levels that can be defined by the network.

3. The BE (best effort) class is appropriate for best-effort data such as e-mail, Web browsing, and so on. (The DiffServ BE class should not be confused with IEEE 802.16e BE SF.)

For backhaul QoS transport, traffic of SF types is mapped to the appropriate class. If the number of the QoS classes over the backhaul is enough, this framework provides the required flexibility for the mobile operator to offer the services. For example, the mobile operator can map the preferred BE SF to an AF class and the basic BE SF to the DiffServ BE class. Figure 10.9 shows the mapping between the SFs and the DiffServ classes on the DL. (A similar concept works for the UL.) As shown in the example in Figure 10.9, the preferred BE SF is mapped to a higher priority class compared to the basic BE SF.

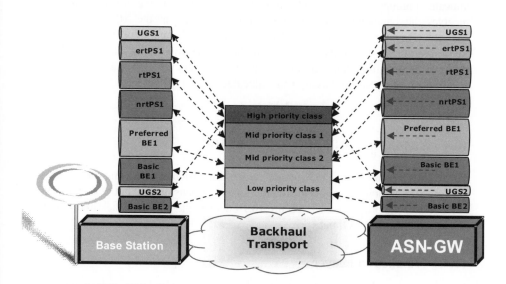

Figure 10.9. An example of mapping SFs to the backhaul QoS classes.

10.3.1.3 WiMAX QoS Over the Core Transport

Since the core transport is overprovisioned for the mobile operator traffic and has abundant bandwidth, there is no need for QoS over the core transport. Therefore, this segment follows the BE model for QoS since, due to large bandwidth of the core transport, there is usually no packet blocking or only very short packet delays.

10.3.1.4 WiMAX QoS Over the ASN-GW

The role of ASN-GW in the QoS framework on UL and DL is quite different. In the UL direction, the packets are coming from backhaul transport with smaller bandwidth to the core network with very large bandwidth. Therefore, there is no need for any QoS in the ASN-GW in the UL direction. However, in the DL direction, the packets coming from the core transport enter the backhaul transport with limited bandwidth after leaving the ASN-GW. Therefore, the ASN-GW has to perform the following QoS functions: SF-based classifications and SF-based rate-capping. A mobile operator is usually interested in having device-based as well as subscriber-based rate-capping. The device- and subscriber-based rate-capping capabilities will be explained later in this chapter.

10.3.1.5 WiMAX QoS Inside the Device

A WiMAX device has QoS functions only in the UL direction. In the DL direction, the device receives WiMAX MAC PDUs over the air interface and forwards the PDU to the correct application. In the UL direction, the device classifies the packets into SFs and makes bandwidth requests from the scheduler at the serving base station on a per-SF basis. The scheduler in response makes a scheduling decision and allocates a number of subchannels on the UL to the MSs. By the time the scheduler receives the bandwidth requests, the status of buffers at the MSs might have changed. Therefore, the scheduler does not make subchannel allocations on a per-SF basis and leaves it for the MSs to make the best decision using the most updated information on buffer status. In the UL direction, the device keeps track of the QoS requirement of its SFs and assigns the allocated subchannels for the MSs to different SFs.

10.4 WiMAX QoS AND PROTOCOL STACK

The WiMAX end-to-end QoS picture is two-dimensional; one dimension covers QoS in different logical segments and the second dimension is about QoS in different layers of the protocol stack. Figure 10.10 depicts the protocol stack in different segments of the WiMAX network.

10.4.1 Application Layer

The entire group consisting of application layer, presentation layer, and session layer in the OSI model is represented by the application layer in Figure 10.10. The ap-

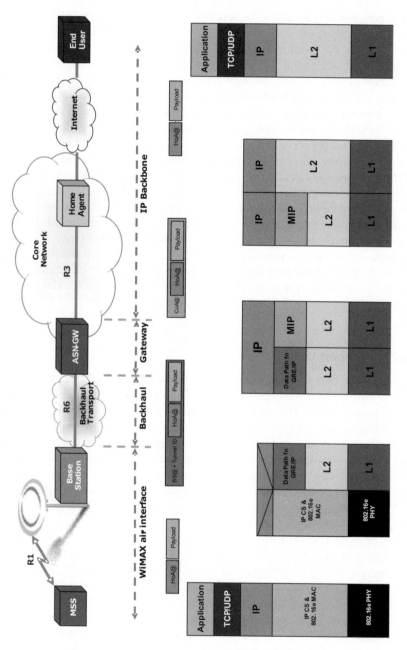

Figure 10.10. Protocol Stack in different segments of the WiMAX network.

plication layer provides services for users and it usually appears only on the two end sides. The application layer has its own QoS optimizations, including compression, decompression, error concealment, and possible rate adaptations. At the transmitter, a compressor may reduce the application layer bit rate. Rate distortion theory introduces a trade-off between the application layer bit rate and the service quality; reducing the application layer bit rate may degrade the service quality. Intelligent encoders are capable of significantly reducing the application layer bit rate while maintaining the service quality. On the other hand, reducing the bit rate and shortening the application block size not only results in lower latency and jitter, which is important for some applications (e.g., VoIP), but it can also dramatically increase the cell-edge performance and the system capacity. There is this misapprehension among some people that the lower the application layer bit rate the better system performance; this is not always true, since every technology has a minimum MAC layer frame size and at some level, lowering application block size brings no advantage to system performance. At the receiver side, the de-jitter buffer minimizes the jitter at the receiver the time warping module improves the service quality for a given average jitter, and the error concealment module maximizes the service quality for a given packet error rate. In some scenarios (e.g., VoIP and multimedia applications) a media gateway (MGW) sitting behind the home agent (HA) inside the core network operates at the application layer for transcoding. However, this does not change the big picture shown in Figure 10.10. The application layer determines all the QoS requirements associated with a service (Table 10.2).

10.4.2 Transport Layer

The transport layer is a set of protocols responsible for encapsulating the application layer blocks into a datagram suitable for delivering them on the network. The transport layer could be TCP or UDP depending on application QoS requirements and is usually sitting at the two end points. For instance, most of the data applications run over TCP since they are loss tolerant and latency insensitive and TCP packet retransmission and rate adaptation capabilities are helpful. However, TCP packet retransmissions can degrade quality of a real-time application (e.g., VoIP). Applications such as VoIP do not need any rate adaptations. Real-time applications (e.g., VoIP) are usually loss tolerant and latency sensitive and RTP/UDP is the appropriate transport protocol for them.

The Real-time Transport Protocol (RTP) defines a standardized packet format for delivering audio and video over the Internet. RTP does not have a standard TCP or UDP port on which it communicates and the only standard it obeys is that UDP communications are done via an even port. The transport layer adds a 20-byte header to the packet as the TCP header size is 20 bytes, whereas UDP (8 Bytes) and RTP (12 Bytes) combined add the same header size. The Internet Assigned Numbers Authority (IANA) specifies the assignment of the port numbers to the transport protocol type (TCP or UDP) and the applications. WiMAX uses the source and destination port numbers in the QoS classifiers to identify the application type.

10.4.3 Network (IP) Layer

The network (IP) layer performs packet routing in the network and potentially plays an important role in the end-to-end QoS picture. IP routing in the WiMAX network is performed over the mobile-IP (MIP) tunnel and the GRE tunnel. All the packets heading to the mobile are routed through the home agent (HA) and the MIP tunnel in the core network to the ASN-GW. In mobile WiMAX with L3 backhaul, a generic routing encapsulation (GRE) tunnel extends the link layer to deliver the packets between the ASN-GW and the base station. Over the GRE tunnel, the packets are first encapsulated with a GRE header and then encapsulated again with IP as the delivery protocol. The link layer extension on L2 backhaul transport can be achieved using Ethernet virtual circuits (EVC).

WiMAX uses the source and the destination IP addresses, and the DiffServ code point (DSCP) in the IP header as the QoS classifiers. This enables an operator to offer different service profiles to different subscribers. WiMAX identifies packets associated with a specific subscriber by checking the IP address assigned to him/her in the IP header and then maps the packets to the appropriate SFs or service classes based on his/her QoS profile. Also, WiMAX allows an operator to offer superior service to its business partners, such as a specific search engine or video server, and encourage its customers to get service from the partners. WiMAX can identify packets associated with a business partner by checking the server IP address (range) in the IP header of the packets. WiMAX also uses the DSCP field of the IP header as the classification rules. Therefore, WiMAX uses the five tuple: source and destination IP addresses, source and destination port numbers, and DSCP to appropriately classify the QoS that each packet belongs to. In differentiated service, the DiffServ class is identified by the DSCP field and a behavior aggregate and policies associated with each DiffServ class are determined by a per-hop behavior (PHB) specified for that class. In the case of Layer 2 link layer extension over the backhaul transport, IEEE 802.1p is used instead of DiffServ to mark the packets and deliver QoS over the backhaul. The IP layer adds a header of 20 bytes to the packet.

10.4.4 Header Compression

The transport layer header size (TCP or RTP/UDP) is 20 bytes and the IP layer header size is 20 bytes for IPv4 and 40 bytes for IPv6 headers. This adds 40–60 bytes of header to the packet. However, applications such as VoIP with EVRC encoder generate packets as short as 22, 10, 5, or 2 bytes, which are much shorter than the headers. In order to mitigate the effect of large overheads, header compression mechanisms such as robust header compression (ROHC) and packet header suppression (PHS) have been developed to compress RTP, UDP, TCP, and IP headers. ROHC and PHS compress the large headers to a few bytes (1–3 bytes with IPv4 and 3–5 bytes with IPv6) and also recover the original header between the mobile device and ASN-GW. Header compression is preformed at the convergence sublayer.

Header compression does not improve the QoS for most data applications since these applications usually generate few long packets, but it dramatically improves the QoS of real-time applications such as VoIP; header compression reduces packet latency of real-time applications, enhances cell-edge performance, and increases system efficiency and throughput.

The medium access control (MAC) and physical (PHY) layers are covered in other sections and, therefore, are not covered here.

10.4.5 IP Packet Fragmentation Issues

The maximum transmission unit (MTU) is a link-layer restriction on the maximum number of bytes of data in a single transmission. Every transmission medium has a limit on the maximum size of a frame (MTU) it can transmit. As IP datagrams are encapsulated in frames, the size of the IP datagram is also restricted. If the size of an IP datagram is greater than this limit, then it must be fragmented. The Internet Protocol allows IP fragmentation so that a single IP datagram can be broken up into two or more IP datagrams of size small enough to pass over a link with a smaller MTU than the original datagram size (RFC 791 and RFC 815 discuss the procedure of IP fragmentation, transmission, and reassembly).

Based on NWG stage 3, the default MTU size to/from the mobile is limited to 1400 bytes. Theoretically, an IP datagram can be as large as 65,495 bytes, but such a large value is never used. Typically, IP MTU is 1500 bytes. Therefore, an IP packet in a WiMAX network may be fragmented using IP fragmentation. In other words, large IP packets may or may not be fragmented at the ASN-GW on the downlink or at a device on the uplink, based on IP fragmentation. Let us look at the possible issues in the following.

In one case, the DF (don't fragment) bit might be set to 1 by the sender (origin host). This means IP packets should not be fragmented even though their size is greater than the next-hop MTU. Consequently, if the IP packet size is greater than the MTU with DF = 1, then an ASN-GW in the DL direction does not fragment the IP packet. Instead the ASN-GW discards the packet and sends an ICMP error message to the origin host as expected per standards. The ASN-GW generates the ICMP error messages in order to get the sender host to reduce its packet size to less than MTU size. If there is no firewall at the sender host, the ICMP error message works well for reducing packet size. However, if there is a firewall, the ICMP error messages are blocked, that is, there is no way to get the sender to be aware that the packet size is too big and, hence, reduce the packet size unless the sender removes the firewall. In the case that there is a firewall blocking ICMP error messages, a solution is to have the ASN-GW do fragmentation if the packet size is greater than MTU size, even at the DF = 1 setting. This enables packets greater than MTU size to go through the ASN-GW and be reassembled successfully at the device.

In the other case, the IP layer creates multiple new IP datagrams whose length satisfies the requirements of the WiMAX network. The IP header from the original IP datagram is copied to the multiple new datagrams. However, IP fragmentation does not copy the transport layer header. In other words, the QoS classification rules that require both IP addresses as well as port numbers cannot properly classify the fragments not carrying the transport layer header. This might cause some issues.

For instance, it is possible that a long video packet gets fragmented; one fragment goes through an rtPS SF while the other fragment travels along the BE SF. Thus, the fragments will be delivered to the other end point at different times and, effectively, latency of the video packet will be determined by the BE SF and not the rtPS SF (Figure 10.11).

A possible workaround for this situation could be using the DSCP field of the IP header instead of the port numbers in the classification rules. However, a long-term solution is performing the IP fragmentation in the CS layer instead of the IP layer, so that the CS layer can properly map all the IP fragments into the appropriate connections.

10.5 WiMAX QoS FRAMEWORK

Section 3 and Section 4 discussed the bearer plane in WiMAX end-to-end QoS in different segments and different layers. However, they only partially covered the end-to-end QoS picture for delivering different services to the end subscribers. Provisioning and signaling are also two important components of the end-to-end QoS picture. In the end-to-end QoS picture, an operator creates a QoS profile for a subscriber and stores it in a database (SPR). The QoS profile embraces a list of the authorized services and the corresponding accounting mechanism. At the provisioning instance before launching a service, a subscriber management system (e.g., AAA or a policy server) signals to appropriately configure the network elements along the bearer path to support the service. Then the network is prepared to run the service.

The WiMAX Network Working Group (NWG) specifies two QoS frameworks in two releases:

1. Preprovisioned (static) QoS framework (NWG Release 1.2). The QoS profiles are preprovisioned inside "AAA" database and the authorized SFs are created only during the network entry phase and after a successful authentica-

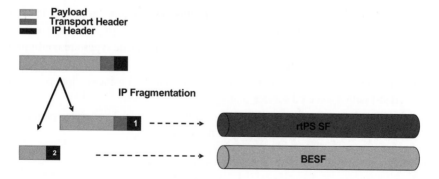

Figure 10.11. IP fragmentation might cause fragmented datagrams to go along inappropriate SFs.

tion process. In the preprovisioned QoS framework, new SFs cannot be created or the existing SFs cannot be modified or deleted after the network entry phase; all the SFs will be deleted during network exit.

2. Dynamic QoS framework (NWG Release 1.5). The SFs may be created, modified, or deleted dynamically at any time after network entry. An application function (AF) requests initiation or termination of a service to a policy function (PF) that triggers a SF creation, modification, or deletion after successful authorization by evaluating the request against the provisioned information.

IEEE802.16e supports all the mechanisms required for dynamical SF creation, modification, or deletion triggered either by the network or the mobile station at any time. However, both preprovisioned and dynamic QoS frameworks allow only network-triggered SF operation, although there have been some proposals presented to the WiMAX Forum for MS-initiated SF operations. Generally, the differences between the two QoS frameworks are the trigger mechanism and the trigger time of the SF operations.

10.5.1 Preprovisioned QoS Framework

As mentioned, with a preprovisioned QoS framework a set of SFs are created as a part of a successful subscriber network entry process and before any IP traffic. Indeed, during the authentication procedure of the network entry, the subscriber's QoS profile is downloaded from the AAA into the SF authorization (SFA) function at the ASN-GW. Then the SFA signals the SF management (SFM) function at the base station to initiate the creation, admission, and activation of the preprovisioned SFs. Figure 10.12 depicts the functional elements in the WiMAX preprovisioned QoS framework in a nonroaming case.

Briefly, the high-level steps involved in provisioning a subscriber in the preprovision (static) QoS framework can be described as follows. The steps before network entry by a subscriber's device are:

- QoS parameters for different service profile IDs are preprovisioned in radio access network elements.
- QoS for different service flow types and different traffic types are configured in backhaul and core network elements.
- The subscriber's device is provisioned through the business support system (BSS) with a selected service plan.
- The subscriber's service profile is stored in a database. The service plan is associated with service profile IDs for network access.

Steps after network entry of the subscriber's device are:

- The subscriber's device information and the associated service profile IDs are sent from the database to the AAA.

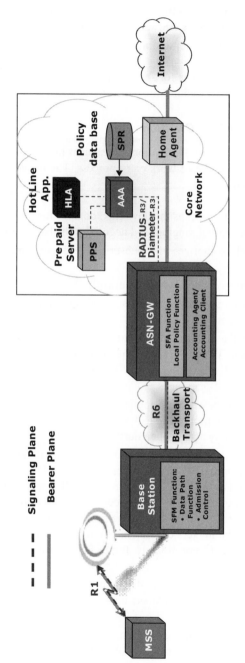

Figure 10.12. Preprovisioned QoS framework functional elements.

- After the subscriber is authenticated, the AAA sends the service profile IDs to the ASN-GW.
- The ASN-GW translates the service profile IDs into QoS parameters and creates service flows to the BS, sending relevant parameters to the BS.
- The BS performs call admission control before granting the service flows.

In the following, call flows for service flow creation, service flow deletion, and call admission control procedures are described in more detail.

10.5.1.1 Service Flow Creation

Figure 10.13 illustrates the SF creation mechanism in the WiMAX preprovisioned QoS framework during the network entry process. The network entry process consists of the following steps:

1. The mobile station, scans the downlink, obtains downlink and uplink parameters, performs initial ranging, and negotiates the basic capabilities.
2. The network authorizes the subscriber and performs the key exchange process.
3. The subscriber registers with the network.
4. Next, the AAA authorizes the subscriber's QoS profile and pushes them into the SFA function at the ASN-GW.
5. At this step, the SFA function at the ASN-GW knows the SFs authorized for the subscriber. The SFA obtains the QoS parameters associated with the SFs and requests SF creation from the SFM at the base station using the RR-REQ (resource request) messages.
6. The SFM at the base station applies the call admission control (CAC) process and admits the appropriate SFs based on its local resource management. The CAC process may also happen during handovers.
7. At this stage, the subscriber still does not have an IP address. WiMAX uses a SF known as initial SF (ISF) for performing the DHCP process between the mobile station as the DHCP client and the DHCP server. The ISF can be used later for data transmission, which requires the ISF QoS parameters to be selected appropriately according to the subscriber's profile. On the other hand, the ISF QoS parameters can be chosen only for a simple DHCP process that will be modified appropriately for traffic transmission after a successful DHCP process.
8. After the ISF establishment, the mobile acquires its IP address using the DHCP process. This IP address may be used during data transmission for traffic classification.
9. Finally, the SFM at the base station will establish all the admitted SFs associated with the subscriber using DSA-REQ (dynamic service addition) DSA-RSP, and DSA-ACK messages over the R1 interface.

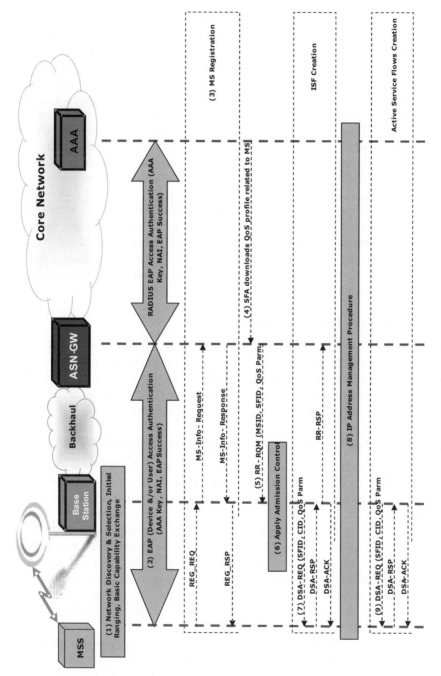

Figure 10.13. Service flow creation process.

Step 9 might be performed before Step 8; in other words, the SFM may establish all the SFs at the same time.

After the network entry, the network is provisioned to run the services and deliver them to the subscriber. Under the preprovisioned QoS framework, all the admitted SFs of a subscriber exist (except idle mode), regardless of traffic flowing. For all the SF types except UGS, this does not mean that a SF not running traffic will consume network resources. In other words, a network element assigns resources to a SF if and only if there are packets ready for transmission. For the UGS SF type in the uplink direction, the network is committed to assign resources to a subscriber regardless of whether the subscriber has packets for transmission or not. The subscriber sends padding (a packet with all 0xFF as payload) in the cases in which the subscriber has no packets for transmission over an uplink UGS SF. Thus, the UGS SFs may waste network resources under the preprovisioned QoS framework.

The admission control in the preprovisioned QoS framework is also a challenging task. Although the other four types of SFs do not consume resources without traffic flowing, the network is committed to deliver the guarantees to the traffic running over an already established SF. However, the CAC has absolutely no a priori information during the network entry about when the subscriber starts an application. This issue will be discussed later.

10.5.1.2   Service Flow Deletion

Within the preprovisioned framework, SF deletion can only happen during the network exit procedure. In other words, all the SFs associated with the MS are deleted during the network exit procedure when the MS is deregistered with the ASN; the initial SF (ISF) will be the last SF to be deleted. Figure 10.14 illustrates the SF deletion mechanism in the WiMAX preprovisioned QoS framework.

10.5.1.3   Call Admission Control (CAC) Challenge in WiMAX 1.0

Call admission control (CAC) is one the essential role that the MAC common part sublayer (MAC CPS) performs in QoS support over the air interface. During network entry or handover process, the SFM at the base station applies the CAC process and admits the appropriate SFs based on its local resource management. The CAC admits or rejects a call and also performs bandwidth reservation. This means that the CAC is responsible for checking the reservation requirements against the available resources in order to investigate whether there are sufficient resources available to fulfill the reservation and to support the requested service. After the CAC admits a subscriber, the network establishes a set of SFs with the right QoS attributes based on the reservation request. These admitted SFs are allowed to carry traffic at any time.

However, the CAC has no a priori information during the network entry about when the subscriber starts an application or whether a SF will carry traffic at all. It is possible that an admitted SF cannot carry traffic at a specific time since there might not enough network resources available at that time. In other words, the pre-

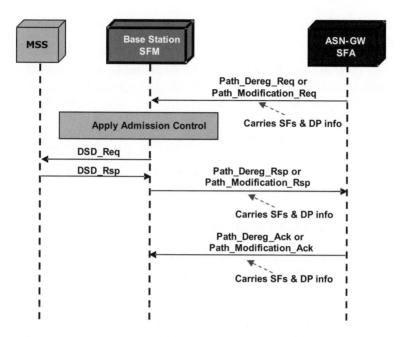

Figure 10.14. Service flow deletion in the preprovisioned QoS framework.

provisioned QoS framework cannot make precise and guaranteed CAC decisions during the network entry or handover procedure.

The details of the CAC mechanism are not specified by the standard. Considering limited network resources and the static nature of the QoS framework in WiMAX 1.0 for establishing SFs, developing an efficient CAC mechanism is very challenging. Such an efficient CAC mechanism can avoid any bandwidth or subscriber overprovisioning, and it can make a precise bandwidth reservation decision during the CAC process. The efficient CAC is an important factor in distinguishing vendors' products since it is a resourceful utilization of network resources and it will provide a better Cap-Ex model for the operator.

10.5.1.4 QoS and Accounting in WiMAX

WiMAX NWG specifies QoS-based accounting, which allows an operator to charge the subscribers based on the service sessions used by them, and not based on connection as traditional accounting does. The WiMAX network can support multiple services for one subscriber simultaneously, with appropriate QoS levels, and charge the subscriber for each service separately.

10.5.2 Preprovisioned QoS Framework Shortcomings

The preprovisioned QoS framework establishes the SFs during the network entry procedure; however, the network has no a priori information about when a subscriber launches a service. This causes a number of shortcomings for the QoS framework.

The preprovisioned QoS framework results in bandwidth overprovisioning. Since the CAC decision is made during SF establishment before running traffic, bandwidth overprovisioning is required. To clarify the issue, let us assume that a subscriber can run the following services:

- ertPS service flow for delay-sensitive services such as VoIP and online gaming
- rtPS service flow for rate-sensitive services such as video/audio streaming
- BE service flow for services such as ftp

A subscriber may not run many applications simultaneously. In other words, WiMAX CAC provisions more bandwidth than required. Thus, during the network planning phase, more bandwidth has to be provisioned (overprovisioned) for the network. Let us briefly review major shortcomings of the preprovisioned QoS framework:

- Scalability issues of the preprovisioned QoS framework. Active SFs are those carrying traffic at a given time. Usually, a subscriber does not run all the applications simultaneously. At a given time, less than 50% of the created SFs are active. Practically, WiMAX network elements (base station and ASN-GW) have SF scalability limitations and the maximum number of SFs that WiMAX network elements can create is limited. Hence, in the preprovisioned QoS framework WiMAX network elements' SF scalability reduces to less than 50.

- Supporting turbo and normal modes is cost-effective in the dynamic QoS framework. Turbo and normal modes provide an on/off switch for the subscriber to increase to higher access speeds and the network bills the subscriber accordingly.

- Call admission control (CAC) in preprovisioned QoS framework issues. CAC admits or rejects a call and also performs bandwidth reservation. The details of CAC are not specified by the standard; however, during the network entry, CAC has no a priori information of if/when an SF will carry traffic. An admitted SF is allowed to carry traffic at any time even if there are not enough resources available. Thus, the preprovisioned QoS framework cannot make precise CAC decisions during the network entry phase.

- UGS SF inefficiencies within preprovisioned QoS framework. UGS SF delivers CBR service to subscribers. However, with the preprovisioned QoS framework, the network has to allocate resources to UL UGS SFs irrespective of whether a UGS SF carries traffic or not. In other words, a UGS SF consumes resources after the network entry until the mobile exits the network. This results in resource inefficiencies, especially on the uplink.

10.5.3 Dynamic QoS Framework

The WiMAX dynamic QoS framework specified under NWG R1.5 enables dynamic QoS selection and enforcement and offers flexible charging models such as service-based as well as flow-based accounting. With the dynamic QoS framework, the network can create, modify, or delete SFs based on application need at any time; this resolves the shortcomings of the preprovisioned QoS framework. However, the dynamic QoS framework requires new trigger mechanisms for SF authorization and management.

Basically, dynamic QoS framework provides:

- Significant control of network resources, which prevents waste of resources while it provides an enjoyable subscriber experience.
- A much more efficient CAC mechanism, which avoids any bandwidth over-provisioning as the CAC can make a precise bandwidth reservation decision during the admission control process. This will provide a better Cap-Ex model for the operator.
- Detection, metering, and control of services to prevent revenue leakage and optimize service revenue.
- Dynamic creation of traffic classification rules, which can be unique to a particular traffic stream in response to some network event.
- Dynamically applying policy, charging, and access control based on service, time of day, location, rate plan, and roaming status to optimize revenue network resource use.
- Significantly improving the ASN equipment SF scalability as the SFs are created only when needed by applications, and SFs not running applications do not exist. This results in a better long-term Cap-Ex model.
- Enabling efficient support of normal and turbo modes as the SF attributes can be modified to switch between the two modes.

The next section explains the WiMAX policy control and charging (PCC) framework to enable the dynamic QoS framework.

10.6 WiMAX POLICY CONTROL AND CHARGING (PCC) FRAMEWORK

10.6.1 Policy Control and Charging (PCC) Framework

As a result of transition from best-effort data service to advanced multimedia services platforms, policy has become an increasingly important area for operators as they evolve their networks into service- and application-oriented platforms. Operators are interested in offering new services and applications such as VoIP and streaming media, bandwidth-oriented tiering of their subscribers, as well as new tools to further secure their networks, such content filtering. Moreover, charging is another important aspect of a policy. By offering new services and applications, the operator needs to be able to charge for these services as not all bits are created equal. The PCC framework provides capabilities for operators in order to control and determine who gets what, when, where, how, and how much of a service delivery.

The policy and charging control (PCC) framework has been developed in 3GPP as an access-agnostic framework to be used by different IP-connectivity access networks (IP-Cans). Figure 10.15 demonstrates the architecture of the 3GPP PCC framework, which contains the following functions:

PCRF (Policy and Charging Rules Function). Provides control over service dataflow detection, gating, QoS, and flow-based charging in the policy and charging enforcement function of the access gateway, and consists of policy control decision and flow-based charging-control functionalities.

AF (Application Function—P-CSCF in IMS). Offers applications requiring dynamic policy and/or charging control, and communicates with the PCRF information regarding the dynamic service session status, that is, activation/ deactivation of applications at the subscriber terminal and the QoS requirements of the application.

SPR (Subscription Policy Repository). Contains the subscription information needed by the PCRF for making decisions regarding subscription-based policies.

PCEF (Policy and Charging Enforcement Function). Responsible for SF detection, policy enforcement, and flow-based charging. PCEF is located in the RAN

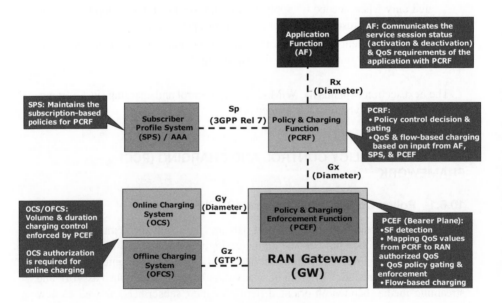

Figure 10.15. 3GPP PCC framework.

GW and implements the policy control by gating and enforcement of the QoS attributes. PCEF maps the QoS class identifier values, provided by the PCRF, to the specific QoS attributes of the access network and enforces the authorized QoS according to the rules received from the PCRF. The PCEF also enforces charging control and reports the data volume and duration measurements to the offline and online charging systems (OFCS and OCS, respectively). For online charging, an authorization from the OCS is required prior to resource usage.

10.6.2 WiMAX PCC Architecture

The 3GPP PCC framework it is not directly applicable to WiMAX as there are some requirements specific to the WiMAX network that require additional functions compared to what is provided by 3GPP Release 7 PCC. The WiMAX release 1.5 PCC architecture requires new functions in the CSN (connectivity service network) and not in the ASN (access services network). The WiMAX network reference model separates the end-to-end WiMAX networks into two parts (ASN and CSN). The ASN piece covers WiMAX base stations, ASN-GW, layer 2 connectivity, AAA, DHCP, FA, and PCC enforcement. The CSN provides IP connectivity services to WiMAX subscriber(s) and usually includes AAA, DHCP, HA, and PCC control-based entitlements.

The differences between the 3GPP PCC and WiMAX PCC architectures are as follows:

- WiMAX PCC requires PCEF relocation support. WiMAX architecture provides no guarantees that the traffic plane for an IP session passes through a network entity during the lifetime of the IP session as a subscriber moves across the WiMAX RAN. However, in 3GPP architecture, the traffic plane for each IP session passes through a single GGSN for the lifetime of the IP session, so PCEF in GGSN does not relocate.
- WiMAX PCC architecture requires a multiple policy enforcement function (PCEF) in ASN-GW (A-PCEF) as well as capability of optional policy enforcement in the core network (C-PCEF) on the bearer plane (possibly at the HA). However, with 3GPP there is a single PCEF at the GGSN.
- The 3GPP PCC framework follows both network-initiated (PUSH) as well as MS-initiated (PULL) models for QoS activation. However, the WiMAX PCC framework follows only the network-initiated (PUSH) model.
- The subscriber authentication in the 3GPP PCC framework is performed at the GGSN, and in the WiMAX PCC framework it is done at the PCRF.

Figure 10.16 illustrates the WiMAX PCC architecture.

10.6.3 WiMAX PCC and Dynamic QoS Framework

The WiMAX PCC framework enables a network to dynamically create, modify, or delete SFs based on application requirements. The benefits of the WiMAX dynamic QoS framework are:

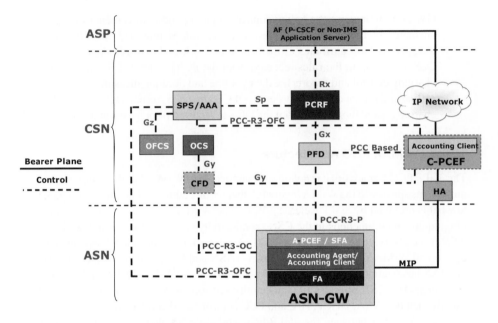

Figure 10.16. WiMAX PCC framework.

1. Provides better capability for providing on-demand services such as video-on-demand, IPTV, on-demand advertising, and music-on-demand, which requires large bandwidth and QoS guarantees. Based on policies for handling on-demand services with advance reservation, a mobile operator can improve the usage of the network resources and enhance the QoS provisioning.

2. By means of end-to-end dynamic QoS, a mobile operator is able to provide a more reliable platform to develop VoIP, video telephony, and mobile IPTV services, and to make various new services available more quickly.

3. To maintain subscriber experience for IMS-enabled services over WiMAX, the charging and QoS policies, as determined dynamically at the IMS core, need to be communicated to the WiMAX network elements where they can be suitably enforced.

Within the dynamic QoS framework, there are two types of SF management:

1. Preprovisioned SF management triggered by the network entry:
   - Initial SF (ISF) creation and establishment
   - IP address allocations and MIP tunnel establishment
   - (Possible) ISF modification and other preprovisioned SFs creation
   - IP-CAN session establishment

2. Dynamic SF management triggered by applications could happen at any time and requires the following steps:

- AF: Service activation
- PCRF: Session binding, PCC rule authorization
- A-PCEF: Bearer binding, QoS mapping, charging rule enforcement
- SFA and SFM: SF creation/modification using RR_Req/Rsp, Path_Reg/Modification_Req/Rsp over R4 & R6
- SFM: Admission control
- SFM: SF creation/modification using DSA/DSC_Req/Rsp over R1

10.6.3.1 PCC Terms

The following terms are used in the PCC framework:

IP-CAN: IP Connectivity Access Network. The IP-CAN type indicates the different underlying access technology such as WiMAX, 3GPP, 3GPP2, xDSL, or DOCSIS.

WiMAX IP-CAN Bearer: A pair of SFs with defined QoS parameters

WiMAX IP-CAN Bearer Binding: Association between PCC rules and WiMAX SFs

WiMAX IP-CAN Session: Association between an MS and an IP network through an IP address

WiMAX IP-CAN Session Binding: Association of AF session and applicable PCC rules to a WiMAX IP-CAN session

10.6.3.2 Mapping of PCC and WiMAX Flows

An application function (AF) session and its associated service dataflows correspond to a WiMAX service dataflow identified by the SDFID that provides data service to subscriber. An AF session consists of a number of packet dataflows, each flow corresponding to a service. A 3GPP service dataflow is defined by PCC rules that correspond to a WiMAX packet dataflow identified by PDFID. A packet data flow is bound to a WiMAX SF (two if the packet dataflow is bidirectional). The preprovisioned SFs of WiMAX are defined as packet dataflows. Figure 10.17 illustrates the mapping of WiMAX flows to PCC flows.

10.6.3.3 Policy Control in WiMAX PCC

At the core of policy control in WiMAX PCC (Figure 10.18), PCRF is a policy rules engine. The main functionality of PCRF is to operate on processes conditions. It triggers in order to make a decision and then performs the appropriate actions based on the conditions. Two major event triggers that can be processed by the PCRF are:

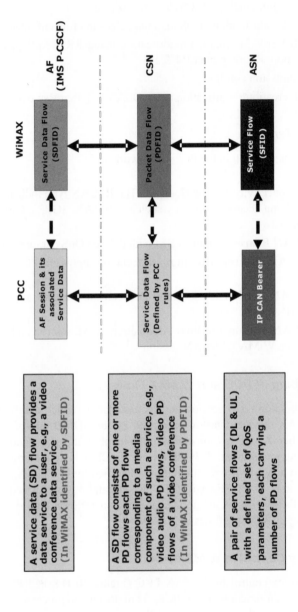

Figure 10.17. Mapping of PCC and WiMAX flows.

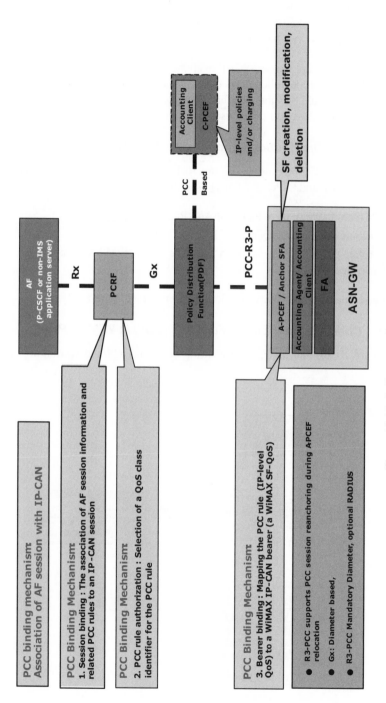

Figure 10.18. Policy control in WiMAX PCC.

1. Establishment of the IP-CAN
2. Modification of the IP-CAN session (i.e., initiation, modification, and termination of an IP-CAN bearer)

After receiving an event trigger, PCRF evaluates a series of conditions in order to determine the appropriate policy decisions that should be made. Elements in the series of conditions could be subscriber entitlement, roaming condition, network congestion, device identifier, subscriber usage, or time of day.

10.6.3.4   Policy Distribution Function (PDF) in WiMAX PCC

PDF is a WiMAX PCC framework (Figure 10.18) specific logical function between PCRF and A-PCEF and C-PCEF (if any) used to:

- Hide the relocation of A-PCEF/SFA to PCRF
- Distribute policies between A-PCEF and C-PCEF
- Translate the 3GPP Release 7 Gx interface to the WiMAX-specific PCC-R3-P interface

PDF could be a stand-alone network element or integrated with PCRF.

10.6.3.5   Charging Distribution Function (CDF) in WiMAX PCC

CDF (Figure 10.20) is a logical entity sitting between OCS and accounting client used to:

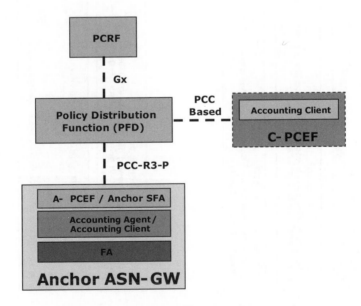

Figure 10.19. PDF in the WiMAX PCC framework.

- Hide the relocation of A-PCEF/SFA to OCS by decoupling Gy from PCC-R3-OC
- Terminate Gy for OCS and translate Gy to PCC-R3-OC
- Distribute the credit rules between ASN-GW and CSN (C-PCEF)

10.6.3.6 PCC Binding Mechanism

PCC rules contain several type of information related to QoS, charging, and identifiers including:

- PCC QoS parameters mapping functions in AF, PCRF, and A-PCEF
- A-PCEF binds PCC rules with WiMAX SFs
- WiMAX QoS mapping in A-SFA
- WiMAX QoS parameters are provided to S-SFA and SFM functions in QoS-Info TLV
- PCRF defines authorized QoS using on service information from AF (e.g., SDP) and/or subscription information received from SPR

Within the PCC framework and with service-based QoS, a bearer resource in the access network is used based on negotiation between what the application requests and what the network can support (Figure 10.21). The AF maps the application-level signaling to service dataflows passed to the PCRF, and a request for reservation is made from the ASN-GW. The ASN-GW establishes an association with the PCRF, so that the authorized QoS parameters are directly pushed to the ASN-GW upon their

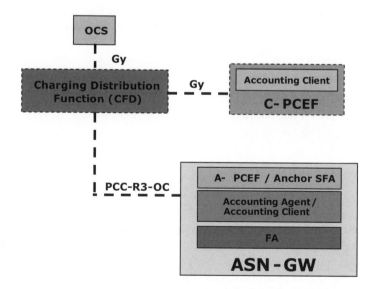

Figure 10.20. CDF in the WiMAX PCC.

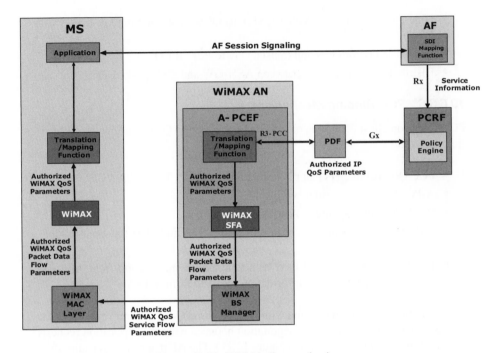

Figure 10.21. PCC binding mechanism.

receipt from the AF. The ASN-GW compares the requested QoS to the authorized QoS and creates a gate for each packet flow. Finally, the ASN-GW informs the application that bearer resources have been granted, allowing the service to flow.

Authorized QoS on Gx (PCC rules) contains the following information:

- QoS Class identifier
- Maximum request BW on UL
- Maximum request BW on DL
- Guaranteed bit rate on UL
- Guaranteed bit rate on DL

10.6.4 WiMAX PCC and IMS

Through the PCC, the integration of QoS policy and charging rules can be realized in the IMS network. As an enhanced QoS policy-control framework, the PCC can implement the QoS policy control, not only according to the service information, but also the charging rate of the SF and the subscription data obtained from the SPR, such as current location, subscriber profile, and date and time of the day.

Session-based QoS control is defined by the policy and charging (PCC) architecture. According to different applications of subscribers, as well as the local control policy, the IMS can control the IP network resources occupied by a specific application. For instance, the IMS can control and manage the bandwidth allocated to the application and define its priority.

During setup of a session, the application requests the network for related media parameters (e.g., codec, media type, and bandwidth) via the SDP contained in the SIP. The P-CSCF forwards the SDP parameters to the PCRF through the Rx interface. The PCRF authorizes the related media parameters according to the subscriber's profile and the local policy. After authorization, PCRF forwards the related IP QoS control parameters through the PDF (Gx interface) to the A-PCEF ASN-GW (PCC-R3-P interface) and C-PCEF (PCC-based interface) at the core network. The PCEFs analyze the source and the destination IP addresses and the source and the destination port numbers to filter the IP flow. The IMS services can be provided by the same CSN service provider or by a third-party service provider with a contractual relationship with the WiMAX network service provider.

The WiMAX PCC provides the end-to-end IMS solution through the integration of QoS control and charging rules and provides flexible differentiated services for the mobile operators.

10.7 IMPROVING THE WiMAX QoS FRAMEWORK

Mobile WiMAX technology offers wireless Internet with comprehensive network control to address operator requirements, and the capabilities to provide advanced services (Figure 10.22). As described in Section 6, dynamic QoS along with the PCC architecture provide a framework that is able to present advanced services according to today's business needs; also Section 5.2 addresses shortcomings of the preprovisioned QoS framework.

However, there are a number of features that could potentially improve the WiMAX QoS framework. These features assist operators to better control and manage their WiMAX networks and also to provide enhanced QoS for their services. Device-based and mobile-station-based rate capping is an example that is described in more detail here.

10.7.1 Device- and Mobile-Station-Based Rate Capping

One of the enhanced QoS features is support of per-MS rate capping in addition to WiMAX-specified SF-based rate capping. The MS rate capping can be applied at the device or subscriber level, and we simply refer both of them as MS rate capping. MS rate capping enables operators to intelligently manage and control the available network resources on both a per-SF and per-MS basis in order to facilitate tiered service differentiation from a qualitative view. It allows operators to allocate their WiMAX network resources across their subscribers more efficiently and more meaningfully.

In a WiMAX network, rate-capping capability can be performed at two levels: SF- and MS-based. Operators are interested in rate capping of several individual

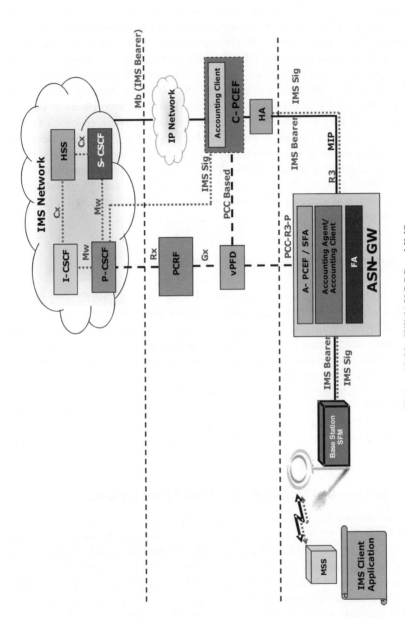

Figure 10.22. WiMAX PCC and IMS.

services (e.g., video streaming, instant messaging, and video,) at the first stage, and then rate capping on a per-MS basis at the second stage. In the case in which the subscriber's authorized services are not using all of their allocated bandwidth at the same time, which is usually the case, this strategy allows a chosen group of services (i.e., a group of preferred SFs) to take advantage of the spare capacity by expanding up to the maximum the resources allocated for that MS. Of course, a proper traffic scheduler is required between the two levels of rate-capping modules (Figure 10.23).

Each subscriber will be assigned a number of SFs on the uplink (UL) and on the downlink (DL) in a WiMAX network. Each one of these SFs will be rate capped individually based on the SFs profiles; this rate-capping capability is in the WiMAX specifications. Additionally, the aggregate traffic associated with a mobile station (MS) can be appropriately rate capped. This means a scheduler can be placed before aggreg`ting the traffic of all SFs. In the first level, SFs will be rate capped individually; then, a priority scheduler forwards the packets of these SFs to the next level that performs the MS based rate capping.

Both operators and their subscribers benefit from this rate-capping strategy. This rate-capping strategy allows operators to provide very strong, tiered-level services for their subscribers and charge them accordingly. The subscribers are guaranteed to receive the aggregate bit rate that they have been charged for if the network is not too congested and the subscriber is not at the cell edge.

10.8 SUMMARY

WiMAX wireless technology supports very robust QoS capabilities for delivering different types of applications and services to the end user. WiMAX MAC can handle IP traffic, Ethernet, and ATM natively, and it is specified to support future transport protocols not yet invented. WiMAX allows the links to be dynamically

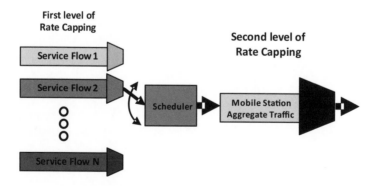

Figure 10.23. Two levels of rate capping: SF-based rate capping and mobile station aggregate-traffic-based rate capping.

configured, which smoothes the balancing act between raw capacity and quality on the fly. This capability improves WiMAX capacity and spectral efficiency. QoS is becoming even more important as new, rich devices are emerging that are capable of running multimedia applications such as video.

The WiMAX technology's rich end-to-end QoS framework allows operators to offer many of the existing services as well as the emerging services to their subscribers. This makes it possible for the wireless operator to attract more subscribers and also charge them based on the services delivered. In conclusion, both wireless operators as well as wireless subscribers benefit from the WiMAX QoS framework: for the operators, the revenue is increased, and for the subscribers may choose the services they desire and pay for what they use.

10.9 REFERENCES

[1] Air Interface for Fixed and Mobile Broadband Wireless Access Systems, IEEE 802.16e-2005.

[2] WiMAX Forum™ Mobile System Profile Release 1.0. Approved Specification (Revision 1.4.0: 2007-05-02), WiMAX Forum, 2007.

[3] WiMAX Forum Network Architecture Stage 2 and 3—Release 1.0 (Revision 1.2), WiMAX Forum, 2009.

CHAPTER 11

MOBILE WiMAX INTEGRATION WITH 3GPP AND 3GPP2 NETWORKS

POUYA TAAGHOL, PERETZ FEDER, AND RAMANA ISUKAPALLI

11.1 Introduction

In the evolution of wireless networks, it is essential that the new wireless technologies reuse the existing networks and provide interworking with legacy systems. Like any other wireless technology, 4G will also be deployed in phases and it will be important to fill the initial 4G coverage gaps with legacy 2G/3G access technologies to provide a ubiquitous and seamless user experience. The integration of emerging 4G access technologies (such as Mobile WiMAX) with the existing 2G/3G access networks (e.g., CDMA/EVDO, GPRS/EDGE, and UMTS/HSPA) into a common system architecture can be the first step in the migration toward mobile broadband networks that can provide users with the best connection at any place and at any time. Such integration allows the operators to reuse existing back-end systems, simplifies operation aspects (network management, customer acquisition, and converged billing), speeds up network deployment, enables access of independent services, and enables seamless intertechnology handover. However, such integration requires addressing all the interoperability issues that emerge when different access technologies are combined, such as the provisioning of a common authentication, authorization, and accounting (AAA) scheme, end-to-end quality of service (QoS), and a vertical mobility management mechanism.

The main requirement for achieving seamless integration of WiMAX and 3GPP/2 access networks is to minimize the handover interruption and preserve the QoS as the mobile station (MS) moves between mobile WiMAX and 3GPP/2 access technologies. The objective is to make the transition from one access network

WiMAX Technology and Network Evolution. Edited by Kamran Etemad and Ming-Yee Lai
Copyright © 2010 the Institute of Electrical and Electronics Engineers, Inc.

to another as transparent as possible to the user, that is, to offer a *seamless mobility* experience. With seamless mobility, users may exploit the availability of several access technologies to best meet their charging and QoS requirements, for example, by automatically selecting the "most appropriate" access network type based on some preconfigured preference settings and operator policies. At the same time, operators may exploit seamless mobility in order to offer compelling value-added, ubiquitous services as well as improve their network capacity and availability of services.

Seamless mobility can be achieved by enabling the MS to conduct seamless handovers across mobile WiMAX and 3GPP/2 access networks, that is, seamlessly transfer and then continue their ongoing sessions from one access network to another. A seamless handover is typically characterized by two performance requirements: (1) the handover latency should be no more than a few hundreds of milliseconds and (2) the QoS provided by the source and target access systems should be nearly identical (to sustain the same communication experience). These two performance requirements are not trivial to satisfy when mobile WiMAX and 3GPP/2 access networks are combined in a common or possibly single architecture. In order to offer seamless handover, several issues need to be addressed. One of them is how fast the service dataflow (i.e., the stream of data packets associated with an ongoing service) can be switched from the old path through the old access network to the new path through the new access network. Use of common or similar link-layer access network procedures (e.g., authentication, security, and mobility procedures) in both the source and the target access network can considerably expedite the handovers. One of the challenges associated with the integration of Mobile WiMAX and 3GPP/2 access networks arises from their differences in terms of AAA procedures, QoS mechanisms, and mobility protocols. Such challenges and associated solutions are the main focus of this chapter.

Interworking can be categorized into two approaches—a nomadic approach and a full-mobility approach. In the nomadic approach, session continuity between different access technologies is not required. That is, data sessions that exist in the networks of one access technology will not be carried over to the other technology when the user switches between the two. Hence, IP sessions that exist in the first network are terminated in the source system before the user enters the target system (the end user would, in fact, notice a service disruption in this kind of intertechnology handoff). However, the market demand for wireless broadband with full mobility is increasing with the emergence of mobile internet devices. Hence, it becomes essential to support interworking with full mobility and seamless session continuity as the user crosses boundaries of WiMAX coverage areas. In the full-mobility approach, the users will be able to maintain their IP sessions and have service continuity without experiencing any significant degradation in their services (like voice over IP or video on demand), other than the possible difference in access technology performance.

In this chapter, we present the current status of WiMAX interworking with 3GPP/2 systems and address the technical issues around the integration of WiMAX into the existing 2G/3G networks. Our goal is twofold. First, we describe and explain the key issues that emerge from the effort to integrate Mobile WiMAX in a mobile network architecture that must also support the legacy 3GPP/2 radio access technologies (i.e., UMTS/HSPA and CDMA/EVDO) to enable the nomadic interworking. Second, we present solutions that enable the MS (with limited RF coexistence capabilities) to attach to and seamlessly traverse between WiMAX and 3GPP/2 access networks using full-mobility interworking. For each of the 3GPP and 3GPP2 systems, we first provide a brief overview of the network architectures and then we address the various aspects of interworking, such as integrated AAA mechanisms, consistent QoS support, seamless vertical handover, and interoperator roaming. The material assumes readers' familiarity with the basic WiMAX network architecture and IMS and PCC integration, which are described in previous chapters.

11.2   WiMAX–3GPP INTERWORKING

As part of Release 8 of the 3GPP specifications, the 3GPP group has been studying and specifying an evolved packet core (EPC) under the system architecture evolution (SAE) work item. The 3GPP EPC has built-in support for interworking with the trusted non-3GPP systems using IETF protocols such as AAA and Mobile IP. In other words, the EPC and WiMAX CSN are converged in many aspects, making the interworking simpler. However, interworking with Release-8 3GPP means that the operator has to adopt the new 3GPP EPC architecture and upgrade its core network accordingly. The WiMAX Forum NWG has recently started a work item to specify an overlay solution that allows interworking with pre-Release 8 3GPP systems without requiring an upgrade to the legacy core network. This chapter presents 3GPP interworking solutions with pre-Release-8 EPC and pre-Release-8 GPRS/EDGE core networks.

11.2.1   Interworking with 3GPP Release-8

11.2.1.1   Architecture of 3GPP Release-8 EPC

Release-8 3GPP introduces the evolved packet system (EPS) concept which is composed of a new radio access network, called evolved UTRAN (E-UTRAN), and a new all-IP core network, called evolved packet core (EPC). The evolved packet core can be considered an evolution of the legacy GPRS architecture with additional features to improve performance, support the E-UTRAN access network, and support integration with non-3GPP radio technologies such as WLAN and WiMAX. As shown in Figure 11.1, access to the evolved packet core is supported not only via 3GPP-specific access networks (e.g., E-UTRAN, UTRAN, and GERAN) but also via non-3GPP access networks, such as WLAN and WiMAX technologies.

   In order to demonstrate how a Mobile WiMAX access network can seamlessly be integrated into a 3GPP network, we first provide a brief overview of the evolved 3GPP network architecture shown in Figure 11.1.

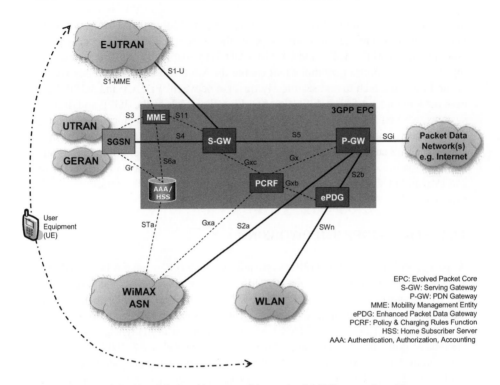

Figure 11.1. Simplified architecture of the evolved 3GPP network architecture.

As shown in Figure 11.1, a number of diverse access networks such as WLAN, WiMAX, GERAN, UTRAN, and E-UTRAN are connected to a common core network (the evolved packet core—EPC) based on IP technology through different interfaces. All 3GPP-specific access technologies are connected through the serving gateway (S-GW), whereas all non-3GPP-specific access technologies are typically connected through the packet data network gateway (P-GW) or the evolved packet data gateway (ePDG), which provide extra security functionality for untrusted access technologies (such as legacy WLANs with no strong built-in security features). The serving gateway (S-GW) acts as the mobility anchor for mobility within 3GPP-specific access technologies, and also relays traffic between the legacy serving GPRS support node (SGSN) and the PDN gateway (P-GW). In the case of E-UTRAN, the S-GW is directly connected through the S1-U interface, whereas SGSN is the intermediate node when GERAN/UTRAN is used. It is important to mention that a mobility management entity (MME) is also incorporated in the architecture for handling control functions such as authentication, security, and mobility in idle mode.

For access to EPC through WLAN and WiMAX, different data paths are used. A WiMAX ASN is directly connected to a P-GW through the S2a interface as a trusted access network. On the other hand, a WLAN is accessible through the evolved packet data gateway (ePDG) as an untrusted access network. All data paths from the access networks are combined at the P-GW, which incorporates functionality such as packet filtering, lawful interception, charging, UE IP address allocation, and routes traffic over SGi to an external packet data network or to the operator's packet data network (for accessing IP services provided by the operator). Apart from the network entities handling data traffic, the EPC also contains network control entities like the home subscriber server (HSS) for keeping user subscription information, and the authentication, authorization, and accounting (AAA) server for determining the identity and privileges of a user, tracking his/her activities, and enforcing charging and QoS policies through a PCC architecture.

11.2.1.2 WiMAX Integration with 3GPP EPC

The main requirement for achieving seamless integration of WiMAX and 3GPP access technologies is to provide an appropriate authentication infrastructure, optimize the handover interruption, preserve the quality of service (QoS) as the mobile device moves between mobile WiMAX and 3GPP access technologies, and enable interoperator roaming. With seamless mobility, users may exploit both access technologies to best meet their charging and QoS requirements, for example, by automatically selecting the most appropriate access network based on operator policies. At the same time, operators may exploit seamless mobility in order to offer compelling value-added services as well as improve their network capacity and availability of services.

11.2.2 WiMAX–3GPP EPC Interworking Architecture

The 3GPP evolved packet core exposes generic, IP-based interfaces toward the non-3GPP access networks (e.g. Mobile WiMAX). Figure 11.2 demonstrates the logical interfaces connecting Mobile WiMAX and the 3GPP EPC. We note that this architecture is largely based on the specified 3GPP specifications (in particular, reference [5]) but it also features some enhancements that we introduce in this chapter in order to support the optimized handover mechanism that we propose later in the following sections.

There are four major logical interfaces in the architecture of Figure 11.2:

1. 3GPP STa (equivalent of WiMAX R3-AAA interface [6]). Used for AAA-based authentication of user equipment (UEs) and enforcement of preconfigured QoS.
2. 3GPP Gxa (equivalent of WiMAX R3-PCC-P [6]). Used for enforcement of dynamic QoS and charging rules.
3. 3GPP S2a (equivalent of WiMAX R3-MIP [6]). Used for layer-3 mobility and bearer establishment toward the core network.
4. 3GPP S14/R9: Used for intertechnology network discovery and selection, and for facilitating the optimized WiMAX–3GPP handover (the latter capability is an extension of the S14/R9 logical interface specified in [6]). This in-

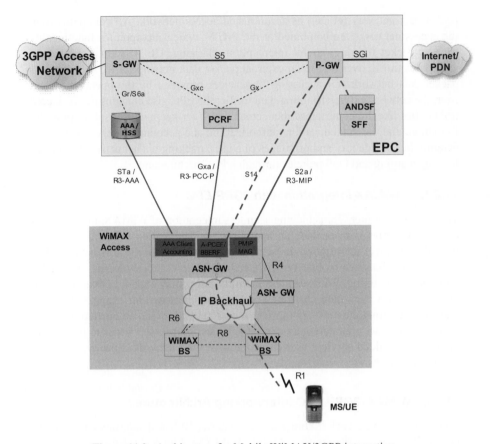

Figure 11.2. Architecture for Mobile WiMAX/3GPP integration.

terface, as well as the functional elements ANDSF [16] and SFF [15] (shown in Figure 11.2), are discussed in the next section.

11.2.3 Access Network Discovery

As the dual-mode mobile terminal (UE/MS) moves across the network, it has to discover other radio technologies available in its vicinity, which could potentially be more preferable than the currently used radio technology. For example, an MS using 3GPP 2G or 3G radio access needs to discover when a Mobile WiMAX access network becomes available and possibly could trigger a handover to the Mobile WiMAX network if it were more preferable to the user and/or the operator, or if the radio signal from its serving 3GPP cell deteriorates significantly. In the most simplistic case, the UE/MS can discover neighbor cells with no assistance from the network by periodically conducting a radio scan in the background. Although this is very simple and does not require any modifications in the network, the problem is that (1) battery consumption can increase considerably, especially when we demand fast discovery, (2) the information discovered about the neighbor cells is only limited, and (3) the UE/MS needs to have two receivers working in parallel (one dedicated to scanning and another for ongoing communications). This drives the need for access network discovery assisted by the network.

The most typical network-assisted discovery solution used today is to have each cell in the network broadcast a list of neighbor cells (of the same or different radio technology) that can serve as candidates for handover. The same concept can be applied to neighbor cell discovery in the integrated WiMAX/3GPP network, provided that all legacy 2G/3G cells are upgraded in order to broadcast information about neighbor WiMAX cells (and vice versa). In addition, when the mobile device is equipped with a single receiver, we must make sure that the radio signal received from neighbor cells is measured without missing any data from the serving cell. To ensure that, the serving base station needs to schedule measurement opportunities to a mobile device, that is short-time windows in which the device can safely leave its serving (say) 3GPP cell and measure neighboring WiMAX radio frequencies. This solution for neighbor cell discovery creates a need for intertechnology measurement scheduling and coordination (unless all UEs/MSs in the network implement dual receivers).

Alternatively, if there is need to minimize the modifications of legacy radio systems, the neighbor cell information may not be broadcast on radio channels but may instead be retrieved by the UEs from a special functional entity in the network. Such an entity has been standardized by 3GPP in order to facilitate discovery of WiMAX cells and is termed the access network discovery and selection function (ANDSF) [16] This function can be considered as a dynamic database that is queried by mobile devices (e.g., with a specific protocol over IP) whenever they need to discover neighbor cells of specific or any radio technology type. It is easy to recognize that such a discovery solution avoids any impact on the radio systems and the associated cost for upgrade. The ANDSF can also be used to provide dynamic operator policies to the mobile devices, which can affect the device's behavior based on some dynamic operator's preferences and rules. In addition, the ANDSF can provide information about neighbor cell attributes such as QoS capabilities, service capabilities, charging rate, and a number of other attributes, which cannot be continuously broadcast on radio channels due to the high radio capacity demand. The mobile device can combine this additional information together with some user-provided data to select the most preferable radio access technology (WiMAX or 3GPP access technology) that satisfies both user preferences and operator policies.

11.2.4 Initial Network Entry

Initial network entry is a process in which the newly arrived mobile station/user equipment (MS/UE) is required to access the network services. Figure 11.3 demonstrates the high-level procedure for the initial WiMAX network entry via the 3GPP EPC.

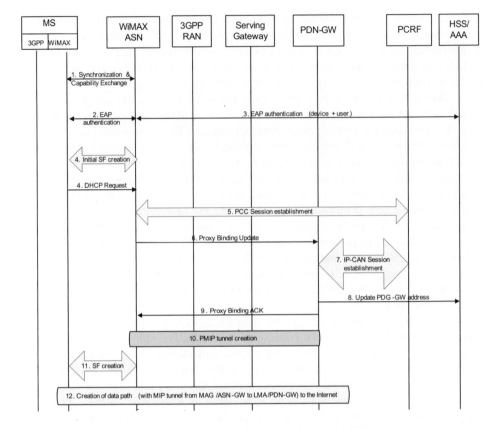

Figure 11.3. WiMAX initial network entry via EPC.

In the first step, the mobile device is required to synchronize itself with the BS and exchange basic capabilities, after which the EAP-AKA procedure is triggered for authenticating the identity presented by the MS/UE. As specified in [5], the user network access identifier (NAI) is constructed per RFC 4282 [20] using the IMSI (international mobile subscriber identify) in the USIM, MNC (mobile network code), and MNC (mobile country code) as "0<IMSI>@mnc<MNC>.mcc<MCC>. 3gppnetwork.org," for EAP-AKA authentication. For example, for EAP-AKA authentication, if the IMSI is "234150999999999 (MCC = 234, MNC = 15), the root NAI then takes the form 0234150999999999@mnc015.mcc234.3gppnetwork.org." After the successful authentication, a three-way handshake is performed (with the PKMv2 protocol) between the MS/UE and BS for exchanging the security keys used over the air interface. Subsequently, the registration process begins, which triggers data path establishment between the BS and the ASN-GW, and then a service flow is set up according to the preprovisioned subscriber profile that is downloaded from the AAA server to the ASN-GW during the authentication phase. At this point, the MS/UE assumes that the layer 2 connection is established and its operating system triggers the Dynamic Host Configuration Protocol (DHCP) procedure to obtain an IP address. In turn, this triggers a bearer establishment between the ASN-GW and the P-GW by means of the Proxy MIP Protocol (accomplished with the proxy binding update and proxy binding update acknowledge messages). During this procedure, the PCRF may be contacted to retrieve policy parameters for the P-GW and ASN-GW (especially if this initial network attachment was a result of intertechnology handover).

As can be noted from the procedure above, the 3GPP evolved packet core enables a MS/UE to attach to core network resources via WiMAX by means of an AAA server, a PMIP interface between the WiMAX ASN-GW and P-GW (called S2a), and, optionally, a PCC infrastructure for providing dynamic policy and charging rules to the elements policing the user data traffic (typically, the ASN-GW and P-GW).

11.2.5 Dynamic QoS and Policy Control

An important issue for providing seamless mobility is to maintain a specific level of QoS consistently across the WiMAX and the 3GPP radio access networks. This involves several considerations such as the QoS mappings and semantics on the two access networks as well as appropriate resource allocations. WiMAX allows creation of multiple service flows, each with its specific quality of service. This is similar to the 3GPP PDP-context activation procedure. In addition to this, for equivalent user experience on both networks, classification rules will have to be kept common across both access networks by coordinating the appropriate QoS rules.

QoS consistency between WiMAX and 3GPP access networks can be provided via the PCC infrastructure. Definition of the PCC framework and procedures for EPC [6] in an access network-agnostic form greatly simplifies the mapping functions required for quality of service. In 3GPP EPC, it is expected that the charging is performed in the EPC and only the 3GPP bearer binding and event reporting function (BBERF) is required from the RAN (the BBERF is just a subset of the functions present in the WiMAX A-PCEF).

11.2.6 Interoperator Roaming

The roaming architecture for WiMAX–3GPP integration is shown in Figure 11.4. It supports interoperator roaming, which is an important feature that enables operators to expand coverage regionally and/or globally. The 3GPP EPC allows for roaming interfaces for exchanging operator's policies (S9), authentication exchanges (SWd), routing of user traffic to the home operator (S8), or routing user traffic to the visited network (this is called local breakout, LBO) for accessing local services or the Internet. For broadband Internet traffic, it is more cost-effective to route the traffic toward the Internet at the visited network (LBO) to avoid costly interoperator data transports. In case user traffic is routed to the home network, the S8 logical interface can support mobility either with the Proxy MIP Protocol or with the GPRS Tunneling Protocol (GTP), as shown in Figure 11.4.

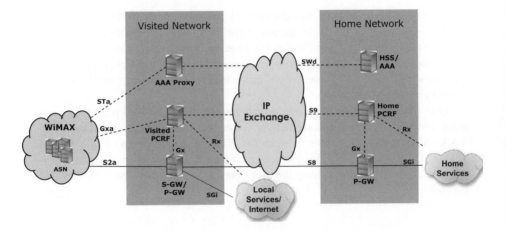

Figure 11.4. Roaming architecture for mobile WiMAX–3GPP integration.

11.2.6.1 Protocol Stack

The protocol stacks of the control and user planes are depicted in Figures 11.5 and 11.6. Figure 11.5 shows the end-to-end user plane where the user traffic is tunneled using the GRE protocol (the S2a GRE/IP tunnel is set up as a result of the PMIP procedure). Figure 11.6 demonstrates the control plane protocol stacks for the STa and S2a interfaces.

Authentication in Mobile WiMAX is carried out using the extensible authentication protocol (EAP) framework shown in Figure 11.7. In the IEEE 802.16e specification [9], the EAP exchanges are transported between the MS and BS within privacy key management (PKMv2) MAC messages.

WiMAX supports two types of authentication: device-level and user-level. The device-level authentication is based on EAP-TLS (see [18]) and allows an operator to verify compliance of a device and its certificate with WiMAX compliance requirements (this feature empowers an operator to adopt a retail device model). The

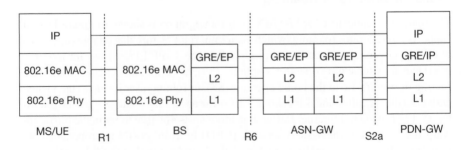

Figure 11.5. End-to-end user plane protocol stack for WiMAX–EPC interworking.

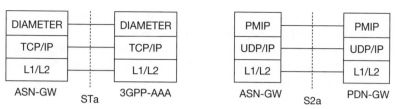

Figure 11.6. Control plane for WiMAX–EPC interworking.

user-level authentication is based on EAP-TTLS [21] or EAP-AKA (see [19]) and it is a way to authenticate the presented user's subscription. In the WiMAX network, the authentication is performed using an AAA framework, with the access protocols of RADIUS or DIAMETER. The ASN-GW hosts the AAA client, and the AAA network allows the home network to authenticate the client. We assume here that IP-address allocation is performed using DHCP, which is the case when mobility is provided with either proxy mobile IP or simple IP.

11.2.7 WiMAX–3GPP Seamless Mobility

As noted before, it is important to provide a seamless mobility experience to the users, that is, enable mobility across mobile WiMAX radio sites and other 3GPP radio sites in a transparent fashion when the two access networks are integrated. From the user's perspective, the seamless experience conceals the heterogenity of the systems and the conceived experience is of an intelligent system capable of manipulating its available resources so as to provide the best service to the user without any intervention by the user. This kind of experience is typically provided, for example, by all kinds of integrated 2G/3G networks, which facilitate the user's mobility across 2G cells and 3G cells in a transparent fashion. From the operator's perspec-

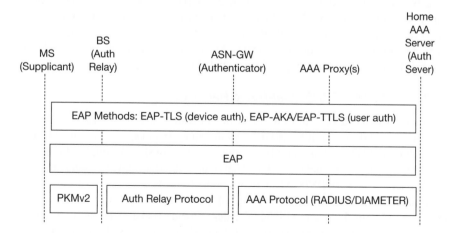

Figure 11.7. Mobile WiMAX authentication framework.

tive, the seamless experience is simply interpreted as improved key performance metrics such as availability, drop rates, and throughput.

There are, however, several issues to be resolved before an integrated mobile WiMAX/3GPP system can provide a seamless mobility experience. Some of these issues are further discussed below.

11.2.7.1   Handover Decision

Assuming that a dual-mode device is using a 3G cell and has discovered (either autonomously or with assistance from the network) its neighbor WiMAX cells, the next question is whether the device (UE/MS) needs to take any actions, that is, whether it needs to initiate a handover to a discovered WiMAX cell. This decision can be made either by the device or by the network. On one hand, it is beneficial when the handover decision is made by the network because the network can redirect the UE/MS to another radio site or frequency that has enough capacity to handle its ongoing communications. The network can also coordinate the mobility of all the dual-mode devices in a way that overall traffic is evenly distributed across all radio resources, congestion is minimized, and the total throughput is maximized. On the other hand, however, the radio network may lack some parameters that impact the handover decision, such as user preferences, the exact type of services active on the UE, and some operator policies pertaining to mobility between mobile WiMAX and 3GPP access networks (e.g., "do not allow handover to 3GPP when a video streaming service is active"). The Rel-8 3GPP specifications mandate that the UE make the decision for handover between 3GPP and mobile WiMAX, which features some key advantages. First, the UE can make the handover decision based on its up-to-date radio measurements, preconfigured user preferences, and all operator mobility policies downloaded from ANDSF. Second, the UE does not need to send any intertechnology radio measurement to the network. Third, the impact on the 2G/3G and mobile WiMAX access networks is minimized because there is no need for a 3G radio access to receive measurement reports about WiMAX cells and make handover decisions for WiMAX. In addition, there is no need for a 3G radio access network to keep track of the available radio resources on the WiMAX side and vice versa. Of course, this approach minimizes the coupling between the 3GPP and mobile WiMAX access networks and the associated upgrade cost.

11.2.7.2   WiMAX–3GPP Dual-Radio Handover

The dual-radio handover call flows in Figure 11.8 show a seamless handoff from a WiMAX network to a 3GPP network. As the name implies, a dual-radio MS has two radios, and both radios can fully operate simultaneously. The dual-radio device can enter a target access network (3GPP network in the figure) while being active in another access network (WiMAX network in the figure).

Initial network entry of an MS to a WiMAX network using an EPC core network in shown in Figure 11.3. The MS then decides to hand off to a 3GPP network, based on the configuration parameters for such handoff. Figure 11.8 shows a make-before-break type of a handoff. That is, the MS completes the network entry to a second (here 3GPP) network before it exits out of the original (WiMAX) network.

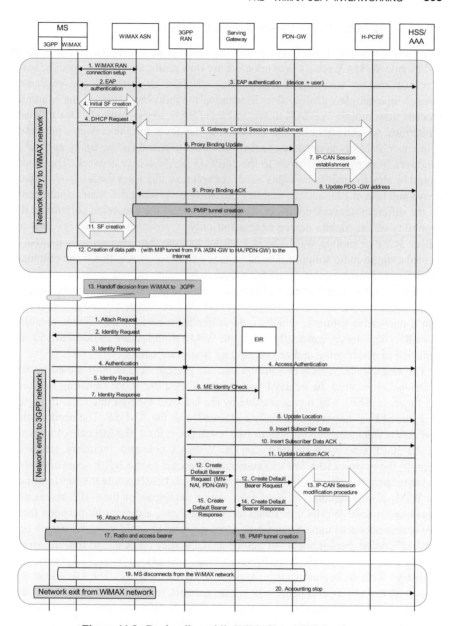

Figure 11.8. Dual-radio mobile WiMAX-to-3GPP handover.

Steps 1–18 show a network entry to the 3GPP network, whereas steps 19 and 20 show the network exit from the WiMAX network. We believe that initially 3GPP/WiMAX dual-radio devices will be used for the dual-mode access but over time the devices will be replaced with a single-radio alternative with which only one transmitter can be active at any given time, hence reducing the battery consumption and the cost of the device.

11.2.7.3 WiMAX–3GPP Single-Radio Handover

After discussing the above issues pertaining to the seamless mobility across 3GPP and mobile WiMAX access networks, we are now ready to present a specific solution that enables seamless mobility. This solution is presented and explained through an example signaling flow illustrating the individual steps of the handover process from mobile WiMAX to the 3GPP UTRAN access network. Due to space limitations, we consider only one direction of handover but the concepts presented below can be easily generalized to the other direction and can be easily applied to handover from mobile WiMAX to the 3GPP GERAN access network. The reader would be able to identify how this handover solution addresses the issues discussed above and how it can achieve seamless mobility by placing different requirements on the different network elements. The key aspect of the presented solution is the capability of the mobile device to transmit only on a single radio access network (either 3GPP or mobile WiMAX) at a particular time. For obvious reasons, this is termed a single-radio solution. The advantage of this solution is that it can eliminate the RF coexistence issues faced by a dual-radio solution and, consequently, improve the handover performance. However, this solution also requires more intelligence in the mobile device and in the network. We considered in the previous section a dual-radio solution, which could suffer from performance issues resulting from RF coexistence, especially when the WiMAX and 3GPP radio networks are deployed in neighboring and/or overlapping frequency bands.

The single-radio handover solution that we present here is built around a new functional element in the evolved packet core, called the 3GPP signaling forwarding function (SFF), which is accessible by the UE over the S14 interface. In Figure 11.2, the SFF is shown as optionally collocated with the ANDSF. The need for this new functional element comes from two restrictions. First, the MS cannot transmit on the 3GPP side while operating on the WiMAX side and, therefore, needs an agent in the network (the SFF) to authorize its access to the 3GPP access network and to prepare the appropriate 3GPP resources on its behalf, while the MS is still on the WiMAX side. Second, we need to minimize the impact on the 3GPP access networks and mobile WiMAX as much as possible; thus, a direct link between these two access networks must be avoided. Also, we need to avoid WiMAX access network scheduled measurements of neighbor 3GPP sites on behalf of the UE, so we assume that the UE conducts 3GPP radio measurements on its second receiver (if available) without any assistance from the WiMAX network. The UE communicates with the SFF over a generic IP access network (e.g., the mobile WiMAX access). From the 3GPP access network perspective, the SFF emulates a simplified radio network controller (RNC) base station controller (BSC) for the case of UTRAN and GERAN access, respectively, whereas from the mobile WiMAX access network perspective the SFF emulates a WiMAX ASN-GW base station. The single-radio handover solution is schematically depicted in Figure 11.9 and is further explained below.

The UE/MS uses a source mobile WiMAX access system and at the core is being served by a P-GW. It is assumed that the mobile WiMAX network supports

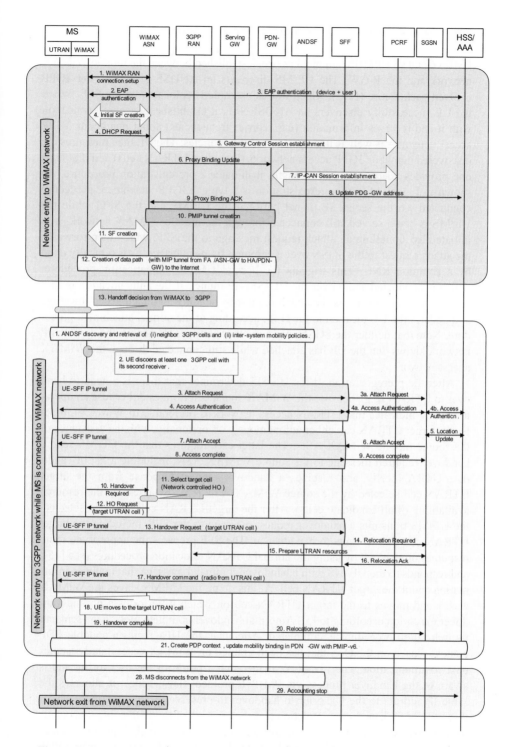

Figure 11.9. Seamless single-radio handover from Mobile WiMAX to the 3GPP UTRAN access network.

PMIPv6 and so a PMIPv6 tunnel has been established between the mobile WiMAX network and the P-GW. The UE/MS discovers an ANDSF in the serving 3GPP evolved packet core (EPC), which in this example is colocated with a SFF. When the UE successfully discovers an ANDSF/SFF, it establishes a secure connection with it and receives information (e.g., carrier frequencies) about neighbour 3GPP access networks as well as intersystem mobility policies. The UE then measures the discovered neighbor 3GPP access networks (including UTRAN cells) and if at least one provides adequate signal strength, it initiates a preregistration procedure. As shown in Figure 11.9, the preregistration is a typical 3GPP attachment procedure conducted via the secure IP tunnel between the UE/MS and the SFF, while the UE/MS is being served still connected through the mobile WiMAX network. It is initiated by tunnelling an attach request message to the SFF, which then forwards the attach request to the SGSN over a common Iu–ps interface, as is typically done by a common RNC. This triggers the normal UMTS authentication procedure, which is conducted again over the UE–SFF IP tunnel. If the authentication is successful, the SGSN accepts the attach request by sending an attach accept message and updates the UE location in the HSS, according to the normal attachment procedure. Note that neither the SGSN, nor any other network element in the core 3GPP network knows that the UE has attached while still connected through the WiMAX access network.

After the preregistration, if the UE/MS determines that there is a need to hand over to UTRAN (e.g., because the WiMAX signal deteriorates or because an operator policy indicates that UTRAN access network is preferable to WiMAX), it selects a target UTRAN cell either autonomously or, optionally, with the assistance of the source mobile WiMAX ASN. The latter case requires that the UE/MS send a handover required message to the source WiMAX ASN, including a list of candidate UTRAN cells, and receives a handover request response from the target UTRAN cell selected by the source WiMAX ASN (e.g., by assessing the resource availability of all candidate cells). After the target UTRAN cell has been selected, the UE/MS transmits a handover request message to the SFF, including the target UTRAN cell that it wants to handover to. The SFF prepares the appropriate radio resources on that cell by using the normal UTRAN relocation procedures (see [15]) and responds to the UE/MS with a handover command message that includes information about the target UTRAN cell. At this point, the UE/MS leaves the WiMAX access and moves to the target UTRAN cell on which it performs a normal handover execution according to [5]. A normal handover complete message is sent and the relocation procedure is completed. After that, the UE creates a suitable PDP context (based on the QoS that has been granted on WiMAX access) and resumes data communication. The PDN-GW is the anchor point for the user traffic and is aware of the handover process via PMIPv6 protocol flags; hence, it allocates the same IP address to the UE prior to handover to preserve session continuity. Note that the UMTS handover completion and the creation of the PDP context accounts for a relatively short interruption in the data transmission, which in practice ranges up to a few hundreds of milliseconds. Note also that after the creation of the PDP context, the PMIPv6 tunnel in the P-GW is relocated from the WiMAX ASN-GW to the SGSN. It is important to note that with a single-radio handover the network controls the resource allocation in the target system including the data bearer paths. Hence, PMIP is more suitable for single-radio solutions compared to client-based MIP protocols.

11.2.8 Interworking with Pre-Release-8 3GPP

This section describes the recently started WiMAX Forum work item addressing core network aspects of interworking between WiMAX and pre-Release 8 3GPP Packet Switched Core Network. This work item will provide architecture for seamless data mobility between WiMAX subscribers and a large base of existing legacy 3GPP packet data subscribers, without impact on deployed 3GPP network elements while supporting both dual and single-radio configurations. At the time of this writing only dual-radio architecture has been defined and that will be addressed here.

11.2.8.1 *Deployment Model*

The following deployment models are possible with this architecture:

- A 3GPP operator also owning a WiMAX network. In this model, the common operator may continue utilizing the 3GPP core authentication infrastructure, home location register (HLR), as the main repository of authentication credentials. The common operator will also deploy a WiMAX AAA. Alternatively, the common operator may provision the dual-mode MS with a second subscription specific to the WiMAX access network and retain this second subscription only at the WiMAX AAA.
- A WiMAX operator who wants to provide 3GPP access at the network edge and, therefore, establishes a partnership or roaming agreement with a 3GPP operator: In this model, each operator will conduct an independent authentication of the MS by their respective systems and the MS will have to be provisioned with an independent dual set of access credentials. Alternatively, the WiMAX operator can use the 3GPP HLR as the main repository of authentication credentials for its subscribers. When the MS attempts to access the WiMAX network, the WiMAX AAA will query the authentication parameters from the 3GPP HLR in order to conduct the WiMAX EAP authentication.

11.2.8.2 *Authentication Mechanism*

Unlike Release-8 3GPP architecture, pre-Release-8 3GPP architecture does not have the EPC network. However, just like EPC interworking, the solution must use the existing authentication infrastructure of both access technologies and optimize the handover interruption period by preauthenticating through one access network before switching to the other. Likewise, the MS needs to preserve the QoS attributes of all the active sessions during handoff.

In order to use the authentication infrastructure, the MS must identify itself with the proper subscription credentials. Depending on the deployment model, the MS may have either two independent sets of credentials, one each for 3GPP and WiMAX networks, or a single active set of credentials with the home 3GPP network. The possibility of a single active set of credentials with the WiMAX network is not considered since it entails a core 3GPP legacy change.

The dual-mode UE/MS may maintain two independent sets of credentials: one set with the pre-Release-8 3GPP HLR for access to the pre-Release-8 3GPP network and another set with the WiMAX AAA for access to the WiMAX network. In the case of a dual set of credentials, each accessed network conducts its own access authentication, for example, SIM/USIM for the 3GPP and EAP-TTLS for the WiMAX access. In the case of a single set of credentials, the MS is authenticated with the 3GPP HLR using SIM/USIM and on the WiMAX side it is authenticated with the WiMAX AAA, which uses the HLR as the source of authentication parameters for conducting the EAP access authentication.

11.2.8.3 Network Reference Model for Interworking Architecture

Figure 11.10 shows the nonroaming interworking architecture reference model. A new pre-Release-8 interworking node (IWK) is introduced, which interfaces with the GGSN through the Gi interface and R3 to the WiMAX CSN.

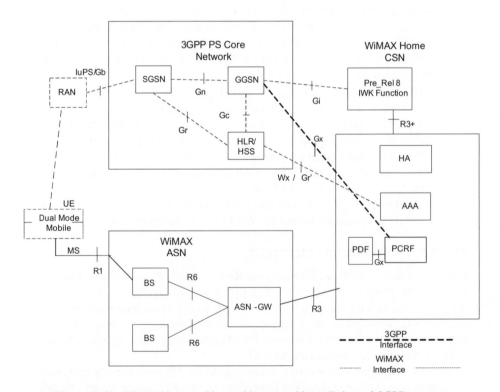

Figure 11.10. WiMAX interworking architecture with pre-Release-8 3GPP systems.

The IWK function is composed of a AAA proxy, proxy MIP client, and a foreign agent (FA). It connects with the GPRS network using the Gi interface and to the CSN using the R3 interface defined in WiMAX. The IWK function learns about the GGSN when its address is provided in the AAA request message address and simply adds a routing address through the GGSN toward the MS over the Gi. The HLR location is a deployment choice and can also be located as part of the home CSN.

11.2.8.4   Handover Impacts

Figure 11.11 provides the call flow describing the dual-radio entry into the WiMAX and 3GPP networks. This shows an initial network entry into a WiMAX network, followed by handoff to a 3GPP network and a subsequent network exit from the WiMAX network. This model shows the case of a single subscription in which the AAA gets the credential from the HLR (this is typically the case in the EAP-AKA authentication method), as shown in step 3a of Figure 11.11. The NAI is the parameter that ties the two sessions together at the same anchoring point (HA). This NAI is based on the IMSI information that is common to the user subscription (single or dual subscriptions) and is forwarded to the HA in WiMAX entry (step 5). Similarly, 3GPP network entry includes a Mobile IP binding update for the dual-mode MS for the same unique NAI. The pre-Release-8 3GPP IWK obtains the IMSI from the GGSN and creates a NAI-based IMSI for the registration in 3GPP (step 7). Once a MIP tunnel is created between the HA and FA in the IWK function, the IWK function routes the bearer traffic to the GGSN and from the GGSN it is delivered to the MS.

Once handoff from the WiMAX network is completed, the MS disconnects itself from the WiMAX network and stops the WiMAX session accounting, as indicated in steps 28 and 29 of Figure 11.11.

11.2.8.5   Policy and Charging Impacts

Policy and charging control (PCC) functionality in a WiMAX–pre-Release-8 3GPP interworking architecture is currently under specification in the WiMAX forum. The goal is to reuse the existing pre-Release-8 3GPP and WiMAX R1.5 PCC Release 7 architectures to provide an integrated PCC framework for both networks. The policy and charging enforcement function (PCEF) entities in each network are located in the GGSN and in the ASN-GW. The PCEF in the GGSN connects to the PCRF over the existing pre-Release-8 3GPP Gx interface, whereas the PCEF in the ASN-GW connects to the PCRF via the PDF over the PCC-R3-P interface, as defined in WiMAX R1.5. pre-Release-8 3GPP PCC assumes a single, fixed PCEF for the lifetime of an IP session, that is, the traffic plane for each IP session passes through a single PCEF and there is no support for relocation of a PCEF during a session. In an interworking architecture, however, the PCEF associated with the user session changes during a handover. Therefore, the PCRF entity has to becomes mobility aware in order to support PCEF relocation across the two networks during handover. If the PCRF and PDF combination becomes mobility aware, they are

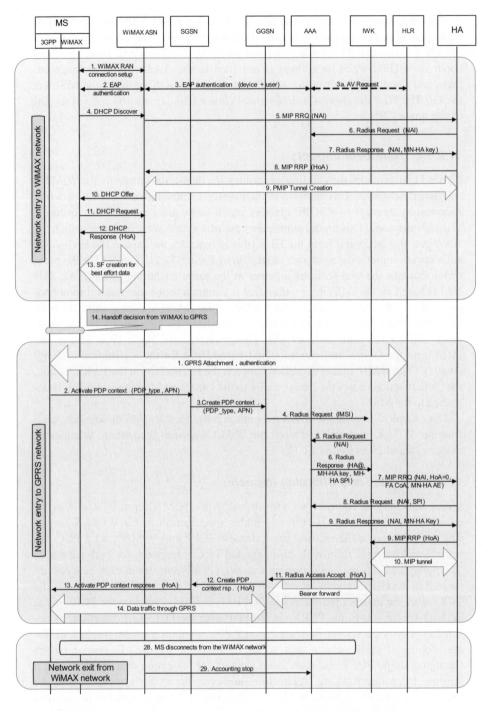

Figure 11.11. WiMAX-to-pre-Release-8 3GPP systems handoff call flow.

able to correlate the Gx DIAMETER session with the GGSN to the PCC-R3-P DI-AMETER session with the ASN-GW. After handoff, the PCRF installs the same policy and charging rules when applicable.

11.3 WIMAX–3GPP2 INTERWORKING

In this section, we discuss the interworking between WiMAX and the 3GPP2 Code Division Multiple Access 2000 (CDMA-2000 1x EV-DO (evolution data optimized), hereafter called EVDO networks. The dual-mode MS/AT used for this interworking behaves as any other WiMAX-only device in a WiMAX network and an EVDO-only device in an EVDO network. That is, it complies with both the WiMAX standards [12] and EVDO standards [14] for network discovery and selection, network entry, authentication, and mobility when it operates in either one of these networks. We address the full-mobility approach and discuss interworking using Mobile IP [11]. We describe a loosely coupled network architecture in which the EVDO core network, with its network elements like home agent (HA) and authentication, authorization, and accounting (AAA) server are shared by both the WiMAX and EVDO access networks. This enables common billing and authentication for both networks using the same accounting and billing servers. This model is ideally suited for service providers who use a phased approach when evolving from third-generation to fourth-generation networks.

11.3.1 The 3GPP2 Network Architecture

The EVDO network architecture model is detailed in reference 14. The access network consists of mainly base stations and a radio network controller (RNC), whereas the core IP network has a packet data serving node (PDSN) that hosts the FA and HA. EVDO specifies a point-to-point protocol (PPP) layer between the MS and the PDSN for both default packet and multiflow packet applications. An EVDO mobile station or access terminal (MS/AT) always performs MIP registration following a PPP negotiation. This implies that the device must trigger a PPP negotiation that makes the link setup time on EVDO a little longer than in WiMAX. Additionally, EVDO has the concept of the unicast access terminal identifier (UATI) session and a PPP session. The lifetime of a UATI session can be longer than a PPP/MIP session, but a PPP session can only exist if the MS already has a UATI session. This is an important aspect because the air-interface session setup (including negotiation on session parameters and protocols) takes about 3 to 5 seconds on average and is considered a time-consuming procedure. Preestablishing an air-interface session and leaving it idle (without an active PPP/MIP session) while the MS is active in a WiMAX network would be a desirable way to reduce handoff time. This has some implications for the dual-mode device behavior; assuming that EVDO coverage is everywhere and WiMAX coverage is spotty, the device could periodically register with the EVDO network to make sure that its UATI session is kept alive to avoid extra handoff time. However, monitoring and transmitting with both technologies consumes more battery power on the MS/AT. This is a trade-off between the device's battery life coupled with additional complexity and handoff performance.

11.3.2 WiMAX–3GPP2 Interworking Architecture

A network architecture that supports full mobility with session and service continuity across WiMAX and EVDO networks is shown in Figure 11.12. The figure shows WiMAX and EVDO access networks sharing one IP core network. This is a loosely coupled model because the WiMAX and EVDO networks have separate and independent data paths to the core network. Each network follows its unique network entry procedures, authentication methods, intratechnology mobility, paging, and so on. As shown in the figure, the WiMAX and EVDO networks are connected to a common IP core network. An end user can use the same application services (for example, video on demand) in either one of the two access networks, since the two access networks have access to the same applications through the common IP core network. The data paths are separate for WiMAX and EVDO access networks. This architecture supports full mobility across the two access networks during intertechnology handoffs by maintaining a MIP tunnel between the HA in the core network and the FA in the access network. Figure 11.12 also

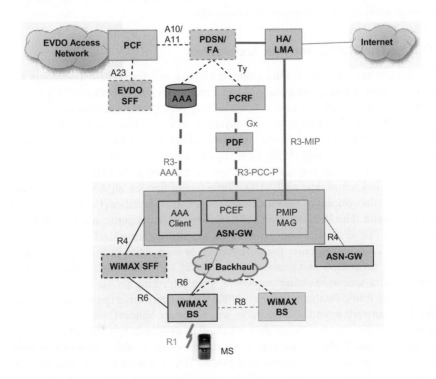

Figure 11.12. WiMAX–3GPP2 interworking network architecture.

shows two components, WiMAX SFF and EVDO SFF, that are used for single-radio MS.

The mobile IP protocol is used to reach any MS using a unique IP address assigned to it in its home network by the HA. The MS roaming in a foreign network registers itself through a foreign agent with the home agent. If the registration is successful, the FA assigns a temporary care-of address (CoA) to the MS. The HA creates a MIP tunnel to the FA. That is, the HA creates an extra IP header for the care-of address of the MS over the IP packets addressed to the MS's home address (HoA). Any correspondent node can still reach the MS by its home address. The packets originating from the CN are routed to the HA and then through the MIP tunnel to the FA and to the MS. Hence, the MS can still be reached by its home address in the foreign network.

The concept of MIP tunnels can be extended to enable interworking during handoff across different access networks. For example, whenever an MS enters a WiMAX network from an EVDO network, the common HA creates a new MIP tunnel with the FA in the WiMAX network. Any incoming packets of any ongoing applications (like video on demand, voice over IP, etc.) to the MS are now directed by the HA through the MIP tunnel with the FA in the WiMAX network. Similarly, whenever the MS enters an EVDO network from a WiMAX network, the common HA creates a MIP tunnel with the FA in the PDSN and tears down the tunnel with the FA in the ASN-GW. The existing IP sessions between the MS/AT and the core are still active. The end result is that there is IP session continuity and, hence, service continuity to the end user during the intertechnology handoffs.

Intertechnology handoffs can be implemented in three different ways. The first of these is break before make, in which the link in the serving access network is torn down before a new link is set up in the target access network. The second is make before break, in which the link in the target access network is made first before it is torn down in the serving access network. The third is make before make with simultaneous bindings, in which the link in the serving and target access networks are maintained for a brief period of time before the link in the serving network is torn down. Make before break with simultaneous bindings provides the best service continuity with minimum packet loss during handoffs, followed by make before break and then break before make. However, there are additional algorithms (e.g. duplicate packets coming along the two links) required for make before break with simultaneous bindings that need to be included for a successful implementation.

11.3.2.1 Client MIP and Proxy MIP Models

The Mobile IP Protocol, used to implement interworking among different access networks, can be classified into two different types: client mobile IP (CMIP), in which the client (MS) implements MIP, and proxy mobile IP (PMIP), in which the client on the device uses DHCP to obtain an IP address and the network implements the MIP on its behalf. WiMAX networks support both CMIP and PMIP, whereas EVDO supports only CMIP. The difference in these two models is how the MS registers itself and obtains its IP address and how the network completes the registration. In the CMIP model, the MS integrates an additional MIP stack. The following example uses MIP Ver. 4. The MS sends a MIP registration request message (RRQ) during network entry. Unlike CMIP, a PMIP-based MS does not implement a MIP protocol stack. Instead, the MS uses a DHCP DISCOVER message and the network (ASN-GW), on behalf of the MS, sends a MIP registration request to the HA. The detailed call flow for CMIP-based handoff is explained in Section 11.3.7, whereas the detailed call flow for PMIP-based handoff is given in Section 4.3.8.

The mobility session is associated by the network-based home agent with the identity of the mobile subscription or NAI of the device. When CMIP mode is used by the MS/AT for both EVDO and WiMAX access networks, the dual-mode MS uses the same NAI in both the WiMAX and EVDO networks. When PMIP mode is used in WiMAX and CMIP in the EVDO network, the dual-mode MS uses a pre-provisioned NAI as an identifier for both CMIP and simple IP accesses, and the PMIP client in the WiMAX network maintains the association of the mobility session for that NAI. The dual-mode device uses, for the entire duration of the IP session, the same NAI that is preprovisioned for access in the EVDO system, even for access authentication in the WiMAX system.

11.3.3 Protocol Stack

Figure 11.13 shows the data plane for both WiMAX and EVDO networks. It shows a dual-mode MS that has two protocol stacks, one for each technology. Note that we show here both protocol stacks, for comparison and ease of understanding. However, the MS is active in only one network at any given time. The figure shows the MIP layer in the protocol stack between the FA and HA. Any packet sent from a correspondent node (external IP device) to the MS's home address is routed to the home agent. The HA knows the care-of address of the MS and has an active MIP tunnel with the FA (in either WiMAX or EVDO network). The HA adds an IP header with the CoA to the original packet and delivers it to the FA through the MIP tunnel. The FA, in turn, strips off the CoA header and delivers the original packet to the MS. The details of this procedure can be found in reference [11].

11.3.4 Integrated Dynamic Policy and QoS

EVDO supports a device-initiated quality of service (QoS) model, whereas WiMAX Release 1.0 supports a network-initiated QoS model.* The QoS classifier is obtained mostly from the device in EVDO (the details are given in [7]). In WiMAX [12], however, the MS does not send any QoS classifier to the radio access network. Any QoS classification information must come from the network. Interworking between EVDO and WiMAX with the same QoS class for any application (like voice over IP) across these two access technologies after handoff requires additional signaling from the network. It is possible to maintain the same QoS class when the handoff happens from WiMAX to EVDO, since the MS/AT in an EVDO network can request the same assigned QoS. However, maintaining such a QoS class when transferring from EVDO to WiMAX would require further algorithms not presently developed.

---

*Mobile initiated service flows are currently being discussed in WiMAX forums.

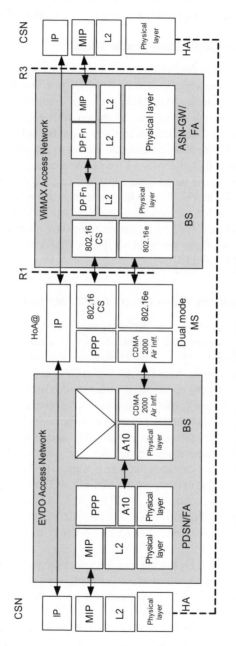

Figure 11.13. Mobile IP data plane for EVDO and WiMAX using a dual-mode MS.

11.3.5 WiMAX–3GPP2 Seamless Mobility

In this section, we discuss the seamless mobility during WiMAX-to-3GPP2 hand-offs and vice versa, using dual-radio and single-radio mobile MS.

11.3.5.1 WiMAX–3GPP2 Dual-Radio Mobility

In this section, we explain the call flows for intertechnology handoffs. As discussed earler, Mobile IP can be implemented in WiMAX in either the client (MS) using CMIP or in the network using PMIP. However, MIP implementation in EVDO uses CMIP only. In this section, we explain all the possible scenarios for handoffs. All the call flows explained in this section apply to the make before break scheme. The call flows for break before make scheme are similar to the corresponding call flows shown below, except that the MS/AT exits the serving access network (e.g., EVDO) before entering the target access network (e.g., WiMAX). There are some open issues with make before break with simultaneous bindings, such as duplicate packets arriving at the MS simultaneously when both networks are active during handoff, that need to be solved by implementations not described in the standards. Here, we show the call flows for WiMAX–EVDO handoffs (and vice versa) using CMIP in both WiMAX and EVDO.

11.3.5.2 Client MIP in WiMAX and 3GPP2 Networks

WiMAX-to-EVDO Handoff. MS performs the WiMAX RAN connection setup using initial ranging, device authentication, and user authentication. The details of these steps can be found in [9, 12, 13]. During the WiMAX connection setup, EAP authentication is performed with the AAA server(s) using the MS's initial NAI. Multiple AAA servers may be involved in cases such as roaming. In such a case, the access request message is directed by the AAA server in the visited network to the home AAA, which eventually authenticates the user. Upon successful access authentication (i.e., EAP success), the home AAA assigns an HA for the MS's session. The ASN-GW initiates initial service flow (SF) creation (for downlink and uplink) for IP connectivity establishment. The HA uses the MN-HA key (received in the step 8 of Figure 11.14) to compute the MN-HA AE key, assigns a home address (HoA) to the MS, and responds with a MIP registration response (RRP) message to the ASN-GW (step 9 in Figure 11.14). A data path is established between the MS and HA through the FA located in the ASN-GW for IP traffic originating or terminating in the MS.

In step 14 of Figure 11.14, the MS decides to switch over to the EVDO network, based on the signal strength and/or any other configured parameters. The details of initial network entry to the EVDO network are given in [7] and [8]. The MS/AT builds a MIP RRQ message using either all-zeroes (or all-ones) indicating dynamic HoA assignment, or by including its own value of the HoA already assigned during the WiMAX initial access. The MS/AT also includes MN-FA challenge extension, MN-AAA authentication extension, and NAI extension as per the X.S0011-D specification [8] . The NAI used in this message has to be the same one used by the MS in the WiMAX network. The MS then sends this message to the PDSN/FA through

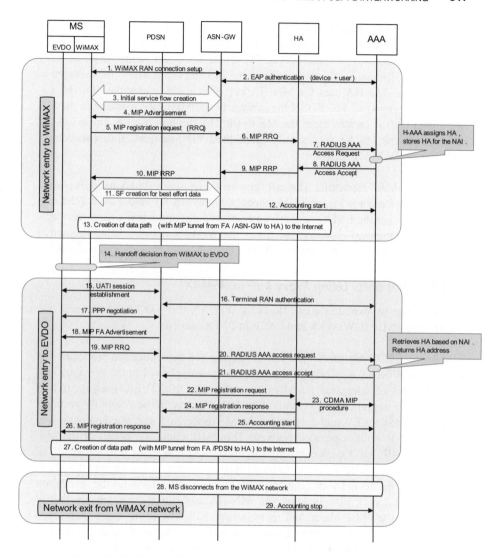

Figure 11.14. Dual-radio WiMAX-to-3GPP2 EVDO handoff (CMIP-based).

the EVDO RAN. Using the NAI value, the home AAA identifies the HA assigned to the MS in the WiMAX network. It returns the same HA address that has been used for this session over the WiMAX network in a RADIUS access accept message to the PDSN. After the user is authenticated by the AAA, the PDSN will forward the MIP RRQ to the HA, using the address of the HA it received in the access accept message. It does not include any access technology extension in the message, to indicate to the HA that the message is associated with a MS/AT in an EVDO network. The HA detects that there is an existing mobility binding for the MS by checking the NAI, but there is no known security association (MN-HA key) to validate the RRQ message. The HA communicates with the AAA and, using the NAI and SPI, obtains the MN-HA key and validates the RRQ message; for more details, see the X.S0011-D specification [8]. Upon successful validation of the RRQ message, the HA assigns the same HoA to the MS and sends MIP RRP message to the PDSN/FA. The HA uses the MN-HA key it received to sign the MN-HA authentication extension. The PDSN then forwards the MIP RRP message to the MS and a data path is created from the MS to the HA through the FA located in the PDSN. Once the MS/AT successfully enters the EVDO network, it disconnects itself from the WiMAX network.

EVDO-to-WiMAX Handoff. The call flow for EVDO-to-WiMAX handoff is similar to the one shown in Figure 11.14, except that the MS/AT enters the EVDO network first. It uses the EVDO network entry procedures and when a handover decision is made, the dual-mode device enters the WiMAX network using the native WiMAX network entry procedures followed by an EVDO network exit procedure.

11.3.5.3   *Handoffs Using Proxy MIP in WiMAX*

In this section, we explain the call flows for WiMAX-to-EVDO handoffs (and vice versa) using PMIP in WiMAX and CMIP in EVDO networks.

WiMAX-to-EVDO Handoff. The initial ranging and authentication procedures of an MS in a WiMAX network are the same as those explained in the previous section. After it is authenticated, the MS sends a "DHCP DISCOVER" message to find out the DHCP servers in the network. The DHCP proxy in the ASN-GW receives the DHCP DISCOVER and determines the MS associated with the data path over which the DHCP message was received. The PMIP client, on behalf of the MS, generates a MIP RRQ message using the initial NAI and sends it to the HA. The message contains PMIP access technology type extension, indicating that the MS is in a WiMAX access network. The MIP RRQ message also includes the revocation support extension. It also includes the MN-HA authentication extension that is computed by the PMIP client, and its associated SPI. The PMIP client then sends the MIP RRQ to the HA. Upon receiving the message, the HA detects that there is no existing mobility binding for the MS (identified by the NAI). After successfully validating the RRQ message (using the services of the AAA), the HA assigns a HoA address to the MS and responds to the PMIP client (ASN-GW) with a MIP RRP message. The initial network entry is completed (as per the call flows shown in Figure 11.15) and a service flow is created between the MS and the WiMAX RAN. A data path is set up between the MS and the HA through the FA in the WiMAX network and IP packets addressed to (or originating from) the MS go through this path.

The MS decides to switch over to the EVDO network based on signal strength and/or any parameters configured in the MS for initiating the handoff. The entry to the EVDO network and the handoff is the same as shown in Figure 11.14.

EVDO-to-WiMAX Handoff. The call flow for EVDO-to-WiMAX handoff is similar to the one shown in Figure 11.15, except that the dual-mode device enters the EVDO network first. The MS/AT then enters a WiMAX network using the native procedures for WiMAX network entry before exiting the EVDO network.

11.3.5.4 WiMAX–3GPP2 Single-Radio Mobility

Single-radio handoff architecture development has recently started at the WiMAX Forum. In a single-radio configuration, a mobile device can only transmit on a single technology at a time and receive on one of the two access technologies. The expect-

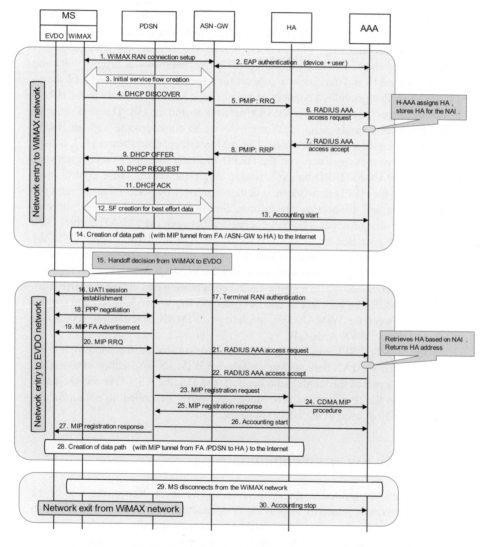

Figure 11.15. Dual-radio WiMAX-to-3GPP2 EVDO handoff (PMIP-based).

ed battery life of such device is, therefore, longer and the device price is lower when compared to the dual-mode, dual-radio configuration devices. The dual-mode mobile device, when connected to the WiMAX technology, sends high rate packet data (HRPD) network entry signaling information to the HRPD signaling forwarding function (SSF) that interfaces to the HRPD RAN. Likewise, when connected to the HRPD access technology, the mobile device sends network-entry signaling messages to the WiMAX SFF that interfaces to the WiMAX ASN. At the time of this writing, the complete WiMAX Forum NWG Stage 2 and Stage 3 single-radio standards documents' call flows have not been finalized. Hence, we are only providing here the network reference model to describe this evolving single-radio work.

Figure 11.12 shows the network reference model (NRM) for WiMAX and EVDO interworking. The MS is a single-radio MS, that is, it has dual-receive (from the base stations of the two disparate networks) and single-transmit (to only one base station) functionality. Interworking among the two access technologies is achieved using the SFF. The SFF is a logical entity that provides layer-3 tunneling support for the MS and may include support for a secure tunnel. Figure 11.12 shows a HRPD-SFF that provides a connection between the MS/AT, actively in the HRPD mode but connected through the WiMAX network to and the HRPD access network (AN). It allows a single-radio, dual-mode device to communicate with an HRPD AN that may be on a private network. The A23 interface, described in [17], is used to carry control messages between the HRPD SFF and the HRPD access network during a WiMAX-to-HRPD handoff using a single-radio MS. These control messages prepare the HRPD network for a device handover by performing the required preregistration and connection procedures. Similarly, Figure 11.12 also shows a WiMAX SFF that provides a layer 3 connection between the MS/AT, actively in a WiMAX preregistration mode but connected through the HRPD network, to the WiMAX ASN. It allows a single-radio, dual-mode device to communicate with the WiMAX ASN, preregister and authenticate in the WiMAX network, and prepare it for the expected handover. In a single-radio configuration and during the HRPD-to-WiMAX handoff, the interface is used to transport the WiMAX MAC and control messages between the WiMAX SFF emulating a WiMAX base station, access network, and the WiMAX ASN-GW.

As the WiMAX SFF emulates a WiMAX base station, during handoff It emulates the source WiMAX base station while a real WiMAX base station assumes the target role. As part of the single radio handoff, the MS, the ASN-GW and the home agent perform the IP allocation procedures previously described, to obtain for the MS the same HoA used in the HRPD network.

11.4   WiMAX–IMS INTERWORKING

IMS functions are limited to the application/service layer. Hence, the role of the WiMAX network is primarily to provide IP connectivity to the user to access IMS services. However, there are some WiMAX-network-specific considerations for IMS integration, such as P-CSCF placement and discovery in roaming scenarios and WiMAX support of IMS-based emergency services, which are discussed here.

11.4.1 WiMAX IMS Architecture

In a nonroaming scenario, as shown in Figure 11.16, the user is always connected to the home network and, therefore, accesses IMS services via the home network (hCSN). IMS services can be provided by the same business entity providing the CSN infrastructure or by a third-party service provider with a contractual relationship with the WiMAX network service provider. For simplicity, IMS entities are assumed to be located in the CSN. In a roaming scenario, users can access IMS services in the home network from a visited ASN/CSN to which they roam. If the visited CSN also supports IMS interworking, the user can access the home IMS services using the proxy call session control function (P-CSCF) located in the visited network. The decision as to which P-CSCF to use is made by the home network and will depend on several factors, including roaming arrangements, home network policies, and subscriber profile. Using the local P-CSCF enables the visited network operator to compute roaming charges based on the IMS services being used over their network. It also enables them to provide local breakout for certain IMS services such as emergency service. The WiMAX IMS architecture [15] assumes that the P-CSCF is colocated in the same network (home or visited) that hosts the entity allocating addresses for the MS (i.e., DHCP server or HA). In a Mobile IP scenario, having HA in the home network and the P-CSCF in the visited network can create an undesirable trombone effect in routing of the signaling traffic when reverse tunneling is enabled.

11.4.2 P-CSCF Discovery

The P-CSCF, the first point of contact with the IMS core network (CN), is a proxying registration responsible for session setup and other messages between the IMS client and the IMS CN. The P-CSCF can reside in the home network or the visited

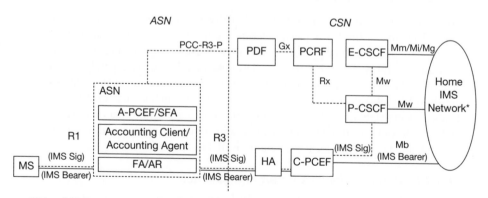

*Other IMS Entities, such as I-CSCF, S-CSCF, HSS, and BGCF, are not shown for brevity.

Figure 11.16. IMS–WiMAX Mobile IP nonroaming architecture.

network in roaming scenarios. This section describes how the MS discovers the address of the P-CSCF in order to register and initiate a session with the IMS CN. Current 3GPP procedures for P-CSCF discovery include a PDP context-based procedure that is GPRS-access-specific and a generic DHCP-based approach that can be used for other access technologies.

In the DHCP-based approach, the MS uses the DHCP SIP server option [13] to indicate that it wishes to receive P-CSCF addresses/FQDNs from a DHCP proxy/server. The WiMAX ASN can support either a DHCP proxy or DHCP relay function to provide the P-CSCF addresses to the MS. The ASN indicates its IP Service capabilities (including DHCP relay/proxy capabilities) to the AAA server during device/user access authentication. The AAA server, based on the ASN capabilities information, home network policies and/or subscriber profile, provides either the P-CSCF address (IP addresses or domain names) or a DHCP server information (if the P-CSCF information can be obtained) in the AAA reply to the ASN. If the P-CSCF address is received, the ASN stores this information and acts as a DHCP proxy enabling the MS to retrieve the P-CSCF addresses directly from the ASN using DHCP procedures. If, however, the ASN receives a DHCP server address, the ASN provides a DHCP relay functionality and relays the DHCP messages from the MS to the DHCP server. The DHCP server provides the P-CSCF information in the DHCP exchange with the MS.

In roaming scenarios, the visited AAA acting as an AAA proxy can append the visited P-CSCF, visited HA, and/or visited DHCP server information to the AAA exchange between the ASN and the home AAA server. The home AAA server makes the final decision on whether the visited or home entities should be used. It then includes the selected entities in the AAA reply to the ASN. The home AAA assigns a P-CSCF and other entities (i.e., DHCP server, DNS server, or HA) to be collocated in the same network (home or visited) to avoid the routing issues described in the previous section. If both visited and home P-CSCF information is provided by the home AAA, the ASN can choose the preferred P-CSCF in either the visited or the home network.

11.4.3 Emergency Services

IMS-based emergency services (ES) call delivery involves the following main steps: detection of the ES request, determining location of the nearest emergency service network, and, finally, routing the call to it. These steps are performed by the IMS entities (P-CSCF and E-CSCF) as described in [13] and are generally independent of the WiMAX network. There are, however, some specific functionalities that the WiMAX network needs to provide to support ES, such as unauthenticated ES access to subscriptionless devices, communicating location information as required to the ES network, prioritization of the ES call with matching QoS and service flow creation, and providing a local breakout if available.

11.4.3.1 Subscriptionless Emergency Service

When a subscriptionless device enters the WiMAX network, an optional initial network entry attempt is made by the device using provisioned credentials, which fails authentication since the device/user has no subscription. However, when the user makes an ES call, the MS attempts an ES-based network entry using an ES-decorated NAI (e.g., {sm=2}username@NSPrealm, where 2 indicates an ES call). The WiMAX network permits ES-only access to the device using this ES-decorated NAI. During this step, the MS is assigned an IP address and discovers a P-CSCF address. The MS then sends the ES VoIP call to the discovered P-CSCF. The rest of the steps with respect to detection and routing of the ES call by the IMS CN are as described in [13].

11.4.3.2 Location Information

If the MS has location information available, it will include this information in the ES request. The location information may consist of network location information, that is, the location identifier and/or the geographical location information. The IMS CN uses this location information to direct the call to the appropriate ES network. If the MS does not have the location information and is not GPS capable, the ES network queries the location server (LS) to find out the accurate location of the MS, which in turn triggers a retrieval procedure for network-based location determination using the WiMAX ASN or MS-based location determination procedures.

11.4.3.3 Prioritization

For reliable ES call delivery, it is important to prioritize the ES session over non-emergency sessions. The IMS CN can use the PCC framework to request the WiMAX IP-CAN to assign the appropriate QoS and service flows for prioritization of the ES session.

11.4.3.4 Local Breakout

A visited network can provide local breakout for emergency service, whereby an emergency IMS session is established in the visited network regardless of whether the MS is registered with IMS in the home network. This function is under development in WiMAX.

11.5 SUMMARY

This chapter discusses WiMAX–3GPP/2 interworking. It presents the network reference model with WiMAX access and 3GPP (Release-8 or pre-Release-8) or 3GPP2 core network. It also shows how a dual-mode MS can operate in the network model with the call flows for network entry, exit, and handoffs. An important aspect of the interworking covered in this chapter is the seamless handoff between WiMAX and 3GPP/2 networks, where a dual mode allows the MS to maintain its IP sessions (thus maintaining the continuity of a VoIP call, streaming video, or any other IP session) during handoffs across WiMAX and 3GPP/2 networks. This chapter also presents an overview of the interworking between WiMAX access and IMS core networks.

11.6 REFERENCES

[1] 3GPP TS 23.003, Numbering, Addressing, and Identification (Release 8), April 2010.

[2] 3GPP TS 23.060, General Packet Radio Service; Service Description; Stage 2; (Release 8), March, 2010.

[3] 3GPP TS 23.203, Policy and Charging Control Architecture (Release 7), December 2009.

[4] 3GPP TS 23.203, Policy and Charging Control Architecture (Release 8), March 2010.

[5] 3GPP TS 23.401, General Packet Radio Service (GPRS) Enhancements for Evolved Universal Terrestrial Radio Access Network (E-UTRAN) access (Release 8), March 2010.

[6] 3GPP TS 23.402, 3GPP System Architecture Evolution: Architecture Enhancements for Non-3GPP Accesses (Release 8), December 2009.

[7] 3GPP2 C.S0024-1 version 1.1: CDMA2000 High Rate Packet Data Air Interface Specification.

[8] 3GPP2 X.S0011-D: CDMA 2000 Wireless IP Network Standard, December 2006.

[9] IEEE 802.16e-2005: Air Interface for Fixed and Mobile Broadband Wireless Access Systems, December 2006.

[10] IETF RFC 5213, Proxy Mobile IPv6, August 2008.

[11] IETF RFC 3344, IP Mobility Support for IPv4, IETF RFC 3344, B. Patil, P. Roberts, C. E. Perkins, August 2002.

[12] WiMAX Forum Network Architecture Stage 2 and 3 Documents—Release 1. (WMF-T33-001/002/003/004-R015v01_Network-Stage-3-Base and WMF-T32-001/002/003/004/005-R015v01_Network-Stage-2), November 2009, http://www.wimaxforum.org.

[13] WiMAX Forum Network Architecture: WMF-T33-101-R015v01 IP Multimedia Subsystem (IMS) Interworking, Release 1.5 version 01. November 21, 2009. http://www.wimaxforum.org.

[14] X.S0011-D, CDMA-2000(r) Wireless IP Network Standard: Introduction, http://www.3gpp2.org/ Public_html/specs/X.S0001-001-D_v1.0_060301.pdf, February 2006.

[15] WiMAX Forum NWG, Working Draft Single Radio Interworking between Non-WiMAX and WiMAX Access Networks, Draft-Txx-sss-R015v01-A, April 2010, http://www.wimaxforum.org.

[16] 3GPP TS 24.302, Access to the 3GPP Evolved Packet Core (EPC) via Non-3GPP Access Network, Stage 3 (Release 8), March 2010.

[17] 3GPP2 A.S0023-0 v1.0: Interoperability Specification (IOS) for High Rate Packet Data (HRPD) Radio Access Network Interfaces and Interworking with World Interoperability for Microwave Access (WiMAX), April 2009.

[18] B. Aboba, D. Simon, R. Hurst,, IETF RFC 5216, The EAP TLS Authentication Protocol, March 2008.

[19] H. Arkko, and J. Haverinen, IETF RFC 4187, Extensible Authentication Protocol Method for Third Generation Authentication and Key Agreement (EAP AKA), January 2006.

[20] B. Aboba, M. Beadles, J. Arkko, and P. Eronen IETF RFC 4282, The Network Access Identifier, December 2005.

[21] P. Funk and S. Blake-Wilson, IETF RFC 5281, Extensible Authentication Protocol Tunneled Transport Layer Security Authenticated Protocol, August 2008.

CHAPTER 12

MULTICAST AND BROADCAST SERVICES IN WiMAX NETWORKS

KAMRAN ETEMAD AND LIMEI WANG

12.1 INTRODUCTION

Multicast and broadcast services (MCBCS) are downlink point-to-multipoint services in which a common set of multimedia data packets are transmitted with coordination from one or more BSs to multiple terminals in the designated MCBCS service area. This feature in concept is similar to the MBMS features in 3GPP systems, for example, HSPA and LTE, and BCMCS features in 3GPP2 systems, for example, in EV-DO Rev A, in which the resources within a radio channel can be shared dynamically between point-to-point unicast services and point-to-multipoint broadcast and multicast services.

In WiMAX, the MCBCS is an optional network capability introduced in air interface System Profile Releases 1.0 and 1.5, with network support defined in WiMAX Network R1.5. The MCBCS, which relies on the over-the-air multicast and broadcast service (MBS) feature in IEEE 802.16-2009, provides end-to-end capability to efficiently deliver a common multimedia content to multiple users using shared radio resources through broadcasting or multicasting. Whereas broadcast service is a push-type service offered to MSs without requiring any request from the user, a multicast service is only offered to users who have explicitly requested and joined a multicast group associated with and authorized to receive such multicast service.

This section describes the MCBCS concepts and design principles in WiMAX covering both air-link aspects and end-to-end network architecture. We start with introducing basic requirements and concepts, followed by air-link MAC/PHY specifications of MBS in IEEE 802.16-2009, and then we describe the network architecture.

12.2 BASIC TERMS, REQUIREMENTS, AND USE CASES

WiMAX MCBCS may be deployed to deliver local, regional, or national multimedia content to users across a large network in which cell sizes, frequency reuse, and users' mobility may widely vary. The MCBCS content may be delivered as real-time streaming or nonreal-time and prescheduled transmissions.

Multicast services may be static, wherein the data transmission and its mode are insensitive to the number and location of target users and their mobility, or it may be dynamic, in which case the data transmission and the transmission mode can change based on number and location of target users and their mobility within and across BSs. For example, data is not transmitted unless there are some users joined the service and expect to receive the data and, when transmitted, the data may be unicast if the number of active users is small.

The broadcast service typically is offered as a free or value-added service to all users who are generally authorized to access the network. This mechanism may be used, for example, to broadcast traffic alerts and local advertisements to users. The multicast services, on the other hand, are typically subscription-based and offered with various charging structures to users who have explicitly or implicitly requested and agreed to receive such services. This mode is more suited for delivering real-time and nonreal time streaming of multimedia content such as videos, news, and music.

Figure 12.1 shows the content distribution overview in which application service providers generate and distribute multimedia packets to a MCBCS controller/server in the network service provider's (NSP's) domain, which then routes the packets to appropriate access networks, manages service provisioning, and so on. The MCBCS content channels are typically transmitted to a geographic area called the MCBCS transmission zone, which may consist of one of more MBS zones. Each MBS zone defined in IEEE 802.16-2009 consists of one or more BSs. Transmission zones are configured independently based on target service areas for each content channel or groups of channels, for example, at local, regional, and national levels. The transmission zone, therefore, may be overlapping for different programs/contents. In this example, the MCBCS controller/server distributes two-channel contents to the WiMAX network. MCBCS channel #1 will deliver to the transmission zones containing MBS zone #1 and #2. MCBCS channel #2 will deliver to the transmission zone which contains MBS zones #1, #2, and #3. The data transport between the MCBCS controller/server to the MBS zone is through IP multicasting.

The network should be able to deliver MCBCS data to all subscribed MSs concurrently with their unicast services and even when the MS is in its power saving (sleep and idle) modes. The MS should also receive incoming call/service notification while receiving MCBCS programs. Once subscribed and provisioned with target channels, the MS should only need minimal or no interaction with BS/network

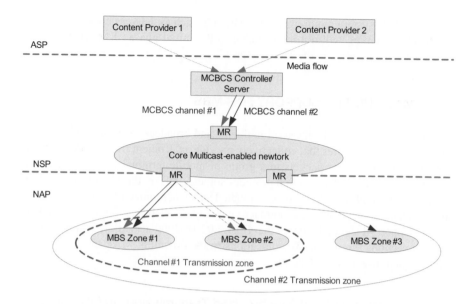

Figure 12.1. Content distribution overview.

for channel switching among those channels while moving within a MBS zone. Even across MBS zones, the network may provide the MS with MBS configuration information of neighboring zones such that the MS does not need to request this information upon entry to the new zone.

The MCBCS user should also be able to receive or reject his or her subscribed MCBCS program while at home or roaming. The network provides the means for program discovery and initial acquisition of service information, including a program/content guide, as well as MCBCS access and acquisition information. The system should also support MS- or network-initiated activation of additional MCBCS channels/services while delivering the current MCBCS and/or unicast data.

The WiMAX system provides the necessary end-to-end QoS for real-time streaming of MCBCS applications such as audio and video, and it also supports multiple priority levels for MCBCS flows, so that if contention for resources arises, a higher priority MCBCS flow will have precedence over lower priority flows. The system can trigger the collection of the MCBCS statistics, such as lost frame, assigned resources, and achieved bit rate to deliver the collected statistics to OAM&P functional elements. MCBCS also supports time-based accounting as well as volume-based accounting.

The WiMAX MCBCS framework includes procedures for authentication and authorization of the users for designated subscribed MCBCS programs to protect the service. Although IEEE 802.16-2009 does support layer-2 encryption for MBS data, the initial release of MCBCS relies primarily on application-level encryption of content.

Some of the system requirements for MCBCS such as delivery confirmation and verification roaming will not be available in the initial phase of MCBCS but are planned to be addressed in the second phase.

12.3 MAC AND PHY SUPPORT FOR MBS

In IEEE 802.16 standards, the term multicast and broadcast services (MBS) refers to a collection of features that collectively provide an efficient framework for transport of multimedia data packets to a group of mobile terminals, using a common multicast connection. This section provides some details on the MBS framework in IEEE 802.16-2009, with focus on coordinated and synchronized MBS, service flow management, and synchronization/macrodiversity, as well as mobility and power-saving support. It should be noted that IEEE 802.16-2009 also allows simple multicast connections through traffic allocations in the downlink MAP using multicast CIDs without use of MBS features.

12.3.1 Coordinated and Synchronized Transmissions in MBS Zones

Each BS supporting MBS belongs to a certain MBS zone with a unique MBS zone ID. An MBS zone is defined as a group of BS's in which the same CID and security association (SA) are used for transmitting the MBS service flow(s). The MBS is offered in the downlink only and may be deployed with coordination or optional synchronization among the BSs of the corresponding MBS zone. The coordinated MBS involves only frame-level synchronization of the MAC PDU's transmission across all BSs in the same MBS zone. On the other hand, time and frequency synchronization of MBS burst transmissions, when feasible, would allow the MS to use macrospace diversity combining and achieve higher SINR and, therefore, throughput.

Each BS may also provide the MS with MBS content locally within its coverage and independently of other BSs. Such single BS provision of MBS is simply a configuration in which an MBS zone consists of one BS only.

To ensure proper multicast operation on networks of BSs employing MBS, the MCIDs used for common MBS contents and services must be the same for all BSs within the same MBS zone. This allows the MS that has already registered with a service to be seamlessly synchronized with MBS transmissions within an MBS zone without communicating in the UL or reregistering with other BSs within that MBS zone.

More specifically, for all BSs within the same MBS zone, the following coordination is always assured:

- Identical classification of MBS packets and their mapping to SDUs
- Identical mapping of SDUs into the MBS bursts
- Same SDU transmissions in the same frame

- Same SDU fragment sequence number and fragmentation size across frame transmissions

This level of coordination in the MBS zone, which may also be considered as frame-level synchronization, enables the MS to continue receiving MBS transmissions from any BS within that zone, even when the MS is in power saving (sleep and idle) mode without need for the MS to register with the BS from which it receives the transmission. This frame-level synchronization may be used even in non-single-frequency networks (non-SFN) in which frequency reuse is higher than 1.

MBS transmissions may optionally be frequency- and time-synchronized across all BSs within an MBS zone to enable MSs to use macrospace diversity combining and receive the MBS transmission at higher CINR and, therefore, improved data rates and/or reliability levels. When macrodiversity is enabled, the MBS burst positions and dimensions as well as PHY parameters must also be the same across all BSs within the same MBS zone. More specifically, the macrodiversity mode requires that all BSs within the same MBS zone have:

- Same modulation and coding, that is, DIUC, of MBS bursts
- Same mapping of SDUs to PDU and same mapping of PDUs to bursts
- Identical order of bursts in the MBS region
- Same MBS MAP construction

The specific means of synchronizing MBS transmissions, including classification, fragmentation, and scheduling at a centralized node called the MBS sync controller is outside the scope of the IEEE 802.16-2009 standard but is defined in the WiMAX Forum.

12.3.2   MBS Service Flow Management

Similar to unicast services in IEEE 802.16-2009, the MBS service flows are managed through a DSx messaging procedure that is used to create, change, and delete a service flow for each MS. During a service flow initiation, the base station sends a DSA-REQ message to the MS (see Chapter 2). The DSx messages carry important service-flow information such as QoS, SFID (service-flow identifier), MCID (multicast connection identifier) if the network decides to multicast the data, or a CID (channel identifier) if the network decides to unicast the data in the air to the MS. The MS will then send a response message to the BS and the BS will acknowledge the success of service-flow creation.

For MBS, in which service typically involves multiple selectable content channels, the standards also define a more optimized DSx message structure so that multiple connections and service flows can be managed, using a compound group DSx. For example, using group DSx, if the MS is interested in multiple services, it can create all of the service flows in one complete DSx exchange, which reduces system overhead and latency. From the MS perspective, the SFID assigned by the anchor authenticator is unique and the MCID is also unique per MBS zone, that is, each MCBCS content program in a MBS zone will be assigned a MCID. This MCID is common to all the MSs for that content in the MBS zone.

DSA also provides the MS with the MBS zone ID for the subscribed service flows to indicate a service area through which a MCID and SA for a broadcast and multicast service flow are valid. A BS supporting MBS includes a list of MBS zone identifier(s) to which it belongs in the DCD message. One BS may belong to multiple MBS zones of the same or different sizes and coverage. A BS may not support MBS or may not be a member of any MBS zone, in which case it will set the MBS zone ID to zero.

It should be mentioned that in the WiMAX Forum there is an optional framework to set up and update MBS connections without using the DSx method, through some application-layer signaling. In this case, the interaction between MS and BS/network is different, as will be described later in this chapter.

12.3.3   Tracking and Processing MBS Data Bursts

Following the successful creation of the MBS service flow(s), typically using a DSA procedure, the MS continues monitoring the broadcast the DL-MAPs and looks for an information element called MBS-MAP-IE (see Figure 12.2). Once the MS receives the MBS-MAP-IE, it verifies the associated MBS zone ID. If the MBS zone ID matches the one the MS received in the DSx message, the MS will then use the pointer within MBS-MAP-IE to locate the corresponding MBS permutation zone and the corresponding physical-layer parameters. All MBS transmissions are sent in their designated MBS permutation zone, the location of which is indicated in the MBS MAP IE. See Chapter 2 for concepts of subcarrier permutation zones.

The MBS permutation zone starts with a management message call, MBS-MAP, which includes one or more information elements called MBS-DATA-IE (see Figure 12.2). Through MBS-DATA-IEs, the MS verifies if the MCIDs listed in MBS-DATA-IE are among those MCIDs it was allocated though the DSx messaging. MBS-MAP points to its next occurrence and also the location of MBS bursts, which are typically 2–5 frames after the current frame. These pointers serve as a daisy-chain primate MAP, which allows the MS to follow the MBS control and data transmissions without reading every frame and interacting with the BS.

MBS data bursts may contain different content channels, each mapping to different MCIDs. The standard allows parsing and selective discard/processing of content channels based on their corresponding MCIDs. This selective processing of MBS data helps lower the memory requirements and battery consumption of mobile devices.

It is also possible to map multiple content channels to the same MCID, in which case content channels may be differentiated with different logical channel IDs (LCIDs). In this case, selective processing is also possible but LCID is not supported in the Mobile WiMAX system profile.

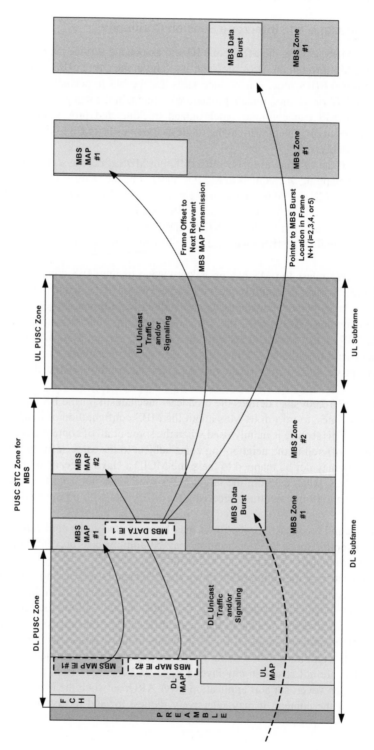

Figure 12.2. MBS signaling and data tracking in IEEE 802.16 MAC.

12.3.4   Intrazone and Interzone Session Continuity

For a given MBS service, the MBS zone ID and associated service parameters provided to the MS through the DSA message are valid in the entire MBS zone. Therefore, within that MBS zone, it is not necessary for the MS to perform the network reentry, handoff, or location update procedure to maintain the reception of MCBCS service. Such MS can continue synchronized tracking and processing of MBS bursts without leaving the power-saving mode as it moves from BS to BS within the same MBS zone.

However, if the MS moves to a BS in a different MBS zone, the service-flow parameters and mapping of channels to MCIDs may not be the same as in the previous MBS zone. The MS can detect that it is in a new zone by detecting the change from the MBS zone identifier list that the target BS's DCD (downlink channel descriptor) advertises or through the MOB_NBR_ADV message of the serving BS. If the list does not contain the MBS zone ID that the MS currently has, it means the MS is in a new zone.

In this case, unless the MS has prior knowledge of the MBS configuration in the target MBS zone, it should request the target BS for MBS update, that is, new service flow management encodings for the multicast and broadcast service flows. For such interzone MBS update, the MS should perform allocation-update procedure. The BS at the serving ASN will then map the service flow with the new MCID and send it to the MS. The MS may also perform network entry to trigger an MBS update, which may result in temporary interruption of the data across zones.

IEEE 802.16-2009 also defines the broadcast mechanism for the BS, in the border of MBS zones, to inform its MSs about the MBS configuration of neighboring MBS zones. This feature is mainly used when the some or all of content channels in one zone are offered in the neighboring zone, subject to the same security association, but they may not be mapped to the same MCIDs. Having received this information, the MS, when entering the target MBS zone, does not need to request an MBS update and it may continue receiving the MBS data with no interaction with the BS.

Multicast and broadcast service flows may be encrypted at the application layer or MAC or both. Upper-layer encryption may be employed to prevent nonauthorized access to multicast and broadcast content.

Also, considering the nature MBS flows, the ARQ and HARQ are not applicable to multicast connections as there is no layer 1 or layer 2 feedback from the MS. Although not included in the system profile, IEEE 802.16-2009, allows the MBS to be used with time diversity similar to that used in HARQ retransmissions, whereby some HARQ parameters are used for MBS bursts to allow proper sequencing and time-diversity combining when MBS bursts are repeatedly transmitted. Such retransmissions are not based on any layer 1 or layer 2 feedback from the MS. The system does, however, support application-layer ARQ, so when the MS fails to receive parts of the content, it may selectively request for and receive the missing parts.

12.4 MCBCS NETWORK ARCHITECTURE

There are two approaches to supporting and managing multicast broadcast services in WiMAX networks, namely, the DSx-based approach and application-layer approach. The DSx-based method is the mandatory framework as it uses the standard IEEE 802.16-2009 service-flow management procedure, whereas using an application-layer message to carry layer 2 parameters is an optional approach that may be simpler in some deployment models. Considering that the DSx-based framework is the baseline and mandatory in WiMAX, we will mainly focus on DSx-based architecture and design. The application-layer approach will be introduced at the end of this chapter for reference.

12.4.1 Baseline MCBCS Network Reference Architecture

The key functional elements in the reference network architecture for WiMAX MCBCS based on DSx are shown in Figure 12.3 and described below. The function in { } indicates optional approach.

- *MCBCS Content Server.* The MCBCS content server is used to store the contents of MCBCS services and provide the contents to the MS. How to manage the MCBCS content server as well as the interface between MCBCS content server and MCBCS controller/server is out of scope of the WiMAX specification. The operator may use existing or third-party content servers.

MCBCS functions in the CSN are as follows:

- *MCBCS Controller/Server.* The MCBCS controller/server is a logical network entity that manages the MCBCS services in the CSN domain. It includes service-guide construction and distribution, and application-layer security management to ensure that the content can be viewed only by the authorized users and to manage the program and IP multicast groups. The MCBCS controller/server can be in any physical form such as BCAST, with the requirement that the entity implement the adaptation layer to conform to the WiMAX R3 interface.
- *AAA/PF.* Similar to unicast, the AAA/PF authenticates and authorizes the MS to access particular MCBCS services.
- *Subscriber Profile Database.* The subscriber profile database is a database to store all the user profiles, whether or not this profile database is only for MCBCS service or also for unicast is a deployment issue and, thus, out of scope of the WiMAX specification.

MCBCS entities and functions in NAP are as follows:

- *MBS Proxy.* MBS proxy is an important function in a MBS zone in the NAP.

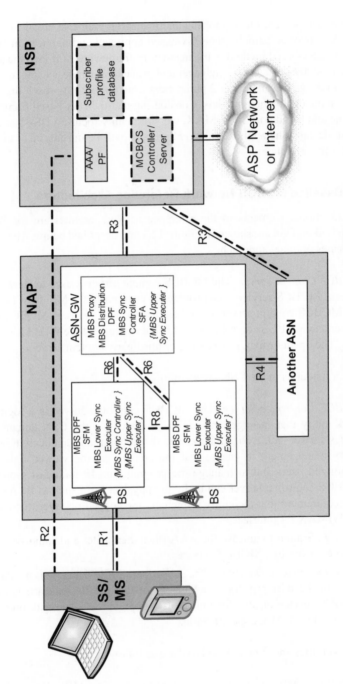

Figure 12.3. MCBCS DSx approach to network architecture.

It acts as a coordinator and session manager to manage the session between ASN and CSN in a MBS zone and also triggers the MBS distribution DPF for the data-path management. It is a policy-enforcement point that also allocates radio parameters such as MCID to a MBS service flow. Depending on the operator policy, the MBS proxy may manage one or more MBS zones, but for a given MBS zone, there is only one MBS proxy.

- *SFA.* The function of service flow authorization (SFA) in MCBCS service is the same as for unicast service and it is per-MS. The only difference is that the SFA may communicate with the MBS proxy to get updated MBS service parameter information.

- *MBS Distribution DPF.* The MBS distribution DPF function is located at ASN-GW and it is colocated with MBS proxy to manage the bearer path for the MCBCS service for the entire MBS zone. It creates, modifies, and deletes the GRE tunnel and the data path. The primary MBS distribution DPF is an anchor point of a MBS zone in the NAP to receive data from the CSN. After the MBS distribution DPF receives data from the CSN, it classifies the data based on the classification rules and assigns a GRE sequence number to the data packet. If volume-based accounting is required, the MBS distribution DPF will also act as an accounting agent to count the volume of received data. It coordinates with the MBS sync function to support MBS synchronization.

- *MBS DPF.* MBS DPF is a function entity located at the BS to support the MBS data-path management.

- *SFM.* The function of service flow management (SFM) in MCBCS service is the same as for unicast service. It is used to create, modify, and delete service flow through the R1 interface with the MS by the DSx procedure.

- *MBS Sync Function.* The MBS sync function is used to coordinate and synchronize the MBS contents in the downlink transmission over the WiMAX network. It includes the MBS sync controller function and MBS sync executer function. The MBS sync controller function must be colocated with the MBS proxy or optionally in the BS. If the MBS sync controller is located at the BS, it is called a super BS. The MBS sync controller function is a centralized control function in a MBS zone to interact with MBS distribution DPF to specify the synchronization rules and deliver the rules to the MBS sync executer. The MBS sync executer then constructs the MAC PDU, burst, and, finally the PHY frame. The MBS sync executer has two parts. First part (upper sync executer) is used to construct the MAC PDU and burst, which must be located at BS optionally at the GW. The second part (lower sync executer) is used to construct the final PHY frame and transmit to the air. It is also responsible for broadcasting the MBS_MAP_IE and MBS_MAP messages.

- *MBS Client.* The MBS client function is located at the MS and it complies with the IEEE 802.16-2009 standard to receive the data and control information through the air interface. The MBS client function also supports the IP multicast based on the IPv4 and IPv6 stack, and all other upper-layer functions that ensure that the MS can receive and display the MCBCS contents properly.

12.4.2   Service Initialization and Provisioning

The MCBCS service initialization involves the static or dynamic association between the MBS proxy at the ASN and the MCBCS controller/server at the CSN. This association will help the MCBCS controller/server to know which MBS zone and corresponding MBS proxy is interested in receiving a particular MBS service. The static association is outside the scope of the WiMAX specification. The dynamic association can be triggered in the network by the very first MS's request for MCBCS service. Figure 12.4 shows the details of the association triggered by the very first MS that made the subscription. MCBCS service is a preprovisioned service.

Once the MS is authorized during the network initial entry phase, the AAA sends an access accept message to the anchor authenticator, which will then decide which MBS proxy in the corresponding MBS zone it needs to forward the MBS policy to. The anchor authenticator sends a MCBCS_RR_Request message with the MCBCS controller/server IP address to this specified MBS proxy, and the MBS proxy will associate itself with the MCBCS controller/server for this service. After the association is complete, the MBS proxy sends a MCBCS_RR_Response message back to the anchor authenticator.

Service provisioning is described through MS-side state transition procedures, which are followed independently by each MS, as well as the network-side state transition procedures, which are almost independently performed on the network side. In general, MS service provisioning contains server discovery, subscription, join, and leave procedures, whereas network-side service provisioning includes service announcement, session start, data transfer, session update, and session stop procedures. These procedures are further described in the following and shown in Figure 12.5.

12.4.2.1   Network Service Provisioning

Network Service provisioning procedures include the following states:

- Service announcement. Allows the user to be informed of the available service through distribution of a service guide.

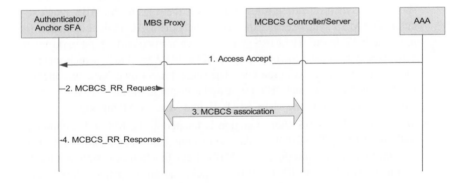

Figure 12.4.   MCBCS association.

- Session start. This is a notification procedure to the access network when the MCBCS controller/server is ready to transmit data. Once the access network receives session start notification, if the data path has not been created yet, it establishes the data path, or it may change the state of the reserved radio resource from reserved to active.
- Session update. This refers to the optional process of modifying the data path if the service QoS or priority is changed.
- Session Stop. Once the data transmission is completed, the MCBCS controller/server will notify the access network through the session stop procedure. During the session stop procedure, it will trigger MS to leave this service, and then delete the MCBCS service flow from the data path. It may also delete the data path if the last service flow on that data path is deleted. For more details, see Section 12.4.2.3.

12.4.2.2 MS-Side Provisioning

MS-side service provisioning involves procedures performed by each individual MS as follows. Server discovery is the very first step that the MS performs to obtain the service guide and discover the relevant MCBCS controller/server from which MS can get services. If the MCBCS controller/server information is already preconfigured in the MS, the MS can skip server discovery; otherwise, the MS may get the server information through DHCP procedures.

Subscription is the next step to establish a relationship with the MCBCS service provider that has the service subscription record on its user profile. Subscription can be online and/or offline, but its details are outside the scope of WiMAX specifications. Depending on the operator policy, some MCBCS services such as free broadcast programs or emergency services may not require subscription.

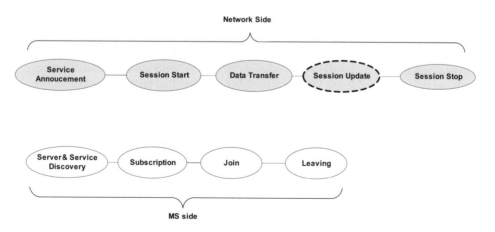

Figure 12.5. MCBCS service state transition.

Joining the MCBCS may be network initiated for the preprovisioned MCBCS service or be MS-initiated and dynamic for nonprovisioned MCBCS services. Network-initiated joining assumes that the MS has already subscribed to the MCBCS services before its network entry and the service is either free or flat rate. In the case of dynamic joining, the MS can subscribe to the MCBCS service online, after successful network authorization, and immediately trigger the service joining procedure. Alternatively, an MS that has already subscribed to the service may join the service at a later time, decided on by the user, when time- or volume-based charging is applied. For preprovisioned MCBCS services upon successful network entry, the AAA pushes the MS service profile to the anchor SFA. The anchor SFA communicates with the MBS proxy to get updated MBS service information as well as access parameters for the service and then assigns SFIDs and triggers service flow creation for that MS. Similar to unicast, the SFM at the BS initiates the DSA procedure based on IEEE 802.16-2009 to create the service flow for the MCBCS service with the MS.

Leaving procedure is similar to the joining procedure except that leaving is used to trigger deletion of a service flow when MS is not interested in the service any more. In such case, MS sends a leaving message to the network, and the network triggers DSD to delete the service flow. Both joining and leaving are for multicast services only.

12.4.2.3 Data Path Management

The content provider can provide one or more media flows to the MCBCS controller/server. The MCBCS controller/server, based on the local policy, may merge multiple media flows into one MCBCS content. How the MCBCS controller/server merges the media flows is outside the scope of the WiMAX specification.

Figure 12.6 shows the relationship of the MCBS identifiers from MCBCS controller/server to the ASN. Each MBS content from the MCBCS controller/server is identified by a channel ID that is 1:1 mapped to the MBS service flow identified by the transmission zone ID with the PDF ID. The PDF ID is unique per transmission zone and it is common to all the MSs that are interested in that content. Thus, this identification will not be changed when the MS moves from one MBS zone to another. One or more MBS service flows may be carried by a data path identified by a data path identifier (DPID) at the ASN. Within each MBS zone, the MBS service flows is 1:1 mapped to the MCIDs at the MAC layer so that the MS can do selective discarding based on the MCIDs that it is interested in. The data path protocol stack is shown in Figure 12.7.

12.4.2.3 Data Path Creation, Modification, and Deletion

The MCBCS data path is used to deliver a MCBCS service for all the MSs in a given MBS zone of a transmission zone; it is shared by all the MSs. The MCBCS data path can be preconfigured as well as dynamically set up based on network trigger events. In the case of a preconfigured MCBCS data path, the association between the MCBCS controller/server and the MBS proxy is statically established. The data

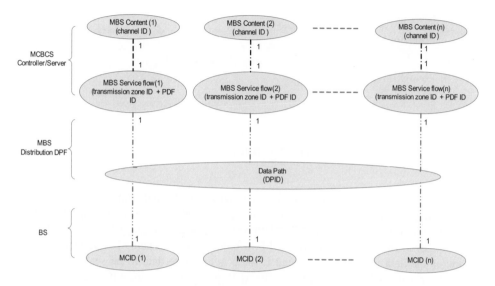

Figure 12.6. Relationship between MCBCS identifiers.

path already exists with preconfigured QoS, even if there is no MS interested in any MBS contents and no data is transmitting on that data path.

When dynamically setting up the data path, the data path is not preestablished. It is either triggered by the first MS that is interested in a MCBCS service or it can be triggered by a session-start signal from the CSN, which indicates that data is ready to transmit. If that data path is triggered by the session start request message from the MCBCS controller/server to the MBS proxy, the MBS proxy will send a trigger to the MBS distribution DPF to start the creation of a data path with all the MBS DPF at the BS in the MBS zone. See figure 12.8.

The data path can be modified if the QoS is changed, the service priority has

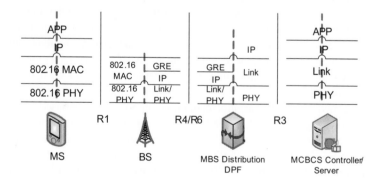

Figure 12.7. MCBCS data path protocol stack.

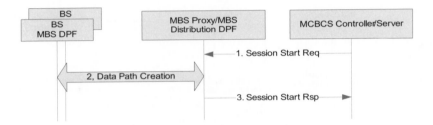

Figure 12.8. Data path creation.

been changed, or if there is a service flow added or deleted from this data path. The MBS distribution DPF function can send a Path_Mod_Req message to the MBS DPF at the BS, and the MBS DPF can response back using Path_Mod_Rsp messages to the MBS distribution DPF to modify the data path. The service flow and/or the data path can be deleted if there is no more data to be sent from the CSN for that MCBCS service flow. The MBS proxy triggers the MBS distribution DPF to delete the corresponding MCBCS service flow on the data path. If it is the only service flow on the data path, the MBS distribution DPF will also delete the data path by sending a Path_Dereg_Req message to the MBS DPF and the MBS DPF response a Path_Dereg_Rsp message to complete the deletion of the data path.

12.4.3   Charging and Accounting for MCBCS

The MBS accounting agent, which is colocated with the MBS distribution DPF, is used to count the actual bytes received for the service for NAP statistics and also for supporting volume-based accounting. The accounting client, which is very similar to unicast accounting client, can be collocated with an anchor SFA at the NAP or can be in the CSN. It collects the charging and accounting information via the support of accounting agents based on the service identification, such as program ID, DPF ID, transmission zone ID, service type, and transport type, as well as subscriber identification. A MCBCS program ID is used to identify a program package that the user subscribed for. It is a logical concept from a user point of view. A program package can contain one or more channels and contents, and each channel/content may have different starting/ending time, QoS, and so on. A MCBCS's service is logical from the point of view of data transportation and it is identified by a DPF ID that is unique in a transmission zone.

There are two types of charging/accounting support: service-based accounting and access-based accounting. Service-based accounting is used to identify the type of accounting records for a given MCBCS content, such as video streaming or file downloading. Access-based accounting is used to identify the amount of bearer resources that are consumed by a MCBCS service. Access-based accounting can be used for NAP statistical and monitoring purposes. It can also be used to charge a network service provider (NSP) or a subscriber, using time-based accounting or volume-based accounting. The time and volume can be based on the service start/end time and the entire volume during that session. It also can be based on the joining/leaving action that the MS performs. For a broadcast service, since there is no explicit join/leave action, the time- and volume-based accounting, if applied, will be based on the entire session for that service.

12.4.4 MBS Data Synchronization

MBS data synchronization utilizes what has been defined in IEEE 802.16-2009 to enable the MS to continue receiving MBS transmissions from any BS within a MBS zone even when MS is in power saving (sleep and idle) mode without need for the MS to register with the BS from which it receives the transmission. The synchronization can be inside of a MBS zone or can happen between zones. In a MBS zone, IEEE 802.16-2009 defines frame level synchronization and macro diversity, which we introduced in the previous section. For inter-MBS zone synchronization, the specification of frame offset in the corresponding MBS_MAP message will help the MS identify the new burst location in the target MBS zone. The framework for supporting this MBS data synchronization in general is shown in Figure 12.9.

The MBS distribution DPF function classifies the incoming data received from the CSN and sends the data to the MBS sync controller to generate the synchronization rules based on the level of synchronization (frame level or macro diversity). The MBS sync controller sends the rules, including timing information, related to the data to the MBS upper sync executer. After the upper sync executer receives the data from the MBS distribution DPF as well as the synchronization rules, the MBS upper sync executer, based on the rules, first constructs the MBS MAC PDUs, then groups those MBS MAC PDUs into a MAC burst. The MBS lower sync executer takes the MAC burst and constructs the PHY burst based on the synchronization instruction and parameters. It then puts different PHY bursts into a permutation zone in the downlink frame and sends it to the air.

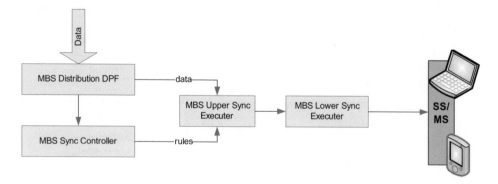

Figure 12.9. MCBCS data synchronization framework.

12.5 MCBCS APPLICATION-LAYER APPROACH

MCBCS application-layer approach is an alternative, optional MCBCS framework that may be used besides the mandatory DSx approach. In this framework, the MAC layer parameters such as MCID are not delivered through MAC-layer signaling message but through application-layer signaling, that is, through a service guide or during the subscription. This section will give an overview of this optional approach, with focus on the differences between application layer and DSx approaches.

12.5.1 MCBCS Application-Layer Approach Network Architecture

The application-layer approach network architecture is very similar to DSx network architecture except that there are no MBS sync functions, including the MBS sync controller and MBS sync executer. Figure 12.10 shows the application-layer approach to MCBCS network architecture.

There are a few other differences between the Application-layer approach and the DSx approach. In the application-layer approach, The MCBCS server at the CSN has all the functions that supported by the DSx MCBCS controller/server. In addition, the MCBCS server for application-layer approach also manages the MCID and MBS zone IDs for each program/channel.

The MBS agent is a function at each BS responsible for supporting the data path bearer management and constructing the final PHY frame based on the information configured by the NAP and the sync information included in each SDU. It is also responsible for transmitting the MBS_MAP_IE, MBS_MAP messages.

The MBS client in the application-layer approach has additional requirements that the MBS client must be able to decode the application-layer signaling containing the MAC layer parameters and push those parameters from the application layer

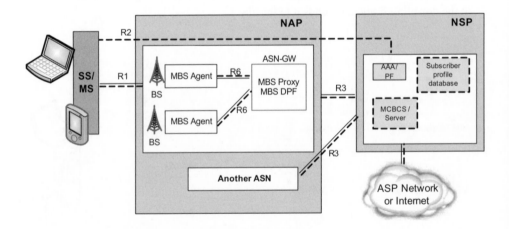

Figure 12.10. MCBCS reference model in the application layer framework.

down to the MAC layer inside the MS. When the MS detects that the zone has changed, the MAC layer should signal to the application layer to get the corresponding MCID in that new zone if the MS did not get the new zone MCID information or it did not push all the zone's MCIDs to the MS MAC layer.

The data path for the application-layer approach is similar to the DSx approach. The data delivered from the CSN the ASN is through a multicast routing; inside a MBS zone, it is through another multicast routing as the BS joins the MR in the MBS zone. Similar to the DSx approach, the application-layer also uses the GRE tunnel, except that it uses data path type 2, which is defined in the WiMAX R1.3. The granularity for the data path in a MBS zone is {MBS zone ID, MCID}. The data path creation procedure is the same as the DSx multicast transport approach except that the data path creation can only be triggered by session start procedure.

There is no sync controller in the application-layer approach as all the radio resources and parameters are preconfigured and there is no dedicated synchronization control signaling from the ASN-GW to the BS. The packaging synchronization information such as timing information is added on top of MAC SDU at ASN-GW as a GRE payload; thus, the sync information is through the data plane along with data delivery.

The application-layer approach has a similar view about service provision, except for a few differences: The online delivery of the service guide must go through the WiMAX R2 reference point and be coded with XML and transport via HTTP/TCP. The service guide may contain the mapping table of the channel content multicast IP address with the MBS zones and MCIDs. The online subscription phase will authenticate and authorize the user and, thus, deliver the mapping tables to the MS and the accounting and charging will be started after subscription. The delivery confirmation is through the R2 reference point and the retransmission is through unicast delivery.

The application-layer approach supports subscription-based, time-based, and volume-based accounting. The differences for time-based and volume-based accounting compared with the DSx approach is that in the application-layer approach, since there is no network layer communication between MS and GW, the charging starts once the subscription is started and will be stopped when the subscription is cancelled.

12.6 SUMMARY

The MCBCS feature is an optional feature in Mobile WiMAX that uses the IEEE 802.16 MBS air-interface mechanism to provide flexible and efficient mechanisms for sending common contents to multiple users using shared radio resources. It supports various usage scenarios and business models. The WiMAX MCBCS enhances MS reception by supporting data synchronizations inside of a MBS zone and, optionally, between MBS zones to enable macrodiversity combining and, therefore, higher data rates and session continuity across the wide-area MBS zones. The daisy-chain feature also improves MS mobility and power saving.

12.7 REFERENCES

[1] Part 16: Air Interface for Broadband Wireless Access Systems, May 2009, http://wirelessman.org/pubs/80216-2009.html.

[2] WiMAX Forum® Network Architecture Detailed Protocols and Procedures Base Specification, November 2009, http://members.wimaxforum.org/apps/org/workgroup/nwg/document.php?document_id=49034.

[3] WiMAX Forum® Network Architecture System Requirements, Network Protocols and Architecture for Multicast Broadcast Services Dynamic Service Flow Based, MCBCS—DSx, November 2009, http://members.wimaxforum.org/apps/org/workgroup/nwg/document.php?document_id=49047.

[4] WiMAX Forum® Network Architecture System Requirements, Network Protocols and Architecture for Multicast Broadcast Services Dynamic Service Flow Based, MCBCS Subsystems Common Sections, November 2009, http://members.wimaxforum.org/apps/org/workgroup/nwg/document.php?document_id=49046.

[5] WiMAX Forum® Network Architecture System Requirements, Network Protocols and Architecture for Multicast Broadcast Services Dynamic Service Flow Based, MCBCS Application Layer Approach, November 2009, http://members.wimaxforum.org/apps/org/workgroup/nwg/document.php?document_id=49048.

CHAPTER 13

LOCATION-BASED SERVICES IN WiMAX NETWORKS

WAYNE BALLANTYNE, MUTHAIAH VENKATACHALAM,
AND KAMRAN ETEMAD

13.1 INTRODUCTION

Location-based services have become a requisite feature for most wireless networks. There are two primary use cases: (1) emergency services access (E911), in which the user's location is reported to the public safety answering point (PSAP), from which emergency responders are dispatched; and (2) personal and corporate applications such as pedestrian/vehicular navigation, personnel or asset tracking, and points of interest lookup. Within this space of applications, location accuracies ranging from < 5 m to hundreds of meters may be acceptable, depending on the specific use case.

Traditional wireline 911 calls rely on a simple database-lookup approach to determine a caller's location during a 911 call. There is a one-to-one mapping between the caller's phone number and the user's location, which is transmitted to the PSAP in civic address format, that is, the caller's phone number serves as the "primary key" in the database lookup.

These lookup methods can apply to fixed wireless equipment such as CPE (customer premises equipment), which is usually operated at the user's household. Here, however, the end user may have the option to update his/her location manually, in the database, in order to allow for the possibility of moving the CPE device to a different location. This was the outcome of FCC order 05-116 [1], which applied to VoIP providers such as Vonage.

In other cases, even for mobile wireless devices, the LBS application location accuracy requirement may be on the order of kilometers. Thus, a cell ID-based location would be sufficient, since WiMAX cell site radii are expected to be in the

500–2000 m range for 2.5 and 3.5 GHz deployments. In the CSN, there will be a database mapping the WiMAX base station ID to a geodetic location of the base station. (This database will henceforth be referred to as the "base station almanac.") Thus, the BSID serves as the lookup key for the user location.

A significant advantage of database lookup methods is fast TTFF (time to first fix), usually just a few seconds once the connection to the location server is established. However, for mobile terminals in a mobile network this approach cannot be used.

Traditional, autonomous location methods such as conventional GPS have proven to be inadequate in providing location to the applications above in many cases. This is due to the fact that the small-form factors associated with portable devices dictate use of a suboptimal GPS antenna, which is often 5–8 dB poorer in sensitivity than an external patch antenna used for a conventional GPS receiver. In addition, the 30 or more seconds TTFF associated with conventional GPS is too slow for many LBS applications, especially E911. Therefore, we look to the WiMAX wireless network to provide methods to enhance or assist the location determination procedures with respect to TTFF, sensitivity, or both. Moreover, the WiMAX network should provide security features to bar an unauthorized requester from obtaining the location of an MS without permission.

13.2   LBS USAGE MODELS AND DESIGN REQUIREMENTS

The LBS solution in WiMAX networks is targeting various usage and business models. The location may be managed primarily by the mobile station (MS) with minimal support from the network or it may be managed primarily by the network with some assistance from the MS. The usage models may involve periodic or frequent requests for location or they may be event triggered by applications on the MS or the network side. The two key design goals for LBS in WiMAX are QoS and security.

13.2.1   Location Security

Security is a very critical piece of any network architecture that supports location-based services. Any location requester (LR) requesting the location of an MS has to be authenticated and authorized. Furthermore, enough security precautions have to be taken for transporting the MS location across the WiMAX network and beyond in a highly secure fashion.

LR verification is achieved by using the AAA protocol for authentication and authorization of the incoming LR. When a LR requests the MS location, the LR is first subject to authentication and then authorization against the user policy. If either of these fail, then the LR is denied the MS location and an error response is returned back to the LR. If both of these are successful, then the location server (LS) will initiate the query for the MS location within the WiMAX network.

The security transport problem is solved by ensuring that the reference points within the WiMAX network over which the MS location are transferred are fully secure by either pre-establishing or dynamically negotiating security associations between the reference point tunnel end points. Bulk encryption is then used to encrypt and subsequently decrypt the payload that carries the MS location. In order to ensure that the MS location is also securely carried outside of the WiMAX network from the LS to the LR, it is recommended that the LS establish a dynamic security association between the LS and the LR and apply bulk encryption on the payload carrying the MS location on the U1 reference point.

13.2.2 Location QoS

The WiMAX network supports two primary QoS parameters for location determination, which are often at odds with each other. These parameters are the location accuracy and the latency for obtaining the location.

Generally, if a location method takes longer to obtain a fix, it should, in return, provide a more accurate fix for the MS location, and vice versa. If this does not happen, then the need for invoking such a location mechanism (which provides poor latency and poor accuracy at the same time) will diminish. Location accuracy can be further classified into horizontal accuracy and vertical accuracy.

The following sections describe a network reference model for LBS in WiMAX networks as well as a high-level overview of various location determination schemes and models.

13.3 REVIEW OF LOCATION METHODS FOR WIRELESS DEVICES

This section provides an overview of location methods that may be used in WiMAX networks.

13.3.1 Assisted GPS (A-GPS)

As noted above, GPS assistance in the network can significantly improve TTFF as well as GPS sensitivity. For example, the sensitivity of most autonomous GPS receivers is around −142 dBm to −145 dBm, due to the basic BER limitations associated with demodulating BPSK in the GPS CDMA receiver. However, a key feature of A-GPS is that the MS receiver does *not* have to demodulate the GPS SV navigation message; it is received via the wireless network from a remote GPS reference network. The GPS navigation message includes the GPS satellite ephemeris data (which describe the SV orbits and, thus, allow their positions to be calculated), SV clock correction data, ionospheric correction data, and so on. Removing the requirement for data demodulation allows the GPS receiver to achieve acquisition sensitivities less than −155 dBm (assuming an adequate complement of on-chip GPS correlators), a tremendous improvement.

Another key aspect of A-GPS is improvement of TTFF. Since the GPS receiver need not demodulate the navigation message, which takes 30 seconds or more, as soon it has acquired four or more GPS SVs and extracted the pseudoranges, a location fix can be computed. Moreover, other available data in the wireless network can improve the SV acquisition time significantly. The net result is that GPS TTFFs of 5 seconds or less are quite common in strong signal conditions.

GPS signal acquisition for each SV involves searching through a two-dimensional code phase and frequency space, as is well documented in works such as [2]. Frequency uncertainty is due to the unknown GPS receiver reference oscillator offset, as well as the unknown Doppler shift caused by the relative motion of the SV and the MS device. Code phase uncertainty results from the unknown MS clock offset relative to GPS system time; there is also an N msec ambiguity due to the 1 msec repetition interval of the GPS C/A code. To reduce these uncertainties, the following forms of GPS assistance can be provided by the wireless network:

- Frequency acquisition aiding. A typical TCXO (temperature-compensated crystal oscillator) in a cost-sensitive MS device having an A-GPS receiver might be on the order of ± 3.0 to ± 5.0 ppm (parts per million) uncertainty over temperature, aging, and shock (even if make tolerance is calibrated out in the device factory). Thus, without any "frequency assist," the GPS receiver will have to (in the worst case) search this entire uncertainty range. The WiMAX receiver will typically perform an AFC (automatic frequency control) procedure to lock (or measure the offset of) its own TCXO reference to the WiMAX BS. Since the WiMAX BS TX frequency will in most cases be locked to a very precise reference (GPS receiver or rubidium/cesium oscillator), once the MS WiMAX receiver locks to the BS frequency, the BS frequency accuracy will be effectively transferred to the MS, with some additional uncertainty due to finite S/N and AFC resolution. Nonetheless, the net result is that the WiMAX MS receiver TCXO uncertainty can be effectively reduced to ± 0.5 ppm or better. If this TCXO is then also used as the reference for the MS GPS receiver, the frequency uncertainty can be reduced from ± 3.0 ppm or worse to ± .5 ppm. This reduces the frequency space search time (worst case) by a factor of six or more.

- Doppler uncertainty is also another component of frequency uncertainty; however, it is different for each SV, unlike the TCXO uncertainty discussed above, which affects the entire GPS receiver for every SV. This Doppler shift is 0 ppm for GPS SVs directly overhead, and can approach ± 12 ppm for an SV at low elevations. (Since the GPS SV orbital speed is 3.87 km/sec, the contribution of the MS device user's own speed relative to the earth will be minor, for a pedestrian or vehicular-based user.) These Doppler offsets can be readily calculated if the initial approximate user position is known, since the GPS SV locations are known via the ephemeris data or prestored GPS almanac. The user position can be approximated by the cell-ID determined location.

- Precise time aiding, to an accuracy of ± 200 μsec or better relative to GPS time, allows the MS GPS receiver to narrow the code-phase uncertainty tremendously. Several commercially available wireless networks, such as

3GPP2 and iDEN (Motorola Integrated Dispatch-Enhanced Network), now offer mechanisms to relate a wireless network signal epoch to GPS time. A broadcast message identifies to the MS the GPS time at said epoch transition. Figure 13.1 illustrates this conceptually.

WiMAX system time is kept in terms of frame numbers, with each frame representing 5 msec in the configuration for WiMAX mobile deployments. A WiMAX epoch consists of 2^{24} frames. At the time of the WiMAX epoch rollover, the GPS-synchronized base station determines the corresponding GPS time, which is typically represented as GPS week number (in the current GPS epoch) and the elapsed seconds count. This information is then broadcast in the LBS-ADV message, GPS time TLV. When an MS recovers this TLV, from knowledge of the current frame number and the current timing within the frame it can compute the GPS time with some small uncertainty, for example, ± 50 μsec, depending on the details of the WiMAX modem and the accuracy with which the BS transferred the GPS time. After this time transfer, WiMAX system time can be consistently translated to GPS system time. Time strobes are generated from the WiMAX modem, one at the beginning or end of every frame occurring every 5 msec (for the mobile WiMAX profile), and these can be used to time assist a GPS receiver in the MS, or transfer time to an external accessory.

Figure 13.2 shows a possible WiMAX MS terminal block diagram, with an embedded GPS receiver present.

The above embodiment assumes that a common TCXO is shared between the GPS receiver and WiMAX modem, as is desirable for cost savings. When the WiMAX modem registers to a WiMAX system with a precise BS TX frequency, as discussed above, it will be able to send AFC reports to the application CPU for pur-

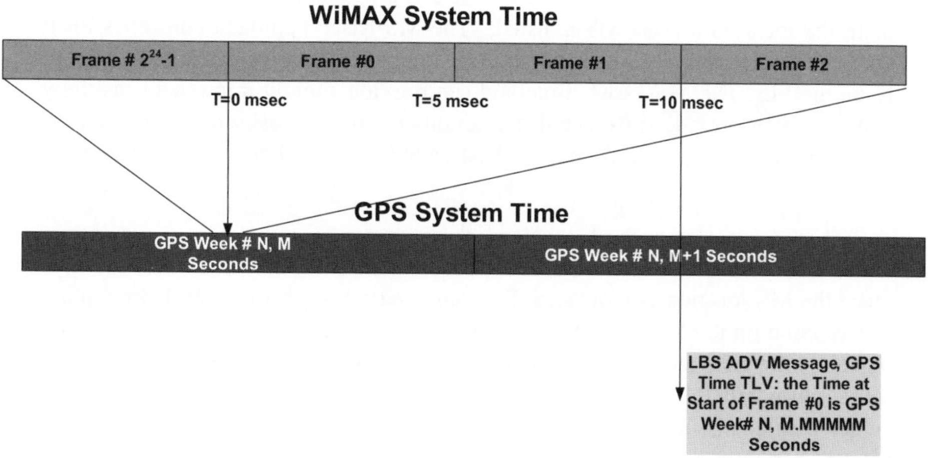

Figure 13.1. Timing aid concept for AGPS.

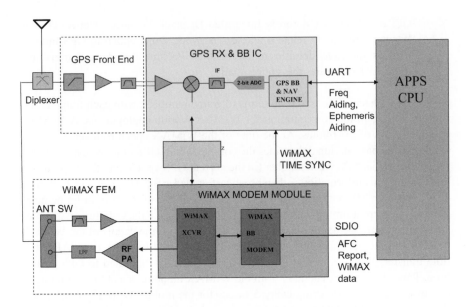

Figure 13.2. Example of GPS and WiMAX modem interaction.

poses of computing the TCXO offset. In addition, per the WiMAX system time transfer method described above, a time strobe can be generated that can relate WiMAX system time to GPS time. During initial GPS signal acquisition, assuming that the WiMAX modem is registered on a WiMAX system, time aiding and frequency aiding can then be transferred to the GPS receiver, using messages from the Apps CPU. This can help speed up GPS TTFF in weak signal by a factor of three or more.

The MS position based on AGPS can be computed on the MS itself (MS-based), or in the location server (MS-assisted). The MS-based approach consumes lower energy per fix, generally speaking, and real-time location tracking can occur directly on the MS. The MS-assisted method can provide slightly improved sensitivity, and may also provide differential corrections or hybrid position computation, as discussed previously, which may be best suited for one-shot fixes, such as in an E911 call.

13.3.1.1 MS-Based AGPS Location

Once the MS location is requested, from an external requestor or an LBS application running on the MS itself, the GPS engine software process is activated. It will power up the GPS receiver and begin the SV acquisition process using the GPS almanac. In parallel, an assist management software layer will retrieve all forms of assists available, such as time and frequency aiding, approximate position, and, most importantly, GPS ephemeris. Once four or more SVs are acquired that meet various internal criteria for carrier-to-noise density (C/No), SV elevation, and geometric distribution, then a user location can be computed as soon as the ephemeris-aiding data is received from the wireless network. The ephemeris data allows the GPS SV locations to be determined, and then the MS location can be computed, usually via a least squares solver [2]. After that, an adaptive tracking loop known as a Kalman filter is applied to produce continual estimates of MS position and velocity, for example, at a 1 Hz rate. The Kalman filter also allows the user position tracking to be maintained for brief periods when the MS experiences a fadeout in GPS signal strength, due to entering a heavily forested or dense urban region. The key aspect here is that the MS-based mode provides for continuous tracking, which is highly desirable for pedestrian or vehicular navigation applications. In addition, in the case in which the LBS application is self-contained on the MS, such as a vehicular navigation application with stored maps and on-screen display, the MS need not contact the network for updated ephemeris-assist data for an hour or more.

13.3.1.2 MS-Assisted Location

With this technique, network assistance data is also provided to the MS GPS receiver to allow rapid acquisition of the GPS SVs. It is, however, *not* required for the MS to receive the SV ephemerides, since the MS position is computed on the LS. The MS acquires four or more GPS SVs, and then uploads the time-stamped pseudoranges to the LS. The LS, with knowledge of the ephemerides and perhaps differential corrections, can then compute the MS location and transfer the fix either back to the MS (if the application runs on the MS) or directly to an external requestor (LR). The MS-assisted approach may provide some sensitivity improvement, as higher computational power is available as well as additional data, and a slightly more accurate fix, but, on the other hand, current MS-based receivers are already approaching –160 dBm in assisted-mode acquisition sensitivity. Moreover, the MS-assisted approach is very cumbersome for tracking applications, and would generate much more two-way data traffic between the LS and the MS if it were used this way. In addition, due to the constant need to upload GPS pseudoranges to the LS, the GPS tracking power on the MS will be much higher than on the MS-based scenario. Thus, the MS-assisted approach is more suited for one-shot fix application scenarios, such as E911 or a personnel tracking application in which the user position would be polled once every 5 minutes or so. In U.S.-based 3GPP2 systems, MS-assisted architectures were deployed initially, but the newer MS devices can support both MS-based and MS-assisted methods, an implicit acknowledgement of the limitations of the MS-assisted approach.

13.3.1.3 Ephemeris Extensions or Long-Term Orbit (LTO) Projections

This relatively new methodology provides for projections of SV ephemeris data for periods as long as seven days or more. The GPS system now provides for a maximum period of applicability of four hours of the SV ephemeris data, with ground-control segment updates to the SVs occurring every hour. With the LTO methodology, complex (often proprietary) algorithms are offered by various vendors to extrapolate the GPS ephemerides for much longer periods, albeit with a loss of SV positional accuracy, and, thus, MS location accuracy. Still, the accuracy degradation is "graceful" and this method could have value for certain scenarios in which network coverage is intermittent or the LS goes down. Logistically, all that is involved from the network standpoint is for the MS to retrieve an LTO data file every few days or so from the LS. This file, perhaps 20–100 K in size, will contain numerous datasets, possibly in compressed format, to represent the ephemeris data over intervals of four hours or less. Since the WiMAX network, of course, supports file downloads, there is really no technology challenge here, and LTO will not be discussed further. In practice, LTO deployment has been slower than anticipated.

13.3.2 Database Lookup Methods

Traditional wireline 911 calls rely on a simple static database lookup approach to determining a caller's location during a 911 call. There is a one-to-one mapping between the caller's phone number and the user's location, which is transmitted to the PSAP (public safety answering point) in civic address format. That is, the caller's phone number serves as the "primary key" in the database lookup.

These database lookup methods also apply to WiMAX devices. For fixed WiMAX equipment such as CPE (customer premises equipment), which is usually operated at the user's household, the user's location could be typically preloaded as the service address, in the carrier's database. Here, however, the end user may have the option to update his/her location manually, in the database, in order to allow for the possibility of moving the CPE device to a different location. (This was the outcome in FCC VoIP 911 order 05-116 [1], which applied to VoIP providers such as Vonage.)

In other cases, even for mobile wireless devices, database lookup methods are also used. For example, a cell ID location consists of reporting the user's position in relation to the serving-cell geometry. This could be the centroid of coverage or the base station location itself. During an E911 call in the United States, a cell ID location is usually used as the initial "Phase I" location for routing the call to the correct PSAP. In other scenarios, the LBS application location accuracy requirement may be on the order of kilometers. Thus, a cell ID-based location would be sufficient, since WiMAX cell site radii are expected to be in the 500–2000 m range for 2.5 and 3.5 GHz deployments. In the CSN, there will be a database mapping the WiMAX base station ID to a geodetic location of the base station. (This database will henceforth be referred to as the base station almanac.) Thus, the BSID serves as the lookup key for the user location. Cell ID can be considered a form of dynamic database lookup since the MS user's serving cell will change as the user roams in the network.

A significant advantage of database lookup methods is fast TTFF, usually just a few seconds once the connection to the LS is established. In WiMAX, a new feature called the LBS-ADV MAC broadcast message allows the MS to know, a priori, the location of the base station, thus avoiding all the lookup latency. This is discussed further in Section 3.4.1

13.3.3 Mobile-Scan-Report-Based Methods

Mobile-Scan report (MSR)-based location methods rely on standard parameters reported by the MS. The basic parameter set that could be used consists of the RSSI, RTD (round-trip delay), and RD (relative delay) fields. These can be reported in the mobile scan response (MOB_SCN-RSP) discussed in Section 6.3.2.3.44 of [3], and then forwarded by the LC to the LS. Alternatively, these parameters can be extracted from the WiMAX modem at the application layer and transmitted to the LS using the R2 interface. Regardless of the transport used, the fix will normally be computed on the LS.

13.3.3.1   D-TDOA (Downlink Time Difference of Arrival)

This method computes user location based on pseudoranges of reference signals received by the MS from multiple base stations. The RTD procedure allows the MS to determine the round-trip time between itself and the anchor BS, so computing the range is given simply by $r1 = c \times RTD/2$, where c is the speed of light. The relative delay parameter for each neighbor cell will give the delay, relative to the anchor BS preamble, of the neighbor cell base stations' preamble arrival time. With these parameters, ranges from the MS to multiple base stations can be computed. With RTD and at least two relative delay parameters, a 2D location for the MS can be derived. Additional RD reports can provide for an overdetermined solution, with RSSI used as a weighting factor. Using the R2 or R3 interfaces to report these MSR parameters to the LS, the LS can compute the user position using the stored BS locations contained in the base station almanac database. Alternatively, using the LBS-ADV message, the MS can compute a D-TDOA position itself, since this message contains the anchor and neighbor BS locations.

13.3.3.2   RF Signature Methods

These involve the construction of an RF signature database, in conjunction with Bayesian statistics, to determine an MS location. Basically, a wireless coverage area is surveyed with a "typical" MS device and, at known locations, parameter datasets are uploaded to a database. These parameters could include BS ID, RSSI, RTD, and relative delay, as well as others that the location vendor deems relevant. Later, to locate an MS in the field, the MS reports its observed parameter set to the LS, which queries the RF signature database to determine the stored location whose corresponding parameter set most closely correlates to that reported.

13.3.3.3   U-TDOA (Uplink Time Difference of Arrival)

This method of location determination is the complement of D-TDOA, in that MS transmissions are used as reference signals for several base stations. Annex L in [3] shows how WiMAX ranging procedures can be used to compute ranges from the MS to multiple base stations.

13.3.3.4 Enhanced Cell ID

In this scheme, the RSSI from multiple BSs and possibly other scan parameters are used, along with the BSID, to determine from a database lookup what subregion of a WiMAX coverage cell the user is actually located in. For example, see reference [8]. In many respects, enhanced cell ID could be considered a form of RF signature method.

13.3.3.5 Hybrid Methods

Although the term hybrid location has many contexts, here we refer to a location method in which GPS and D-TDOA pseudoranges are melded together to produce a user location estimate. This would most likely be used when less than four GPS pseudoranges are available, as may be the case in dense urban environments.

13.3.4 Location Accuracy and Representation

For database lookup methods such as cell ID or enhanced cell ID, WiMAX will provide better accuracy since, for 2.5 and 3.5 GHz deployments, the WiMAX cell radii will be smaller than for a 3GPP or 3GPP2 network, even in rural areas. The primary reason for this is the larger path loss exponent at higher RF frequencies, combined with the system goal of higher data rate throughput. For example, cell radii of 0.5 to 2.0 km will likely be the norm in a system designed for MS devices with max TX power determined by specific absorption rate (SAR) considerations, but often around 23 dBm for handheld devices.

For A-GPS, the location accuracy is determined primarily by the GPS signal strength at the MS, as well as the GDOP (geometric dilution of precision) factor pertaining to the available satellite geometry, so WiMAX will neither enhance nor degrade this. Currently, in strong signal, GPS receivers can realize absolute accuracies in the 5–10 m range, even without differential corrections.

For EOTD methods, the achievable accuracy is a function of several variables. These include WiMAX received RSSI, multipath effects, receiver resolution, and the BS relative timing accuracy. While it is expected that accuracies of 30 m or better can be realized under ideal conditions, there is currently no deployed WiMAX EOTD solution that demonstrates this. One major aspect of this problem is that, presently, most WiMAX receivers are designed to provide resolution and timing estimation sufficient only to maintain the WiMAX system timing lock. A more complex receiver design would likely be required to provide the pseudorange estimation accuracy necessary for WiMAX EOTD location accuracy in the 30 m or less range.

The WiMAX LBS specifications do not explicitly call for a location reporting format. The format used will be a function of the LS for the R2 interface or the external requesting entity in the case of the U1 interface. In other words, the specific representation of user location is out of scope.

13.4 WiMAX NETWORK REFERENCE ARCHITECTURE FOR LBS

The WiMAX network architecture for LBS has been designed in such a way as to minimize the impact on in the WiMAX network [5] has support for use plane, control plane, and hybrid location approaches described in the above section.

The LBS framework in WiMAX accommodates MS-managed location and network-managed location.

In MS-managed location, the network provides the MS with the geographic location of BSs as well as any GPS aiding frequency and timing information over the air. This information may be broadcast/multicast or unicast to the MS. The MS using this information can frequently perform location calculation, for example, by using triangulation schemes, with or without the GPS, at the rates required by the application with no additional support of engagement from the network side. The MS may pass the location to an application resident on the device or a pass it to application servers on the network through upper layer signaling.

For network-managed location, the network provides infrastructure to determine the MS's location based on measurements by the MS in the network or their combination to enhance the accuracy of location before providing that to requesting entities. In this mode, there is need for additional network elements and protocol definitions in the WiMAX network to enable end-to-end LBS.

Figure 13.3 shows the functional reference model for LBS operations in the WiMAX network. The reference points R1–R8 are the standard WiMAX network reference points with specific enhancements for LBS. Furthermore, certain LBS specific functional entities are introduced to enable LBS capabilities in a WiMAX network. These LBS specific functional entities are briefly described in the following.

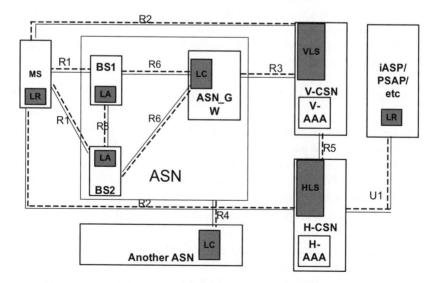

Figure 13.3. Functional reference model for location-based services in a WiMAX network.

- **Location Requester (LR).** The LR is an entity or function that originates a request for location information of a particular MS. The LR may reside outside the WiMAX network [e.g., an external application server (AS) requesting the user location, such as maps.google.com] or in the MS (a navigation application). The LR may also be integrated with or be part of the public safety answering point (PSAP) when location is required for emergency services.
- **Location Server (LS).** The LS in a WiMAX network determines location of the MS. It then provides the determined location to the location requestors as described above. The LS determines the MS location based on one or more mechanisms as described above. The LS is located in the core network of a WiMAX network, commonly called the CSN.
- **Location Controller (LC).** The LC is located in the access network of the WiMAX network, commonly called the ASN. The LC is mainly responsible for coordinating the location measurements of the MS. Upon request from the LS, the LC triggers the location-related measurements and collects all relevant data needed to make a location determination and provides them to the LS.
- **Location Agent (LA).** The LA is located in the BS and is responsible for measuring, collecting, and reporting of location-related data to the LC over the R6 interface. The LA function communicates with the MS over R1 when it triggers or collects location measurements.
- **Mobile Station (MS).** The MS executes various procedures related to LBS. The LR in the MS may request location information from the LS. The MS may request GPS assistance data from the LS over the R2 interface and send back GPS pseudoranges to the LS over R2 as well. In other scenarios, the MS may send WiMAX scan report data to the LS using R2, or to the LA using R1. If the MS determines its own location (with or without assistance from the LS or LA), it may report the location back to the LS over R2 or other external applications directly.

The LBS reference points and interfaces are as follows:

- The R2 reference point is based on OMA SUPL [6] or IETF HELD [7]. This reference point is used to primarily support user-plane location capabilities.
- The R3 reference point uses AAA protocols like RADIUS and DIAMETER as defined in the IETF. This reference point is primarily used to support control plane location capabilities.
- The R4, R6, and R8 protocols use the WiMAX control plane transport that is based on UDP, with the WiMAX headers.
- WiMAX LBS does not mandate any particular protocol for the U1 interface, as this interface is not in the scope of the WiMAX specifications. The WiMAX LS is typically expected to work with LRs that communicate using host of protocols such as Parlay X, MLP, ESP, HELD on the U1 interface.

13.4.1   Support for MS-Managed Location in WiMAX

MS-managed location means that the MS takes necessary measurements, calculates the location, and uses it. Each BS broadcasts its geographic location as well as its neighboring BS's location. This information is broadcast periodically by the serving BS as an L2 LBS-ADV message. The MS uses the LBS-ADV information or measurements and calculates the location on its own (see Figure 13.4). The MS does not need to report its location to the WiMAX network. The MS MAY send its location to applications, for example, for local searches. The exact algorithms used by the MS to determine its location are outside the scope of standards specification. The MS may have a standalone location capability such as an integrated GPS receiver, or it may rely on the geographic location broadcast from the WiMAX network. No LBS-specific functional support is required in either the ASN or the CSN. The MS-managed location procedure is optional for both the MS and the BS.

13.4.2   Support for Control-Plane Methods in the Network Architecture

The WiMAX network fully and efficiently supports the control-plane location methods as described in the previous section. The basic idea behind the control-plane location is that the WiMAX R3 reference point is invoked for the location determination by the LS. When the LS gets a location request from an LR, the LS in turn contacts the correct set of entities in the ASN (also known as the access network) to get the MS location determination going. There are three main types of control-plane location that are supported in WiMAX:

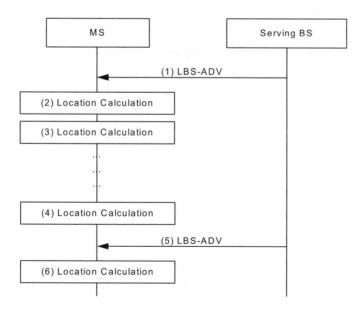

Figure 13.4. MS-managed location.

1. Serving base station identity (BSID)
2. DL signal-based approaches using mobile scanning reports (e.g., EOTD)
3. UL signal-based approaches using base station measurement reports (e.g., UTDOA)

These approaches offer different levels of trade-off between the latency and the accuracy of the location fix. Based on the right trade-off required for the given location request, the appropriate mechanism is chosen by the LS. The LS then contacts the LC in the ASN with the measurement request. The LC then contacts the LAs as needed to get the measurements back to the LC.

It is to be noted that for serving BSID-based location, no measurements are needed. The LC should generally be aware of the MS's serving BSID and will promptly return that back to the LS. The LS typically performs a database lookup to get a rough idea of the coordinates of the serving BSID and will report this back to the LR.

For EOTD and other DL signal-based approaches, the MS will need to measure downlink signals from the serving BS and neighbor BSs and report back to the LA in the serving BS. The serving BS in turn will report the measurements back to the LC and the LC back to LS. The location computation is done in the LS and will be returned back to the LR.

For UTDOA and other UL signal-based approaches, the serving BS will coordinate with the neighbor BSs to measure the uplink signal from the MS at a given point in time and frequency. The measurements from the neighbor BSs are then sent to the serving BS, which aggregates all of the measurements and sends them to the LC. The LC sends them upstream to the LS. The location computation is done in the LS and will be returned back to the LR.

13.4.3   Support for User-Plane Methods in the Network Architecture

User-plane location uses the R2 reference point (instead of the R3 reference point used for the control plane location) directly between the MS and the LS. In other words, when the LS gets a location request, it directly contacts the MS on the R2 reference point and requests either measurements such as mobile scan reports or GPS pseudoranges (in the event the MS has GPS) or the MS location itself (in the event the MS has stand-alone location computation capability).

The LS may also aid the MS with GPS assistance and/or other types of assistance such as the location of the serving and neighbor BSs in order to get a faster fix or to help the MS with the location determination, respectively. Location determination may be done at the LS or the MS depending on the determination mechanism invoked and the number and quality of the measurements available.

WiMAX does not natively define a user-plane location protocol, but instead uses two readily available off-the-shelf protocol stacks to do the job. These two stacks are OMA SUPL and IETF HELD.

13.4.4 Contrast of User-Plane and Control-Plane Methods

User-plane and control-plane location are both used to achieve the same goal—to obtain the MS location. However, there are trade-offs between these two approaches and, depending upon the situation, the LS needs to invoke the right approach.

The big advantage of user-plane location is that the LS can directly get to the MS and signaling is minimized across the various reference points. However, for this to the happen, the MS needs to have obtained an IP address and be fully registered with the LS, and application-layer support is required in the MS. For control-plane location, the LS does not communicate directly with the MS and, hence, there is no hard requirement for the MS to have obtained an IP address. In other words, the control-plane location approach relies more on the L2 connectivity of the MS. However, the signaling costs are generally higher in control-plane location as the signaling will have to traverse multiple reference points before measurements can be obtained.

Another noteworthy point is that user-plane location may not work at all times, especially in the case of GPS, when the MS is indoors. It is well known that GPS does not function well indoors. GPS assistance will help here to some extent, but there could still be a coverage issue if the MS is deep indoors, for example. Control-plane location always relies on native WiMAX signals and those should be readily available indoors if the link budget planning is done correctly prior to the deployment.

13.4.5 Support for Mixed-Plane Methods in the Network Architecture

Apart from the control-plane and user-plane methods, a third approach could be the mixed-plane method. The mixed-plane method is nothing but the LS invoking both control-plane measurements (DL scan reports from the MS, for example) and user-plane measurements (e.g., GPS pseudoranges from the MS) at the same time. The LS can then get the best of both worlds and combine the measurements to get much better accuracy for the location fix. This approach is also fully supported in the WiMAX network.

The trade-off here is that this method costs a whole lot more in terms of latency for the fix and the associated signaling; however, this will translate to much better accuracy for the MS location indoors, where an insufficient number of GPS satellites may be visible.

13.4.6 Support for Accounting

The WiMAX network primarily supports accounting and UDR generation. Different types and granularities of accounting can be activated on the WiMAX network based on the operator requirements. This flexibility in the accounting architecture helps build innovative business models and partnerships between the external LRs and the WiMAX operator.

Typically, there would be an accounting server in the CSN that keeps track of the generated UDRs from the incoming location requests from the external LRs. Whenever a location request is successfully completed by the LS, the LS can then update the accounting server to indicate the same. On a periodic basis, billing and settlement can be done between the LRs and the WiMAX operator based on these UDRs that are available in the accounting server.

13.4.7 Support for Roaming

The WiMAX network provides support for location determination under roaming scenarios. Two types of roaming scenarios are possible. In the first scenario, the location determination is done by the LS in the visited network. The MS location is then passed on to the LS in the home network, which passes it on to the LR.

In the second scenario, the location determination is done by the LS in the home network. The measurements are obtained from the LS in the visited network for doing this.

13.5 SUMMARY

The WiMAX network architecture provides all the required functions for efficiently supporting location-based services with minimal changes to the base WiMAX network. Support is provided for security, QoS, control-plane location, user-plane location, mixed-plane location, accounting, roaming, and other essential network features needed for successfully deploying the location-based services in WiMAX networks.

13.6 REFERENCES

[1] FCC First Order and Notice of Proposed Rulemaking 05-116, VoIP 911 Order, June 2005: http://www.fcc.gov/cgb/voip911order.pdf.

[2] E. Kaplan and C. Hegarty (editors), *Understanding GPS; Principles and Applications,* 2nd ed., Artech House, 2006.

[3] IEEE Std. 802.16-2009, Standard for Local and Metropolitan Area Networks, Part16: Air Interface for Broadband Wireless Access Systems, May 2009.

[4] P. Zarchan and H. Musoff, *Fundamentals of Kalman Filtering, a Practical Approach,* 2nd ed., American Institute of Astronautics and Aeronautics, 2005.

[5] WiMAX Forum Network Architecture, Stage 3: Detailed Protocols and Procedures, Release 1 Version 1.3.0, November 2008.

[6] Secure User Plane Location V 2.0 Enabler Release Package, Open Mobile Alliance™, http://member.openmobilealliance.org/ftp/Public_documents/LOC/Permanent_documents/OMA-ERP-SUPL-V2_0-20080627-C.zip.

[7] HTTP Enabled Location Delivery (HELD), draft-ietf-geopriv-http-location-delivery-13, M. Barnes, J. Winterbottom, M. Thomson, and B. Stark, February 2009.

[8] T. Wigren, Adaptive Enhanced Cell-ID Fingerprinting Localization by Clustering of Precise Position Measurements, *IEEE Transactions on Vehicular Technology,* Sept. 2007.

CHAPTER 14

WiMAX ACCOUNTING

AVI LIOR

14.1 INTRODUCTION

Accounting conjures up a broad set of topics, from the gathering of usage information to how that information is used by the back-office systems (e.g., how bills are generated). A common view is that if a feature or an application has to do with money, it must be an accounting function. Although this view is mostly true, accounting is used for other purposes as well. For example, accounting information is used to audit the performance of the system or to generate periodic monthly performance reports by postprocessing accounting records. Accounting information can be used to feed intrusion detection and avoidance systems, lawful intercept systems, and so on.
The WiMAX Forum specifies the following:

- The accounting information gathered in the access network.
- The accounting record and the protocol for delivering the information to the core.

This chapter examines the basic accounting function, which is commonly referred to as offline accounting or postpaid accounting. WiMAX supports other accounting-related features, such as online accounting or prepaid accounting. Prepaid accounting provides a means for controlling access to services for users who prepay for their service. In WiMAX Release 1.0, prepaid support is described in detail [1].
The Internet Engineering Task Force (IETF) defines two authentication, authorization, and accounting (AAA) protocols; one called remote authentication dial-in user service (RADIUS), as specified by [2,3], and the other called DIAMETER, as

WiMAX Technology and Network Evolution. Edited by Kamran Etemad and Ming-Yee Lai **361** specified by [4]. WiMAX Release 1.0 is based on the RADIUS AAA protocol. Support for the DIAMETER AAA protocol is planned to be introduced in WiMAX Release 1.5. This chapter does not cover accounting using the DIAMETER AAA protocol. Although DIAMETER and RADIUS are significantly different at the protocol level, they are very similar from an operational perspective. Studying the RADIUS-based WiMAX accounting should be easily mapped to understanding WiMAX accounting using the DIAMETER AAA protocol.

14.2   ACCOUNTING ARCHITECTURE

The WiMAX accounting architecture [5] and is similar to the accounting architecture used by 3GPP2 [6]. Both architectures are based on the RADIUS AAA protocol, which is used for authentication and authorization of sessions as well as accounting.

Figure 14-1 shows the AAA architecture for a nonroaming scenario. Accounting records are delivered from the network access provider (NAP) ASN Gateway (ASN GW), commonly referred to as the anchor authenticator (because it is the RADIUS AAA client where the session is first established), to the RADIUS home AAA server in the home network service provider (HNSP) using the AAA RADIUS protocol. Also, if the home agent (HA) is enabled for accounting, then RADIUS-based ac-

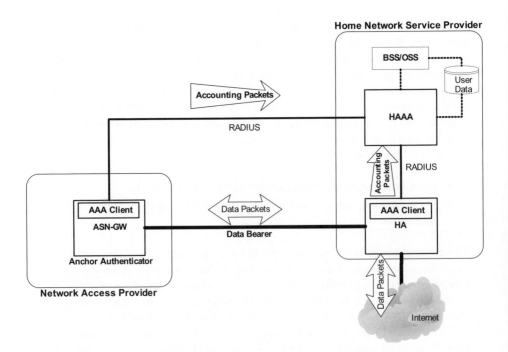

Figure 14.1. Nonroaming AAA architecture.

counting records can be delivered from the HA to the home AAA server (HAAA). After using the information in the accounting records, the HAAA forwards the accounting records to the back-office system(s) or business support system (BSS)/operations support system (OSS). The HAAA may augment or modify the information in the accounting stream with data such as that contained in the user data store. The interface between the HAAA and the BSS/OSS system is not in the scope of WiMAX.

Figure 14-2 shows the architecture-supporting roaming scenario. As the figure shows, the only difference between the roaming and nonroaming architecture is the introduction of the visited network service provider (VNSP), which consists mainly of the visited AAA RADIUS server (VAAA). In the roaming architecture, the HA can exist in either the HNSP or VNSP. Figure 14-2 shows the case in which the HA exists in the HNSP.

When roaming, the accounting records are routed from the network access provider's ASN GW anchor authenticators to the HAAA in the HNSP via the VAAA in the VNSP. The VAAA processes the accounting records and may modify them before forwarding the records to the HAAA in the HNSP. The VAAA may also pass the accounting records to its BSS/OSS system. The BSS/OSS system in the VNSP uses the information in the accounting records for settlement with the HNSP and the NAP. The interface between the VAAA and the BSS/OSS is not in scope of WiMAX.

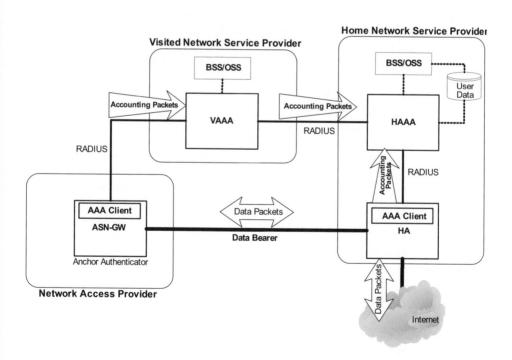

Figure 14.2. Roaming AAA architecture.

Not shown in the above architectural diagrams are AAA intermediaries. These typically exist between the VNSP and the HNSP to facilitate scenarios in which the HNSP does not have a direct business relationship with the VNSP. For example, to support global roaming for its subscribers, an HNSP uses AAA intermediaries called global roaming exchanges to allow its subscribers to roam to VNSPs supported by that global roaming exchange. In this case, AAA accounting records are used for billing settlement between the VNSP, the global roaming exchange, and the HNSP. The accounting packets need to be routed through these entities to the HNSP.

According to the RADIUS specification [3], the delivery of accounting records can be decoupled from the delivery of authentication/authorization records. That is, they do not have to travel through the same path or end up in the same place as the packets used to authenticate and authorize the subscriber. In WiMAX however, accounting records can travel through different servers but they must reach the same HAAA server in the HNSP that authenticated the subscriber. Furthermore, when intermediaries are concerned, the accounting records should pass through the same AAA entities that were involved in the authentication and authorization phase. After all, the authenticating and authorization entities want to get paid. To achieve this in WiMAX, accounting records are routed through the network using the same scheme used to route the authentication/authorization packets.

Also, in RADIUS the accounting packets can be delivered in real-time mode, in which each packet is delivered immediately down the AAA chain as it is received, or in a store-and-forward mode, in which an intermediary node can store accounting packets and then deliver them as a batch to the next hop. WiMAX requires that accounting packets be delivered in real-time to the HNSP. This is because the accounting packets are used as a signal to tell the HAAA server that the session has started or terminated. The HAAA uses these signals to establish and tear down the session state. Also, the accounting packets contain information that is critical for other features to work and thus are required in real-time in the HNSP.

As shown in the previous figures, the AAA client is colocated with the anchor authenticator established during initial network entry. The role of the AAA client is to collect usage information and send it using RADIUS packets to the HNSP via the VNSP in a roaming scenario. Thus, the AAA client needs to be in a position to observe the data. However, in WiMAX the data bearer may relocate and, thus, the AAA client may not always be located in the node on the data path. To facilitate this, WiMAX introduces an accounting agent, which is always located with the data path and, thus, is able to report usage counts back to the AAA client. The accounting agent can be located either in the base station or in the currently serving data path, as shown in Figure 14-3.

14.3 ACCOUNTING CONCEPTS

To understand the data being reported in the accounting stream, we need to establish how WiMAX models users, subscriptions, IP sessions, and so on. Figure 14-4 illustrates this model. The left-hand part of the figure shows the hierarchical rela-

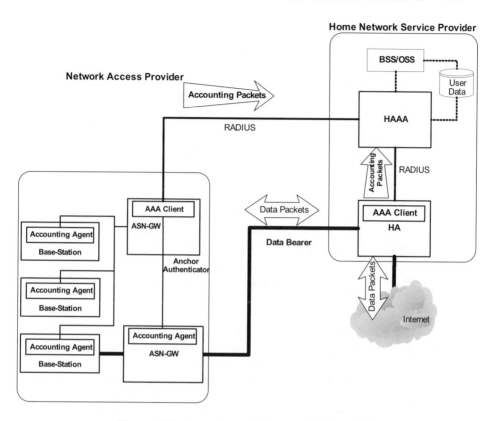

Figure 14.3. Accounting architecture within the ASN.

tionship of the entities that comprise this model, and the right-hand part of the figure contains the identifiers associated with each of the elements of this model. These elements, as identified by the identifiers, appear in the accounting records and other AAA messages.

At the very top of the WiMAX accounting model is a person or a billing entity. Think of this person as the one who receives the monthly bill. External to the home network, the person is identified by the chargeable user identity (CUI) [7]. CUI allows the HNSP to identify the user in such a way as to not reveal the user's true identity and, thus, provide the user with privacy. The CUI is sent in all RADIUS messages and, thus, is available in the accounting stream.

A person may have one or more WiMAX devices represented by subscriptions. A subscription is identified by the network access identifier (NAI) [8]. The NAI consists of several components, two of which are critical to accounting. The home realm (username@homerealm.com) identifies the HNSP and is used to route the AAA messages to the HNSP. The user name identifies the subscription uniquely within the HNSP. Depending on whether the device is using identity hiding or not, the user name portion may change from authentication to authentication. The other

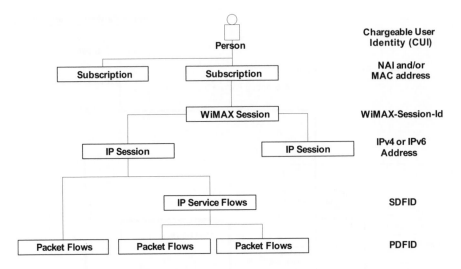

Figure 14.4. WiMAX accounting model.

identifier that may be used to identify the subscription is the media access control (MAC) address of the device. Both the NAI and the MAC address appear in the RADIUS messages, including the accounting packets.

Unfortunately, the model at this level is rather fuzzy and depends on how the operator treats the relationship between a person and his or her devices. Some operators will model the world as one person having exactly one device or one subscription, and some operators model the world so that one person can have one or more WiMAX devices. For example, a person can have a WiMAX-enabled camera, MP3 player, phone, and laptop.

When a WiMAX device connects to the WiMAX network, a WiMAX session is established and the HNSP assigns the session an identifier called the WiMAX session ID, which is unique in the scope of the home realm. The WiMAX session ID remains constant for the lifetime of the WiMAX session, that is, until the device performs the network exit procedures. The WiMAX-Session-Id remains constant even when the MS hands off between network access providers. The WiMAX session ID is reported in the accounting messages, making it the ideal identifier to use for matching all records associated with a given session.

Once the Mobile Station (MS) successfully establishes the WiMAX session, it then acquires one or more IP session(s). Each IP session is identified by the IP address assigned to the MS. Note that the IP address may not be unique between MSs. For example, two WiMAX MSs may be assigned nonroutable IPv4 addresses that are the same as in 10.10.10.1. However, for each device session (identified by the WiMAX Session ID), the IP addresses assigned to the device must be unique.

Each accounting record is associated with a single IP session and, thus, contains a single IP address attribute.

Each of the IP sessions may be further classified into IP service flows. An IP service flow is specified by one or more IP classifiers. An IP classifier specifies the values of the various data elements of the IP header used to determine if a packet is part of an IP service flow. If the values of a given packet match the classifier, then that packet is considered part of the flow. A full explanation of how an IP classifier works is beyond the scope of this book.

When the PD-flow-based accounting mode is used, then the usage associated with each packet flow is reported in its own accounting record that contains the packet data flow identifier (PDFID).

Some services offered by an operator may consist of several IP service flows. For example, a video teleconference service will consist of an IP flow associated with the video stream, an IP flow associated with an audio stream, and, perhaps, an IP flow associated with a chat traffic. To allow the system to easily associate each of these flows with the teleconference service, WiMAX provides a service data flow ID (SDFID). IP service flows belonging to the same service will be assigned the same SDFID. The SDFID appears in accounting messages and helps to correlate accounting records belonging to a particular service.

14.4 ACCOUNTING OPERATIONS

The intent of this section is to introduce the reader to the operations and procedures of accounting and key accounting attributes. Several call flows are used to introduce increasingly complex accounting concepts.

WiMAX supports two types of accounting operations: IP-session-based accounting and flow-based accounting. Both operations are supported by the same procedures. They differ, however, by the information contained in the accounting records and the number of accounting records exchanged.

IP-session-based accounting is the default mode of accounting. That is, all ASN GWs must support IP-session-based accounting. IP-session-based accounting accounts for usage on a per-IP session, that is, the ASN GW reports usage against the traffic delivered from/to the assigned IP address of the MS.

PD-flow-based accounting is optional to support in the ASN GW. An ASN GW that supports PD-flow-based accounting indicates this capability to the HNSP during the network access authentication and authorization procedures. The HNSP can then instruct the ASN GW to report accounting on a flow-by-flow basis. In this mode of operation, the ASN GW sends an accounting record for each flow as defined by PDFID and/or SDFID included in the accounting record. The use of PD-flow-based accounting allows the HNSP to get more detailed usage information and bill the user in a differentiated manner. Flows used for voice can be billed at a higher rate than flows used for browsing the Internet. PD-flow-based accounting generates many more accounting records than IP-session-based accounting.

Next we will look at the accounting procedures using call flows. To simplify the presentation, only key accounting attributes are highlighted and some abstracts are abstracted. For example, the attribute Counters in the call flows is an abstraction for a set of four counters that count uplink and downlink packets and uplink and downlink octets.

14.4.1 Single Accounting Session

The first call flow introduces the most basic accounting procedure. In this call flow, the MS establishes a WiMAX session and an IP session in the ASN. The IP session and WiMAX session are terminated on the same ASN. Figure 14.5 illustrates the process, which consists of the following nine steps:

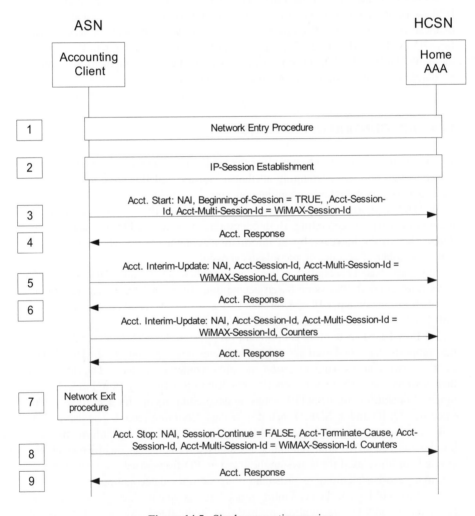

Figure 14.5. Single accounting session.

1. The MS performs the network entry procedure with the home connectivity service network (HCSN) of the HNSP. This procedure involves the home AAA server, which exchanges several messages with the anchor authenticator, where the accounting client is located. During this process, the home AAA establishes a session state for the MS, called a WiMAX session, for which the home AAA assigns an identifier called the WiMAX session ID. The WiMAX session ID is returned back to the ASN. At this point, the MS is authenticated and authorized to establish an IP session.

   During initial network access authentication, the anchor authenticator informs the home AAA which accounting modes its accounting client can support by setting the appropriate values in the WiMAX capability attribute sent to the home AAA. The accounting modes can be IP session-based and/or PD-flow-based. The home AAA will instruct the accounting client which mode it wants to operate under by setting the appropriate values in the WiMAX capability attribute returned to the anchor authenticator.

   Also, during initial network access authentication, the home AAA will indicate to the accounting client (via the anchor authenticator) whether it wants to receive accounting interim updates. Accounting interim updates records are sent periodically to the HNSP to mitigate the loss of accounting information.

2. An IP session is established. The MS is allocated an IP address and the pre-provisioned PD flows received from the home AAA in step 1 are installed.

3. Once the anchor authenticator establishes the IP session, it causes the accounting client to send an accounting request start message to the home AAA. This message will signal that the IP session has started. The message contains the following attributes:

   NAI. The network access identifier is used to identify the subscription and used to route the accounting messages from the accounting client to the home AAA.

   NAS identifier. The network access server identifier of the accounting client used to differentiate between one accounting client and another accounting client. The NAS identifier contains the domain name of the accounting client, for example, nas1.example.com. The IP address of the accounting can also be conveyed using the NAS IP address.

   Accting session ID. This is used to match the accounting start record with subsequent accounting interim-update records and accounting stop records.

   Accting multi session ID. This is used to match all accounting records associated with this MS's WiMAX session. The value is set to the WiMAX session ID received during initial network entry (see step 1).

   Framed IP address. This is the IP address associated with this IP session.

   Beginning of session. This indicates whether the accounting request start corresponds to the start of a WiMAX session (value = TRUE) or that this accounting request start record corresponds to a continuation of a WiMAX session (value = FALSE).

In the case of PD-flow-based accounting, the accounting client sends an accounting request start record for each of the preprovisioned flows that it installed. In this case, each accounting record includes the flow identifiers:

SDFID. Service data flow identifier, which groups two or more PD-flows together.

PDFID. Packet data flow identifier which identifies the PD flow.

Since the session is just starting and there is no usage to report, the accounting request start record does not contain any usage counters.

4. The reception of accounting request start record(s) by the home AAA signals that IP service is being provided to the MS. The home AAA acknowledges receipt by sending accounting response record(s) for each accounting request start message it received.

5. If the home AAA indicated that it wants to receive interim update information (in step 1), then the accounting client sends an accounting request interim update record that includes the subscription identity (NAI), the NAS identifier, accounting session ID, accounting multisession ID, and IP address attributes (described above) and usage counters measured up to this point. In the case of PD-flow based accounting, the accounting client sends an interim update message for each flow containing the SDFID and PDFID and the counters measured for the flow up to this point.

6. For each accounting request interim update received, the home AAA replies with an account response record acknowledging reception of the accounting record.

7. The ASN performs network exit procedures. These can be triggered by the MS or by the network itself.

8. When the IP session is finally terminated, the accounting client sends an accounting request stop message to the home AAA. In the case of PD-flow-based accounting, the accounting client sends an accounting request stop message for each of the flows. Each accounting request stop message contains the subscription identity (NAI), the NAS identifier, accounting session ID, accounting multisession ID, framed IP address and, in the case of PD-flow-based accounting, the SDFID and PDFID identifying the flow. Also, the accounting request stop record contains the final usage counters for the IP session or PD flow.

To indicate that this is the end of the WiMAX session, the accounting client includes the session continue attribute set to FALSE. To indicate the cause of the conclusion of the WiMAX session, the accounting client also includes the accounting terminate cause attribute set to an appropriate value.

9. Upon receiving the accounting request stop record from the accounting client, the home AAA determines that the IP session and, hence, the WiMAX

session has terminated and can now clear any state information that it has maintained for the MS. In the case of PD-flow based accounting, the home AAA receives multiple accounting records, one for each flow. The home AAA will only terminate any state information upon reception of the accounting stop record associated with the initial service flow (PDFID < 1 and 20). The home AAA acknowledges the reception of each accounting request stop message by sending a corresponding accounting response message.

The above example illustrates an accounting message exchanged between the ASN and the home AAA. As can be seen by the call flow, the arrival of the various accounting messages at the home AAA notifies when the IP session starts and when and why it terminated. From the perspective of the home AAA, service to the MS does not start when it authenticates the MS but, rather, when the accounting client sends the accounting request start message, which is sent immediately after the IP session has been established.

Each accounting start-stop pair defines an accounting session. An accounting session records the service the user received during that period of time. The accounting client assigns each accounting session a unique identifier, included in the accounting session ID attribute contained in every accounting record. The home AAA uses the accounting session identifier to match the start record with the stop record and interim update records for that accounting session.

14.4.2 Multiple Accounting Sessions

The above example illustrated a single accounting session that started with the first received IP service and ended when the MS exited the system.

However, in reality the service provided to the MS will change. For example, the MS may be provided with different QOS, or the MS session may cross a tariff boundary (business hours to nonbusiness hours), or the MS may be hotlined or taken out of hotlining. These events would cause the accounting clients to generate multiple accounting sessions.

The next call flow (see Figure 14.6) illustrates the use of multiple accounting sessions. To simplify the call flow, we have omitted the accounting response records generated by the home AAA upon reception of the accounting records from the accounting client. There are seven steps:

1. The network entry procedure is performed and a WiMAX session is established for the MS. The home AAA assigns a unique session identifier for the WiMAX session; in this case the value is 30. The home AAA sends the WiMAX session ID to the anchor authenticator, which will use it in all transactions with the home AAA, including accounting.

2. IP session is established for the MS. The accounting client establishes an accounting session and assigns it a unique accounting session ID. In Figure 14-6, the accounting session ID is 10. The accounting client sends an accounting request start to the home AAA. This record contains the subscription identity

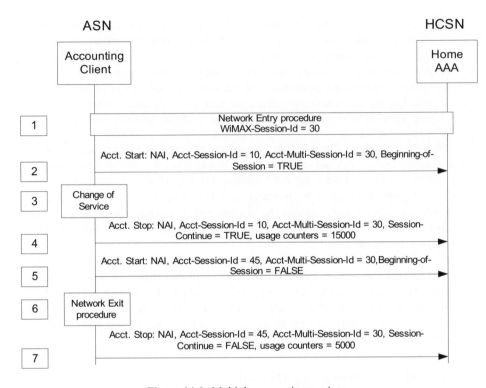

Figure 14.6. Multiple accounting sessions.

(NAI), the NAS identifier, the accounting session ID set to 10, and the accounting multisession ID set to the value of the WiMAX session, in this case 30. The beginning of a session attribute is set to TRUE to signal that the IP session is starting. Also included is the IP address allocated to the IP session of the MS, which is carried in the framed IP address attribute.

In the case of PD flow-based accounting, the accounting client will establish an accounting session with a unique accounting session ID for each PD flow being reported. Multiple accounting request start records, one for each PD flow, will be sent to the home AAA.

3. A change of service is triggered due to some event that would necessitate the home AAA to know about. This is because the operator may want to charge differently for the new service being provided or not charge at all.

4. The accounting client sends an accounting request stop record containing the subscription identity (NAI), accounting session ID attribute set to 10, and accounting multisession ID set to 30. To indicate that the WiMAX session is not terminating, the session continue attribute set to TRUE is included. Also, the usage counters set to 15,000 are included to indicate the total usage under the old service provided to the MS.

In the case of PD-flow-based accounting, an accounting request stop record will be generated for each of the PD flows being reported.

5. The accounting client establishes a new accounting session and assigns it a unique identity of 45. It sends an accounting request start message with the subscription identity (NAI), the NAS identifier, the accounting session ID set to 45, accounting multisession ID set to 30, and beginning of session set to FALSE.

   As in step 2, for PD-flow-based accounting, the accounting client will generate a new accounting session with a unique accounting session identifier for each PD flow, and will send the corresponding accounting request start message to the home AAA.

   The accounting client will clear all of the usage counters.

6. The ASN performs network exit procedures triggered by the MS or by the network itself.

7. When the IP session is finally terminated, the accounting client sends an accounting request stop message to the home AAA. This message contains the subscription identity (NAI), the NAS identifier, the accounting session ID set to 45, accounting multisession ID set to 30, and session continue set to FALSE to mark that this is the termination of the WiMAX session. Also, the usage counters are set to 5000 to indicate how many resources were used for this particular accounting session.

   In the case of PD-flow based accounting, an accounting request stop message is generated for each PD flow.

In the above example, multiple accounting records are received by the home AAA. The accounting session ID is used to pair the accounting request start records and accounting request stop records that demarcate each accounting session.

In the case of PD-flow-based accounting, multiple accounting sessions are established, one for each flow. Each of these accounting sessions has a unique accounting session ID. All the accounting records, regardless of the session, are associated with a common WiMAX session by the accounting multisession ID, set to the value of the WiMAX session identifier received from the AAA during network entry procedures. This value helps the home AAA collect the accounting records for the MS for that WiMAX session.

Since at the start of each accounting session the usage counters are reset, each accounting request stop record contains the usage counters for that particular accounting session. Thus, in the above example, the total usage for the IP session is 15,000 for the first service and 5000 for the last service, for a total of 20,000 units.

14.4.3 Anchor Authenticator Relocation

Another reason for the generation of multiple accounting records is anchor authenticator relocation. From time to time as the MS moves, the anchor authenticator is relocated. When this happens, the accounting client in the old anchor authenticator

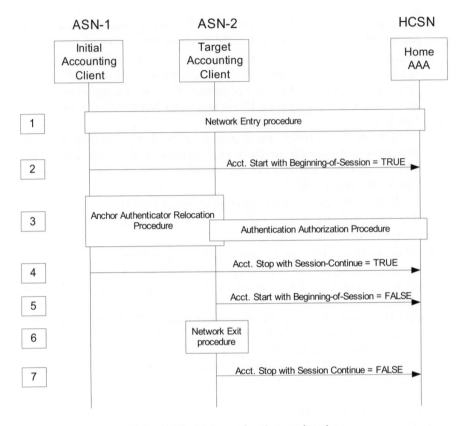

Figure 14.7. Anchor authenticator relocation.

sends accounting stop records to indicate that service has terminated at the old anchor authenticator, and the accounting client in the new anchor authenticator sends out accounting start records to signal the start of service at the new anchor authenticator.

The following call flows illustrate message flows and key values for the anchor authenticator relocation scenario (see Figure 14.7):

1. The MS performs initial network entry into ASN-1. At this point, the home AAA assigns the MS a WiMAX session ID that is sent back to the anchor authenticator.

2. As in the previous scenario, the accounting client sends an accounting request message containing the following information:

   - Accounting session ID set to a value generated by the accounting client for this accounting session

   - Accounting multisession ID set to the WiMAX session ID received during network entry (step 1)

   - Beginning of session set to TRUE, indicating that the session is starting

- User name set to the NAI used in network entry, identifying the subscriber
- Framed-IP address specifying the IP address allocated to the IP session established for the MS
- NAS identifier specifying the identity of the accounting client

As with the other scenarios, in the case of PD-flow-based accounting, an accounting request start message is generated for each PD-flow. Each of the records contains the SDFID and PDFID associated with the flow.

3. The anchor authenticator relocation procedure is triggered. As part of this procedure, the new anchor authenticator authenticates with the home AAA. The home AAA recognizes that this authentication is for an existing MS—same WiMAX session—and returns the WiMAX session ID established during the initial network entry (step 1) to the new anchor authenticator.

4. The initial anchor authenticator in ASN-1 notifies its accounting client that the IP session for the MS has terminated due to a relocation. The accounting client sends out an accounting request stop record containing the following attributes:

   - Accounting session ID set to the same value generated in step 2, matching the value used in the accounting request start message.
   - Accounting multisession ID set to the WiMAX session ID received during network entry (in step 1)
   - Session continue set to TRUE, indicating that the session is not terminating and will continue
   - User name set to the NAI used in network entry, identifying the subscriber
   - Framed IP address specifying the IP address allocated to the IP session established for the MS
   - NAS identifier specifying the identity of the accounting client
   - Usage counters

5. The new anchor authenticator in ASN-2 notifies its accounting client that the IP session for the MS has started. The accounting client sends out an accounting request start record containing the following attributes:

   - Accounting session ID set to a unique value within the scope of this accounting client
   - Accounting multisession ID set to the WiMAX session ID received during the authentication performed in step 3. This value is the same as the value that was assigned by the home AAA to the session during initial network entry procedure (step 1)
   - Beginning of session set to FALSE
   - User name set to the NAI used in the authentication performed in step 3
   - Framed IP address specifying the IP address allocated to the IP session established for the MS
   - NAS identifier specifying the identity of the accounting client in ASN-2

6. The MS performs network exit procedure.

7. When the IP session is finally terminated, the accounting client in ASN-2 sends an accounting request stop message to the home AAA. This message contains:

   • Accounting session ID set to the same value generated in step 5, matching the value used in the accounting request start message

   • Accounting multisession ID set to the WiMAX session ID received during authentication (step 3)

   • Session continue set to FALSE, indicating that the session is terminating

   • User name set to the NAI used in network entry, identifying the subscriber

   • Framed IP address specifying the IP address allocated to the IP session established for the MS

   • NAS identifier specifying the identity of the accounting client

   • Usage counters

In this call flow, it is important to note that the accounting session ID that is used to match accounting records for a given accounting session is locally generated by the accounting client and, thus, it is possible that both accounting clients to use the same value for the accounting session ID. The home AAA must, therefore, use both the NAS identifier and the accounting session ID to distinguish between accounting sessions from one accounting client and another accounting client.

14.5 ACCOUNTING AT THE HOME AGENT

WiMAX supports optional accounting from the home agent. In release 1.0 of WiMAX, accounting reports from the HA are of limited value. An operator may turn on accounting from the home agent to augment the accounting feed received from the ASN, where the accounting feed from the HA may be used to validate the usage counts received from an ASN. In later releases of WiMAX, in which it will be possible that a WiMAX session may utilize multiple HAs, for example, having multiple IP sessions (IPv4 and IPv6) or having local breakout support, then accounting from the HA will be necessary to report how much traffic was handled by each HA.

Operationally, the accounting reporting from the home agent is very similar to the accounting reporting from the ASN. The HA has an accounting client able to generate accounting records and send them to the home AAA. The call flows presented in Figure 14-4 and Figure 14-5 are relevant to accounting record generation of an HA.

The difference between accounting reporting from the HA and accounting reporting from the ASN is in the information being reported. In WiMAX release 1.0, the HA does not receive a preprovisioned PD-flow specification from the home AAA. Therefore, the HA cannot perform PD-flow-based accounting; it can only perform IP session-based accounting.

Since the HA is the anchor point for an IP session, the HA will typically only report one accounting session. That is, for each MS it will generate one accounting request start message, followed by zero or more accounting request interim update messages, and terminating with one accounting request stop message.

In some very rare cases though, the home agent will report multiple accounting sessions, which results in multiple accounting starts and stops. These are triggered by hotlining events when the HA is a hotline device.

14.6 PROCESSING OF ACCOUNTING RECORDS IN THE VISITED NSP

14.6.1 VAAA Processing

The visited AAA of the VNSP receives accounting records from the NAP for each MS that is roaming through the VNSP. The visited AAA forwards the accounting records to the home AAA at the HNSP, and may need to modify them to add the NSP-ID attribute. The VNSP also stores the accounting records to be used for settlement between it and each of the HNSPs and NAPs it has roaming agreements with.

The arrival of the accounting request stop messages with session continue set to FALSE signals the VAAA that the session has terminated. This causes the VAAA to clear the information associated with the session (session state). In the case of PD-flow-based accounting, the AAA waits for the accounting request stop message with session continue set to FALSE that is associated with the Initial Service Flow (ISF) to indicate the termination of the IP session (PDFID will be set to a value of 1 to 20).

14.6.2 BSS/OSS Processing at the Visited NSP

The visited NSP bills each home CSN by duration and/or volume of traffic for all the sessions serviced by the visited NSP during a given time span. If flow-based-accounting is used, the visited NSP may also bill on a per-service flow. For example, the total volume or time utilized for more valued flows, such as used by VoIP services, would be billed at higher rates then the best-effort traffic used for file downloads.

When it comes to settlement time, the VNSP computes the total volume and/or time delivered to customers belonging to a HNSP by processing all the accounting records associated with a particular HNSP. This is done by filtering the accounting records based on the realm of the hone NSP found in the user name attribute in the accounting request stop records. The counts and/or duration are summed up for all the sessions serviced for a particular realm (during a specific time period). In the case of flow-based accounting, the counters and/or time duration are also summed up according to the PDIF and/or SDIF for a given HNSP and for the required time period.

Accounting start records do not contain any usage counters and, thus, are not needed for settlement purposes. Thus, the accounting request start records are removed as soon as the accounting request interim update record (if used) or accounting request stop record is received that is correlated to the accounting request start record by the accounting session ID attributes and NAS ID.

Accounting request interim update records, if available, are used in the case in which an accounting request stop record was not received. Typically, when interim accounting is used, the VNSP keeps the latest accounting request interim update record it received for a particular accounting session (as indicated by the accounting session ID attribute and NAS identifier). When the corresponding accounting request stop record is received, indicating the termination of the accounting session, the accounting request interim update record correlated with the stop record is deleted. If the accounting request stop record is not received after some time has elapsed, then the last received accounting request interim update record is converted to or used as an accounting request stop record.

The number of raw accounting records that the VNSP would need to store can be quite large. The VNSP may cull the accounting records using the above procedures as the records arrive.

Further reduction in accounting record storage is possible. For example, new accounting sessions due to anchor authenticator relocation can be rolled together as long as the new anchor authenticator is within the same NAP (as indicated by the NAP ID, BSID, or NAS identifier). In this case in which the VNSP receives an indication that the session is being handed over from one anchor authenticator to another anchor authenticator, as indicated by the reception of an accounting request stop record with session continue set to FALSE, it can combine all the accounting request stop records with session continue set to TRUE together for the same WiMAX session and the same NAP (NAP ID or the operator ID contained in the BSID attribute are the same).

14.7   PROCESSING OF ACCOUNTING RECORDS IN THE HOME NSP

14.7.1   Home AAA Processing

As shown in the previous call flows, for each MS there could be many accounting sessions, each with their own accounting request start, and, optionally, interim updates and stop records. Upon receiving the accounting records, the home AAA server passes the accounting records to the BSS/OSS systems where they are used to generate customer bills, reconcile billing with roaming partners, and so on.

The arrival of the accounting request start record indicates the start of a new accounting session. In particular, the arrival of the accounting request start record(s) containing the beginning-of-session attribute set to TRUE indicates to the AAA server that the MS has started to receive IP service. At this point, the home AAA also learns the IP address allocated to the IP session of the MS. Until the AAA receives these records, it does not know whether the IP session has started for the MS.

For example, the MS could have powered down, or the allocation of IP session could have failed.

The arrival of the accounting request start records with beginning of session set to FALSE indicates that a new accounting session is starting. These records are not significant to the AAA and are passed to the BSS/OSS.

The arrival of the accounting request stop messages with session continue set to TRUE puts the AAA on notice that something is changing in the session, that the session has not terminated. For example, the session may have transferred from one anchor authenticator to another. The AAA is going to expect to receive an accounting request start with beginning of session set to FALSE with the same WiMAX session ID to mark the beginning of the next accounting session.

The arrival of the accounting request stop messages with session continue set to FALSE signals to the AAA that the session has terminated. This causes the AAA to clear the information associated with the session (session state). In the case of PD-flow-based accounting, the AAA waits for the accounting request stop message with session continue set to FALSE associated with the Initial Service Flow (ISF) to indicate the termination of the IP session (PDFID set to a value of 1 to 20).

In general, the AAA ignores accounting request interim update records and passes them on to the BSS/OSS subsystem.

The AAA passes all accounting records to the BSS/OSS for further processing. How this is done is not standardized in WiMAX. The BSS/OSS does not need all the accounting records. Thus, it is common practice for the AAA to cull the accounting records before passing them to the BSS/OSS. This process is called correlation and mediation.

14.7.2   BSS/OSS Processing

The typical job of the BSS/OSS is to bill the user of a WiMAX device for services received. Also, the BSS/OSS will use the accounting records to settle accounting with the roaming partners (visited NSP).

With respect to settlement with the VNSP, the BSS/OSS follows the same procedures as the VNSP and reconcile its results with the accounting preformed by the VNSP. When the parties disagree, they may resort to looking at the detailed accounting records.

14.7.3   Billing

There are many billing scenarios and the BSS/OSS uses the information available in the accounting request stop and rating functions to create the customer bill. Customers may be billed for total volume or time, or at different rates depending on the type of service used. For example, VoIP usage may be billed on a duration basis. File downloads may be billed on a volume basis, with different rates applied for different times of day.

Since accounting request start records do not contain any usage information, the BSS/OSS uses the accounting request stop records to generate customer bills. Accounting request interim update records, if available, are only used in case an accounting request stop message is missing. In this case, the last accounting request interim update that was received for a particular accounting session (as indicated by accounting session ID and NAS ID attributes) is used as the accounting request stop record.

Each accounting stop record received that corresponds to the billing period is rated based on the octet counters and/or time and converted into a monetary value. When flow-based accounting is used, the PDFID and SDFID as well as granted QoS information, is used to generate the appropriate charge to the user.

14.8   ERROR HANDLING BY THE AAA

In the case in which RADIUS is used to transport accounting packets, the AAA needs to be designed to handle unreliable reception of accounting packets. Recall that RADIUS is based on UDP and, thus, accounting packets can be dropped or received out of order.

The AAA server receiving accounting request records must generate an accounting response record to acknowledge the reception of the accounting request messages. If the accounting client does not receive an accounting response record, it will retransmit the accounting request record.

In the case in which the account response message was lost, the accounting client will retransmit the account request message and the accounting-server will receive a duplicate copy of the accounting request record.

In addition to dealing with duplicate records, the HAAA must also cope with accounting packets that come out of sequence. For example, it is possible for the receiver to get an account request interim update record before receiving an account request start record, or receive an account request stop record before receiving an account request start record.

When the AAA receives an accounting request interim update record for a session for which it did not receive an accounting request start record with beginning of session set to TRUE, then the AAA uses that interim record as an indication that the session has started. Later, if it receives the accounting request start record, it will simply discard it.

Similarly if the AAA receives an accounting request stop record with session continue set to FALSE, indicating that the session is terminating, before it received the accounting request start or Interim Records, then the AAA will simply clear the session state and pass the accounting request stop record to the BSS/OSS systems. The accounting request stop message contains all the information needed to bill the customer. Any missing records that arrive are simply discarded.

14.9   SUMMARY

This chapter presented the protocols and operational procedures of WiMAX accounting. First we looked at the WiMAX architecture where we introduce the network elements that participate in WiMAX Accounting. We then described the WiMAX accounting model. Finally, we presented the accounting operational procedures, which were illustrated using call flows. Accounting is a complex topic which we hope was made easier to understand thus allowing the reader to better comprehend the full WiMAX accounting specifications

14.10 REFERENCES

[1] WiMAX Forum, WiMAX Forum Netowrk Architecture State 3: Detailed Protocols and Procedures. Annex: Prepaid Accounting, T33-002-R010v05, TBD.

[2] Rigney, C., Willens, S., Rubens, A., and W. Simpson, Remote Authentication Dial In User Service (RADIUS), RFC 2865, June 2000.

[3] Rigney, C., RADIUS Accounting, RFC 2866, June 2000.

[4] Calhoun, P., Loughney, J., Guttman, E., Zorn, G., and J. Arkko, Diameter Base Protocol, RFC 3588, September 2003.

[5] WiMAX Forum, WiMAX Forum Network Architecture Stage 2: Architecture Tenets, Reference Model and Reference Points. Part 1, WMF-T32-002-R010v05, TBD.

[6] 3GPP2 X.S0011-005-D v1.0 cdma2000 Wireless IP Network Standard, October 2002. http://www.3gpp2.com/Public_html/specs/P.S0001-B_v1.0.pdf.

[7] Adrangi, F., Lior, A., Korhonen, J., and J. Loughney, Chargeable User Identity, RFC 4372, January 2006.

[8] Aboba, B., Beadles, M., Arkko, J., and P. Eronen, The Network Access Identifier, RFC 4282, December 2005.

CHAPTER 15

WiMAX ROAMING

JOHN DUBOIS AND CHIRAG PATEL

15.1 INTRODUCTION

Roaming enables customers to automatically access their wireless services when travelling outside the geographical coverage area of their home network. This includes Internet, e-mail, voice, video, and other services available on the home network.

When we use our wireless devices on the move, they change access points as we leave the coverage area of one access point and move into coverage of another, usually on the home network. However, if another access point is not available on the home network, the device finds and connects to an access point of a different network made available through a roaming arrangement between two operators.

Roaming is a business and technical relationship, typically between two operators, that enables the subscribers of a home network to connect to and receive services through a second, or visited, network. For example, this includes Internet, e-mail, voice, video, and other services available on the home network.

When a subscriber uses a wireless device while moving, the device will try to attach to different access points when leaving the current coverage area. When the device used by the roaming user is outside the home network coverage area, the device will attempt to connect to an access point of a different network provider as long as a valid roaming arrangement is in place between the visited operator and the home operator.

Roaming provides significant advantages to customers and network operators. Roaming can dramatically expand the coverage area available to customers and en-

WiMAX Technology and Network Evolution. Edited by Kamran Etemad and Ming-Yee Lai
Copyright © 2010 the Institute of Electrical and Electronics Engineers, Inc.

ables an operator to expand its footprint without incurring additional network capital expenditures. Roaming may also provide additional revenue opportunities to network operators.

Roaming can be either national or international. National roaming occurs when the visited network is in the same country as the home network. International, or global roaming, occurs when the visited network is in a different country than the home network. Roaming can also occur between networks using differing technologies, such as WiMAX and Wi-Fi or WiMAX and 3G. This is known as interstandard roaming.

For roaming to take place, two or more operators must connect their networks to provide access to their services and to enable the process of sharing subscriber usage information and to facilitate billing and financial settlement between operators. This can be achieved by operators establishing direct connections between their networks or by connecting their networks through a third-party roaming exchange provider. One advantage of connecting through a third-party provider is that this can enable an operator to connect with many other operators through a single connection. This can allow an operator to expand its footprint quickly and at a lower cost as compared with establishing many direct connections with other operators.

15.2   WiMAX ROAMING BUSINESS DRIVERS AND STAKEHOLDERS

WiMAX is projected to be used by 133 million customers globally by 2012. In today's global market, customers demand competitive prices, higher QoS, and larger service coverage. Roaming is a key element in the growth of WiMAX, as it was in the growth of cellular. To a great degree, the success of GSM can be attributed to roaming. The first GSM Roaming Agreement was signed in 1992 and today there are more than 800 GSM networks connected with each other to provide roaming services to their customers.

Some of the business drivers for WiMAX roaming are:

1. Allows expanded coverage for customers, thus improving user experience.
2. Generates significant revenue.
3. Can result in lower infrastructure costs by sharing networks with other WiMAX operators.

WiMAX roaming will impact various groups of people and organizations. Key stakeholders associated with WiMAX roaming are the end users, WiMAX operators, WiMAX vendors, governments (e.g., FCC), content providers, and various other businesses and organizations:

End Users. End users or customers are impacted the most by roaming. As mentioned earlier, end users expect larger WiMAX coverage areas and roaming will enable end users to utilize their subscription in larger geographical areas.

WiMAX Operators. Operators will be able to provide better coverage to their subscribers and generate extra revenue from both inbound and outbound roaming.

WiMAX Vendors. Vendors will benefit from roaming, since users and operators will require more equipment and services.

Government Organization. Like any other wireless technology, WiMAX roaming will impact various government organizations, such as the FCC, which regulate use of frequency and can impact roaming business practices.

Content Providers. Content providers such as TV broadcasters, and gaming, music, and video content providers will be able to use new channels to deliver their content to end users.

15.3 RELATED STANDARDS AND FORUM ACTIVITIES

Wireless technologies such as WiMAX, GSM, CDMA, and Wi-Fi are capable of roaming. To enable roaming on these technologies, various organizatons such as the WiMAX Forum and the GSM Association (GSMA) develop roaming standards that their members use to enable roaming on their networks.

15.3.1 WiMAX Forum

The WiMAX Forum® is an industry-led, not-for-profit organization formed to certify and promote the compatibility and interoperability of broadband wireless products based upon the harmonized IEEE 802.16/ETSI HiperMAN standard. A WiMAX Forum goal is to accelerate the introduction of these systems into the marketplace. WiMAX Forum Certified™ products are fully interoperable and support broadband fixed, portable, and mobile services. The WiMAX Forum works closely with service providers and regulators to ensure that WiMAX Forum Certified systems meet customer and government requirements.

The WiMAX Forum has various working groups to enhance WiMAX roaming standards and specifications:

Global Roaming Working Group (GRWG). Assures the availability of global roaming service for WiMAX networks in a timely manner as demanded by the marketplace.

Network Working Group (NWG). Creates higher level networking specifications for fixed, nomadic, portable, and mobile WiMAX systems, beyond what is defined in the scope of IEEE 802.16.

Service Provider Working Group (SPWG). Gives service providers a platform for influencing BWA product and spectrum requirements to ensure that their individual market needs are fulfilled.

Regulatory Working Group (RWG). Influences worldwide regulatory agencies to promote WiMAX-friendly, globally harmonized spectrum allocations.

15.4   WiMAX ROAMING MODEL

There are various roaming models that are applicable to WiMAX. These include bilateral roaming, unilateral roaming, hub model, and aggregator models with the following descriptions:

Bilateral Roaming. Bilateral roaming occurs when two operators sign a roaming agreement allowing their subscribers to utilize each other's network for wireless services.

Unilateral Roaming. Unilateral roaming occurs when two operators sign a roaming agreement that allows only one operator's subscribers to utilize both networks for wireless services.

Roaming Hub. Roaming hubs allow operators to sign one roaming agreement with a third-party hub provider and receive the same access and service-level capabilities (voice and data) as with classic bilateral agreements, in addition to data clearing, financial clearing, and fraud protection. The hub model allows an operator to connect with many operators through a single connection with a hub provider. An operator itself can be a hub provider for other operators.

Aggregator. This model is similar to virtual operators, whereby the operator may not own its own network but provides services to its subscriber by signing bilateral and/or unilateral roaming agreements with multiple operators.

The WiMAX Forum has created various documents that can be used to facilitate roaming using the roaming models described above. These are available at www.wimaxroaming.org. These documents include: WiMAX Roaming Agreements 1 and 2 (WRA1 and WRA2) 910, WiMAX Roaming Guidelines [2], WiMAX Roaming Interface Specifications (WRI) [3], and WiMAX Roaming Test Plans [4].

15.5   WiMAX ROAMING AGREEMENT OVERVIEW

Roaming agreements set forth the terms and conditions by which operators agree to provide each other's subscribers with access to their networks. An agreement details which services are to be provided and the rates that the operators will pay for using the other's network. The agreement also includes information relating to QoS, customer care, and billing and financial settlement.

The WiMAX roaming agreement contains three sets of documents:

1. WiMAX Roaming Agreement 1 (WRA1) is the legal agreement between the operators and contains legally binding terms.

2. WiMAX Roaming Agreement 2 (WRA2) is an annex to WRA1 and contains various information operators need to agree on and share with each other to establish and operate roaming services. Each operator completes a WRA2 and provides it to its roaming partner. WRA2 includes the following information: operator contact information, NSP IDs, NAP IDs, authentication type and method supported, frequency and profiles supported, services provided

while roaming, interoperator tariffs, settlement and payment information, and data exchange information.

3. Operators provide their roaming partners with a technical data sheet (TDS) to allow them to configure their various network elements to enable their subscriber to roam onto partner networks.

15.6 WiMAX ROAMING GUIDELINE OVERVIEW

The purpose of this document is to provide guidelines and technical information on how WiMAX networks should be deployed to enable WiMAX roaming. The roaming guidelines provide an overview of WiMAX roaming for the benefit of WiMAX operators, infrastructure and equipment developers, and WiMAX roaming exchange (WRX) providers.

This document describes the roaming architecture, process, and functions, and provides an overview of a technical solution for roaming service between WIMAX Network Service providers (NSPs). The roaming guidelines document is based on the network reference model detailed in the WiMAX Forum network architecture.

The Network reference model from WiMAX Forum Network Architecture Release 1.5 Version 1 [5] is shown in Figure 15.1. It provides reference models for two roaming scenarios as follows:

1. Roaming with home agent located in the visited NSP (see Figure 15.2)
2. Roaming with home agent located in the home NSP (see Figure 15.3)

15.7 WIMAX ROAMING INTERFACE OVERVIEW

The WiMAX roaming interface (WRI) documents are specifications for roaming that provide detailed rules and a description for implementing roaming services. This includes specifications for network discovery, selection, authentication, and authorization. It also includes specifications relating to the exchange of information between operators to track usage, create invoices, and perform financial settlements.

Roaming can be implemented via direct connections between two NSPs or a third-party WiMAX roaming exchange (WRX). Figure 15.4 shows all the functions and interfaces needed for implementing roaming. A NSP can choose to outsource one or more of these functions and interfaces to a WRX.

The WRI encompasses six specification documents described in the following subsections.

15.7.1 AAA Proxy Service

The AAA proxy service document details the functions related to AAA proxy services and the X2 interface. AAA proxy service involves the proxy of RADIUS messages, correlation and aggregation of session records, validating roaming agreements between the roaming partners, and the transfer of aggregated session records

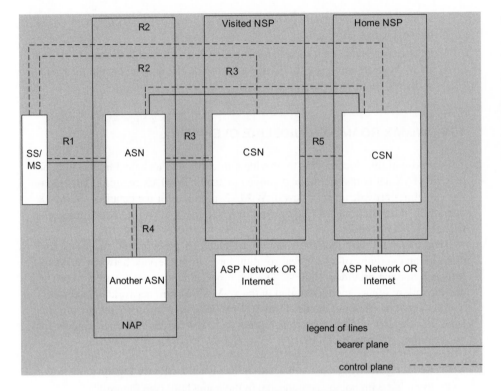

Figure 15.1. WiMAX network reference model with roaming.

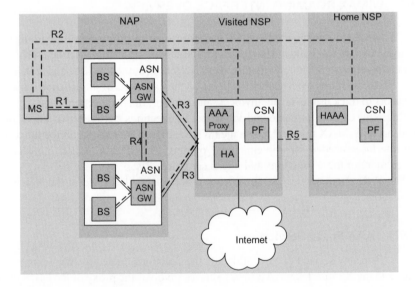

Figure 15.2. Roaming with home agent in visited NSP.

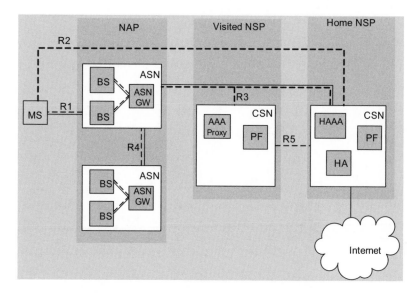

Figure 15.3. Roaming with home agent in home NSP.

to the wholesale rating and fraud management functions. The proxy service validates attributes that are present in the RADIUS records and creates files (aggregated sessions) to be sent to the wholesale rating function via the X2 interface. The AAA proxy service specification also defines the RADIUS attributes, elements, and rules to create an X2 file.

15.7.2   Wholesale Rating

Wholesale rating involves rating of raw aggregated session records. The wholesale rating Document details the wholesale rating function and the X3 Interface.

Wholesale rating is the process of calculating and assigning monetary value to sessions based on files (aggregated sessions) received from the visited network AAA proxy service via the X2 interface. The calculation of monetary value is based on rates and taxes agreed on in the roaming agreement established between two operators. Rated sessions are then used to create files and send them to the clearing function via the X3 interface. The wholesale rating document also defines the elements and rules to create the X3 file.

15.7.3   Clearing

Clearing enables the acceptance and rejection of financial liability via exchange of rated session records. The clearing document details the clearing process and the X4 and X5 interfaces. It also defines the elements and rules to create X4 and X5 files sent via the X4 and X5 interfaces, respectively.

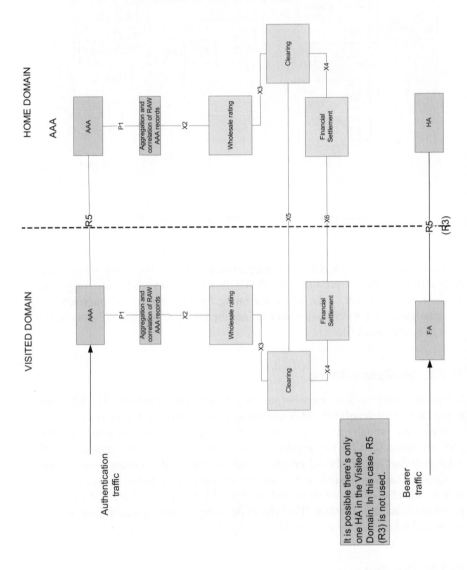

Figure 15.4. WiMAX roaming interface and functional diagram.

The clearing process involves receiving X3 and X5 files, creating and validating X5 files, rejecting and retransmitting corrected X5 files, and providing accounting data to the financial settlement function via the X4 interface by transmitting summarized X5 files (X4 files).

The clearing process occurs in both the home and the visited networks. Each network performs different functions based on whether it is the visited or the home network. The clearing process that occurs in the home network includes receiving X3 files from the home network AAA and X5 files from the visited network AAA. Once files are received, the clearing process in the home network validates the structure and the content of the X5 file. The structure is validated based on the WRI Stage 3 specifications, and content is validated by comparing the sessions and the charges to the X3 files. If the validation fails, the home network AAA creates a reject X5 file and sends it to the visited network. If the validation passes, X4 files are created and sent to the home network via the X4 interface for the financial settlement process.

The clearing function located in the visited network receives X3 files from the visited Wholesale rating function and creates summarized X5 files and transmits them to the home clearing function. It may also receive a rejected X5 file from the home clearing function. When the visited clearing function receives the rejected X5 file, it corrects the errors and retransmits the X5 file. The visited clearing function also creates X4 files and sends them to the visited network for the financial settlement process.

15.7.4 Financial Settlement

Financial settlement involves the actual payment of amounts owed by one operator to another for the use of its network based upon wholesale billing rates established between operators. The financial settlement document describes the financial settlement function and the X6 interface. It also defines the elements and rules needed to create the X6 file.

The financial settlement process occurs in both the visited and the home networks, and each performs a separate financial settlement. The financial settlement process includes the generation of a monthly X6 file based on all the X4 files for the given month generated by the clearing function. Operators then exchange X6 files for validation. If the validation fails, a rejected X6 file is created and sent to the operator roaming partner, which will correct and retransmit it. If the validation passes, then the settlement of accounts is performed by the transferring of monies.

Settlement also includes performing foreign currency exchange and tracking of any outstanding financial obligations.

15.7.5 Interconnection

The interconnection document details financial settlement/clearing, AAA traffic, and bearer traffic data-path connection options, and defines interface requirements to establish roaming. The document provides a detailed description of three options: roaming with no interconnection, roaming with direct interconnections, and roaming with interconnections via a WRX. It defines the interfaces required for direct connections and connections via a WRX.

15.7.6 Fraud Management

The purpose of this document is to define the fraud management process. The interconnect (AAA proxy and aggregation) and data clearing (wholesale rating) processes represent the first opportunity to perform certain types of fraud detection and prevention.

Fraud management will use RADIUS messages and/or rated session records to perform abnormal-usage-pattern detection over a period of time in order to identify and proactively report potential fraudulent activities. This function was not addressed in the first release of the WiMAX Forum roaming interface documents.

15.8 SUMMARY

For customers, roaming provides a broader area in which they can access their services. For a WiMAX operator, roaming represents the extension of its coverage area outside of its home network. Roaming also represents new revenue opportunities for operators as their networks are opened up to a broader user base. WiMAX operators that choose to provide services outside of their home network, or to allow access to their home network by subscribers of other operators, have a variety of business models to choose from. Their choice depends on their business needs. The roaming models described in this document are complementary and will continue to coexist. It is important that WiMAX roaming be available using any of these models.

15.9 REFERENCES

[1] WiMAX Forum Roaming Agreement Template WRA1 and WRA2, Release 1.0 Version 1, September 2008.

[2] WiMAX Forum Roaming Guidelines, Release 1.0 Version 2, June 2009.

[3] WiMAX Forum WiMAX Roaming Interface Stage 2 and Stage 3, Release 1.0 Version 2, 2009.

[4] WiMAX Forum Pre-Commercial Roaming Testing Release 1.0, December 2008.

[5] WiMAX Forum Network Architecture Stage 2 and 3, Release 1.5, Version 1, 2009.

CHAPTER 16

WiMAX NETWORK MANAGEMENT FRAMEWORK

JOEY CHOU

16.1 INTRODUCTION

The trend of the next-generation mobile networks is to separate services from the transport networks. As the result, service-oriented architecture (SOA) has been adopted as the management architecture of the next generation mobile networks [1], including WiMAX. Basically, SOA can be viewed as an architecture that enables the creation of new applications by combining loosely coupled, location-indepen-dent, and reusable services. SOA also assumes that all services are implemented based on the client–server architecture, and support the "find-bind-execute" or "plug-and-play" paradigm to enable rapid service development for meeting the re-quirements of today's dynamic mobile Internet business environment.

However, there are challenges in moving to SOA from the existing OSS (opera-tion support systems) that will require collaboration and intergration of multiple standards. Among them, the most well-known standards are TMN (telecommunica-tions management network) [3] and eTOM (enhanced operations map) [2, 4 ,5].

The concept of TMN was introduced in ITU-T recommendation M-3010 [6] to deal with the complexity of telecommunications management by partitioning the management functions into business management layer (BML), service manage-ment layer (SML), network management layer (NML), element management layer (EML), and network element (NE). TMN also defines the standardized interfaces and protocols used for the exchange of information between layers to enable inter-operability of management functions that may be provided by different vendors.

WiMAX Technology and Network Evolution. Edited by Kamran Etemad and Ming-Yee Lai
Copyright © 2010 the Institute of Electrical and Electronics Engineers, Inc.

eTOM was developed by TMF as a business process framework for categorizing all the business activities that a service provider will use. The focus of the eTOM framework is on the business processes used by service providers, the linkages between these processes, the identification of interfaces, and the use of customer, service, resource, supplier/partner, and other information by multiple processes.

The main differentiator between these two approaches is that TMN is based on the bottom–up approach that has been built on the requirements to manage network equipment and networks; whereas the eTOM is based on the top–down approach that has been built on the need to support processes of the entire service-provider enterprise. Both TMF and eTOM standards will provide functional blocks to help building and validating the SOA model.

This chapter describes the standardization of Mobile WiMAX network management in both WiMAX Forum [7] and IEEE 802.16 [8], which pave the way for OSS/BSS automation and enables multivendor management interoperability. It also covers an emerging technology, self-organizing networks (SON), that aims to increase network performance, coverage, and capacity in the dynamic mobile environment.

16.2 WiMAX FORUM NETWORK MANAGEMENT

Figure 16.1 shows a high-level view of the mobile WiMAX network architecture. The OAM server implements network management applications to provide OAM&P (operation, administration, maintenance, and provisioning) functions to network elements in the RAN (radio access networks) as will as the core networks.

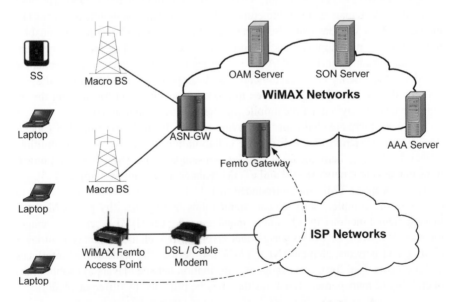

Figure 16.1. Mobile WiMAX network architecture with femto Aps.

The RAN consists of subscriber stations (SS), mobile stations (MS), macro base stations (BS), and WiMAX femto access points (WFAP) that utilizes broadband access technologies such as DSL, fiber, or cable to backhaul WFAP traffic to WiMAX networks via the femto gateway.

WiMAX network management standards are being developed in the WiMAX Forum with the aim to enable multivendor management interoperability between operator's common business support systems (BSS)/operation support systems (OSS) and equipment from different vendors.

Figure 16.2 shows an example of a mobile WiMAX OAM&P reference model that is based on TMN. The TMN approach is used here, since the objective of this model is to assist the definition of the Itf-N interface between network management systems (NMS) in the NML and element management systems (EMS) in the EML. The BML and SML are concerned with the management of all aspects of services that may be directly observed by the subscribers and businesses that are related to the optimization of the return of investment and efficient use of new resources for operators. Field management, order entry processing, customer care, device and service provisioning, trouble management, and billing processing are typical functions in the service and business managements that can be integrated into the business process framework of the eTOM model.

The NMS has the complete view of network operation and provides the functionality to manage a network by coordinating activities across multiple network elements. An EMS is associated with each NE, and plays the manager role to mange NEs on behalf of the NMS. Device management is mainly used to enable the re-

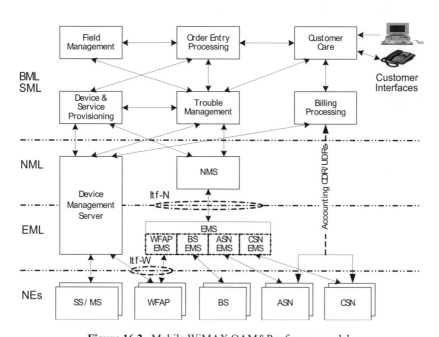

Figure 16.2. Mobile WiMAX OAM&P reference model.

mote management of a huge number of smart mobile devices. It supports device bootstrapping, provisioning, remote diagnostics, and performance metrics measurement and reporting. It is generally believed among industry observers that femtocell technology will not be successful unless WFAP is affordable to the mass market. Therefore, retail model and plug-and-play are required to lower the cost of WFAO deployment. The Itf-W interface between WFAP and EMS/device management is designed to support management interoperability, which is necessary to meet the above requirements.

The standardization of the Itf-N interface enables an operator to use a common NMS to manage NEs from multiple vendors. The integration reference point (IRP) concept, as proposed by 3GPP, is used to define the Itf-N interface. Technologies advance very rapidly by following the trend of Moore's law; however, the management applications tend to stay unchanged over long periods of time. Therefore, IRP methodology utilizes a technology-neutral approach to define the management requirements, interfaces, and network resource modeling that are independent of the technology and the protocol used in the management system implementation.

The IRP methodology consists of three phases, as shown in Figure 16.3:

1. Use Cases and Requirements.The IRP process begins with the requirements phase that defines use cases and requirements for a specific management interface.
2. Information Services (IS). This phase defines the following technology-independent specifications:
   - Interface IRPs, providing the definitions for IRP operations and notifications in a network-agnostic manner

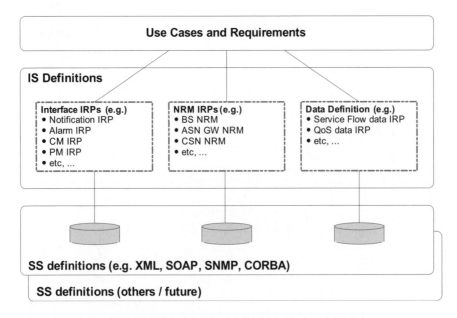

Figure 16.3. IRP definitions phases.

- NRM IRPs, providing the definitions for the network resources models (NRM)
- Data Definition IRPs, providing data definitions associated with network resources defined in the NRM that are to be managed via interface IRPs

3. Solution Set (SS) Definitions. This phase provides the mapping of IS definitions into one or more protocol-specific solution sets (e.g., SNMP, XML, SOAP, and CORBA). This concept provides support for multiple interface technologies as applicable on a per-vendor and/or network-type basis and also enables accommodation of future interface technologies, without the need to redefine requirements and IS-level definitions.

When an operator acquires equipment from multiple vendors, it is desirable for the operator to use common applications in the operator's NML to manage these multivendor networks. Since each NE will carry its own management solution, IRPs are created to assist an operator to integrate multivendor management solutions into their NMS infrastructure. Figure 16.4 is an example of IRP definitions at the EMS northbound interface that contains the following IRPs:

- Alarm IRP to support fault management
- CM IRP to support configuration management
- PM IRP to support performance management

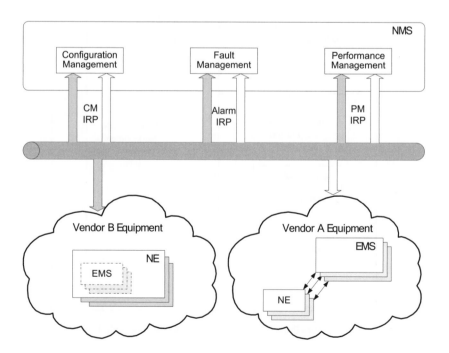

Figure 16.4. IRP for management solutions integration.

The Simple Netework Management Protocol (SNMP) solution set for BS is defined in the IEEE 802.16 network management standard and described in the following section.

16.3 IEEE 802.16 NETWORK MANAGEMENT

Figure 16.5 shows the protocol reference model of the IEEE 802.16 standard that mainly focuses on MAC and PHY layers. The reference model is to be implemented in the BS, SS, and MS, and can be described from data plane and management control plane perspectives. The data plane deals with the transport of user data required for providing services. A CS-SAP (convergence sublayer service access point) is defined to enable layer 3 applications to interface to the MAC layer.

The MAC CS (service-specific convergence sublayer) provides mapping of various layer 3 application data into MAC SDUs (service data units) that are sent to MAC CPS (common part sublayer). Currently, MAC CS provides three convergence functions: ATM (asynchronous transfer mode) CS, packet CS (i.e., Ethernet, IPv4, and

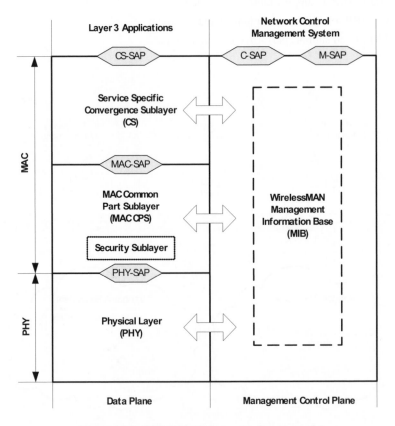

Figure 16.5. IEEE 802.16 Protocol reference model.

IPv6), and generic packet CS (i.e., layer 3 protocol independent). MAC CPS provides media access control functions to allow multiple subscribers to share the common wireless resource. The PHY (physical layer) provides the modem function to covert data into a format that can be transmitted over the physical medium.

The C-SAP (control service access point) and M-SAP (management service access point) expose control plane and management plane functions of the network control management system. The network control management system can access the WirelessMAN MIB (management information MIB) via the M-SAP, and access control plane functions, such as handover control, radio resource management, and mobility management, via the C-SAP. WirelessMAN MIB is designed to support the MAC CS, MAC CPS, security sublayer, and PHY layer.

Figure 16.6 shows the evolution of IEEE 802.16 network management standards. Figure 16.7 shows the WirelessMAN network management reference model that consists of a network management system (NMS), network control system (NCS), and managed nodes, such as BS, SS, and MS. The managed nodes collect and store the managed objects in the form of WirelessMAN MIBs that are made available to the NMS via management protocols such as SNMP.

Table 16-1 shows the WirelessMan ASN.1 MIB modules that have been defined in IEEE 802.16.

Figure 16.8 shows the MIB structure of the WMAN-DEV-MIB module to be implemented in the BS and SS. It defines managed objects relevant to software upgrades and traps. It also includes wmanDecCommonObjects to support event logs for the BS and SS.

Figure 16.9 shows the MIB structure of the WMAN-IF2-BS-MIB module to be implemented in the BS. wmanIf2BsMib is a common MIB module to support both fixed and mobile WirelessMAN. It is structured based on TMN's FACPS model that consists of the following groups:

- wmanIf2BsFm provides the configuration and definitions of traps that are used to report BS alarms, exceptions, and faults.

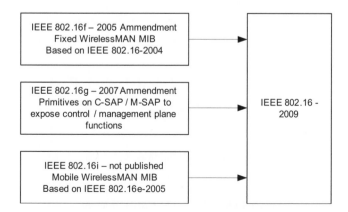

Figure 16.6. IEEE 802.16 network management standard evolution.

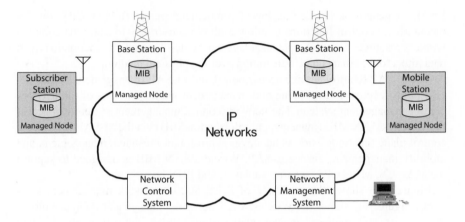

Figure 16.7. WirelessMAN network management reference model.

Table 16.1. WirelessMAN ASN.1 MIB modules

MIB module name	Module identity name	Module identity OID (object identifier)
WMAN-DEV-MIB	wmanDevMib	Iso(1).std(0).iso8802.wman(16).wmanDevMib(1)
WMAN-IF-MIB	wmanIfMib	Iso(1).org(3).dod(6).internet(1).ngmt(2).mib-2(1).transmission(10).wmanIfMib(184)
WMAN-IF2-BS-MIB	wmanIf2BsMib	Iso(1).std(0).iso8802.wman(16).wmanIf2BsMib(1)
WMAN-IF2M-BS-MIB	wmanIf2mBsMib	Iso(1).std(0).iso8802.wman(16).wmanIf2mBsMib(1)
WMAN-IF2F-BS-MIB	wmanIf2fBsMib	Iso(1).std(0).iso8802.wman(16).wmanIf2fBsMib(1)
WMAN-IF2-SS-MIB	wmanIf2SsMib	Iso(1).std(0).iso8802.wman(16).wmanIf2SsMib(1)
WMAN-IF2-TC-MIB	wmanIf2TcMib	Iso(1).std(0).iso8802.wman(16).wmanIf2TcMib(1)

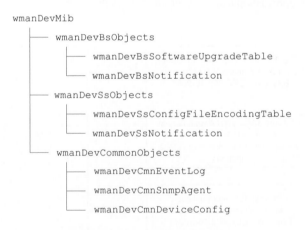

Figure 16.8. wmanDevMib MIB structure.

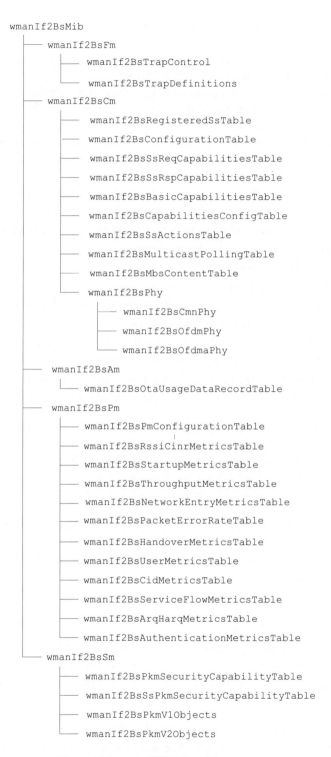

Figure 16.9. wmanIf2BsMib MIB structure.

- wmanIf2BsCm provides configuration capability for MAC and PHY (i.e., common, OFDM, and OFDMA) functions.
- wmanIf2BsAm provides counters to keep track of the number of packets and octets that have been received or transmitted over the air interface, on per-user, connection, or session basis.
- wmanIf2BsPm provides KPI (key performance indicators) counters to characterize the performance over the air interface. It includes the following metrics that are measured on a per-BS-sector basis:
  - Mean and standard deviation UL/DL RSSI and CINR metrics
  - Startup performance metrics (e.g., authentication and ranging success rate)
  - UL/DL average and peak throughput for MAC, PHY, and user layers
  - Average and peak network entry latency
  - UL/DL packet error rate
  - Handover performance (e.g., success, cancel, and reject rates)
  - Average and peak number of users in normal, sleep, and idle modes
  - Average and peak number of user CIDs
  - Average and peak number of service flows in active and provisioned states
  - ARQ/HARQ metrics (e.g., dropped, retransmission, and error rate)
  - Authentication metrics (e.g., HMA, CMA, and short HMA)
- wmanIf2BsSm provides objects to support the security sublayer.

Figure 16.10 shows the MIB structure of the WMAN-IF2M-BS-MIB module to be implemented in the BS. wmanIf2mMib is a mobile WiMAX-specific MIB module that is intended to support mobile WirelessMAN. Therefore, the BS that provides mobile wireless broadband services should implement wmanIf2Mib and wmanIf2mMib. wmanIf2mMib consists of the following groups:

- wmanIf2mBsCm provides the configuration capabilities for supporting mobile broadband services. It supports the following functions:
  - BS–MS capability negotiation, including network service provider provisioning
  - Power saving modes
  - Paging
  - Neighboring BS advertisement
  - Service flows (e.g., QoS profiles, classification, PHS, and ARQ) that are downloaded from the AAA server as an MS enters the network
- wmanIf2mBsPm provides counters to track MS performance (e.g., sleep mode statistics).

Figure 16.11 shows the MIB structure of the WMAN-IF2F-BS-MIB module to be implemented in the BS. wmanIf2fMib is a fixed and nomadic WiMAX-specific MIB module that is intended to support fixed WirelessMAN. Therefore, the BS

```
wmanIf2mBsMib
   ├── wmanIf2mBsCm
   │      ├── wmanIf2mBsConfiguration
   │      │      ├── wmanIf2mBsConfigurationTable
   │      │      ├── wmanIf2mBsSsReqCapabilitiesTable
   │      │      ├── wmanIf2mBsSsRspCapabilitiesTable
   │      │      ├── wmanIf2mBsSsBasicCapabilitiesTable
   │      │      ├── wmanIf2mBsBasicCapabilitiesConfigTable
   │      │      ├── wmanIf2mBsSsCidUpdateTable
   │      │      └── wmanIf2mBsNetworkServiceProviderTable
   │      ├── wmanIf2mBsPowerSavingMode
   │      │      ├── wmanIf2mBsSsPwrSaving2CidMapTable
   │      │      └── wmanIf2mBsSsPowerSavingClassesTable
   │      ├── wmanIf2mBsNeighborAdv
   │      │      ├── wmanIf2mBsNeighborAdvCommonTable
   │      │      ├── wmanIf2mBsNeighborAdertizementTable
   │      │      ├── wmanIf2mBsNeighborOfdmaUcdTable
   │      │      ├── wmanIf2mBsNeighborOfdmaDcdTable
   │      │      └── wmanIf2mBsLbsAdvNeighborBsTable
   │      ├── wmanIf2mBsPaging
   │      │      ├── wmanIf2mBsPagingAdvertizementTable
   │      │      ├── wmanIf2mBsMsPagedTable
   │      │      └── wmanIf2mBsPagingGroupsTable
   │      └── wmanIf2mBsServiceFlow
   │             ├── wmanIf2mBsServiceFlowTable
   │             ├── wmanIf2mBsClassifierRuleTable
   │             ├── wmanIf2mBsPhsRuleTable
   │             ├── wmanIf2mBsQoSProfileTable
   │             └── wmanIf2mBsArqAttributeTable
   └── wmanIf2mBsPm
          ├── wmanIf2mBsSsSleepModeStatisticsTable
          ├── wmanIf2mBsMobileScanRequestTable
          ├── wmanIf2mBsMobileScanResponseTable
          ├── wmanIf2mBsDiversityBsInfoTable
          └── wmanIf2mBsNeighborBsInfoTable
```

Figure 16.10. wmanIf2mBsMib MIB structure.

Figure 16.11. wmanIf2fMib MIB structure.

provides fixed wireless broadband services that should implement wmanIf2Mib and wmanIf2fMib. wmanIf2fMib enables the NMS to provision service flows, including QoS profiles, classification, PHS, and ARQ. This is different from wmanIf2mMib, where the service flow attributes were downloaded from the AAA server as an MS enters the network.

Figure 16.12 shows the MIB structure of the WMAN-IF2-SS-MIB module to be implemented in the SS. wmanIf2fSsMib should be used if the SS supports the SNMP protocol. wmanIf2fSsMin provides SS configuration, traps, and service-flow attributes.

WMAN-IF2-TC-MIB defines textual conventions to be used by other Wireless-MAN MIB modules.

16.4 SELF-ORGANIZING NETWORKS

Cellular networks today still require much manual configuration and optimization to ensure that neighboring cell sites work properly and handoff connections are made successfully. As next-generation mobile networks like WiMAX grow and a

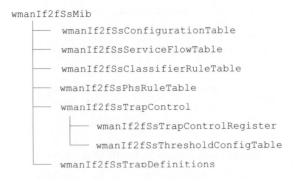

Figure 16.12. wmanIf2SsMib MIB structure.

large number of femtocells get deployed, such manual tasks would greatly burden the operators. Therefore, SONs are designed to enable the radio and network components to interact among themselves, and to configure and tune the mobile system automatically in real time.

The major drivers for self-organizing networks include the following:

- Increasing competition and investment in the mobile Internet market has forced operators to reduce CAPEX and OPEX by minimizing human intervention in both deployment and operational phases.

- Increasing complexity in current mobile networks requires constant provisioning and tuning of huge amounts of parameters in order to optimize network performance, coverage, and capacity.

- The huge amount of spatial–temporal data collected by multiple MSs, WFAPs, and macro BSs at any given time and location will be used not only to understand the wireless environment in providing mobile data services, but to assist in trend analysis, network planning optimization, and automatic centralized management functions.

- Has the ability to detect, mitigate, and recover faults automatically in order to achieve five nines reliability and maintain customer satisfaction.

SONs involve network elements (NEs) in radio access networks (RANs) and core networks to enable automatic configuration, to measure/analyze KPI (key performance indicator) data, and fine-tune network attributes in order to achieve optimal performance. An SON solution is based on client–server architecture, in which the SON server, as shown in Figure 16.1, implements a SON application, based on KPI metrics reported by SON clients such as SS, MS, WFAP, and macro BS. The measurements and objects of KPI metrics will be defined in the management MIB. The KPI objects can be reported to the SON server via the management protocol.

A SON consists primarily of self-configuration and self-optimization.

16.4.1 Self-Configuration

Self-configuration is the process of brining up network elements with minimum craftsperson intervention. Figure 16.13 shows an example neighboring BS self-configuration algorithm in the SON server to automatically determine the neighbor list of a macro BS. When a BS goes online, it will submit its BSID, cell site, sector bearing, and BS attributes to the SON server that implements the following steps:

1. Store online BS attributes, including cell site and sector bearing, into the database
2. Identify the online BS neighbor list, based on cell site and sector bearing
3. Send a message to each neighboring BS in the list to report online BS attributes
4. Send a message to the online BS to report the neighboring BS attributes

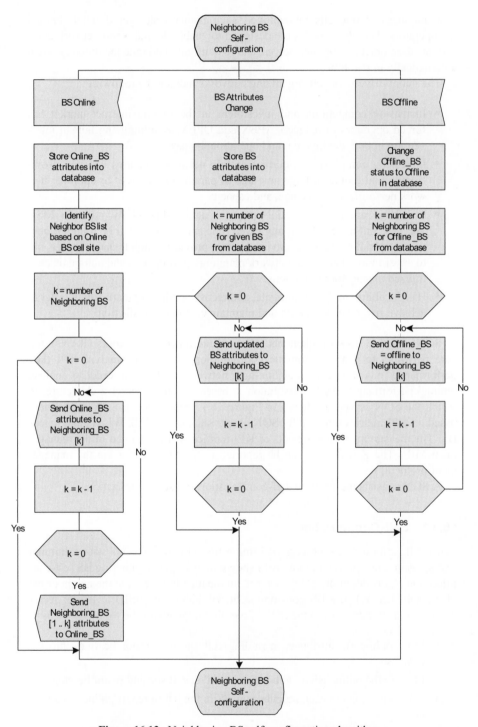

Figure 16.13. Neighboring BS self-configuration algorithm.

When the NMS detects a BS going offline, it will submit its BSID to the SON server that implements the following steps:
1. Update offline BS status in the database to offline
2. Retrieve the offline BS neighbor list from the database
3. Send a message to each neighboring BS in the list to report that the offline BS is offline

When a BS changes its BS attributes, it will submit its BSID and updated BS attributes to the SON server that implements the following steps:

1. Store updated BS attributes into the database
2. Retrieve BS's neighbor list from the database
3. Send a message to each neighboring BS in the list to report the updated BS attributes

Figure 16.14 is a cell planning example that describes how a SON server can identi-

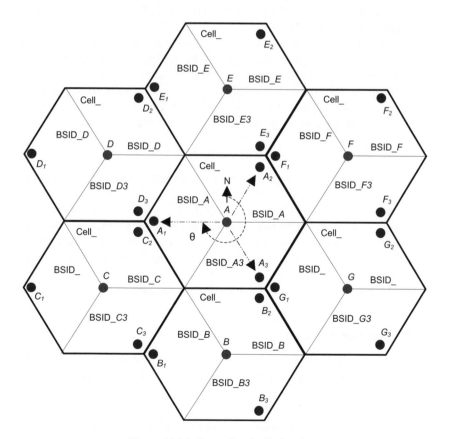

Figure 16.14. Example of cell planning.

fy the neighbors based on cell site and sector bearing. The light shaded dot in the center of each cell indicates the cell center in longitude and latitude. The haversine formula (1) shown below can be used to compute the distance d between two cell centers.

$$R = \text{earth's radius (mean radius} = 6371 \text{ km)}$$

$$\Delta\text{lat} = \text{lat}_2 - \text{lat}_1$$

$$\Delta\text{long} = \text{long}_2 - \text{long}_1$$

$$a = \sin^2(\Delta\text{lat}/2) + \cos(\text{lat}_1) \times \cos(\text{lat}_2) \times \sin^2(\Delta\text{long}/2) \tag{1}$$

$$c = 2 \times \text{atan2}[\sqrt{a}, \sqrt{(1-a)}]$$

$$d = R \times c$$

It is assumed that the online BS is in Cell A with cell center $= A$. By computing the distances between A and all other cells in the serving area, the SON server can identify its neighboring BSs that have the shortest distance to Cell A. For single-sector cellular networks, Cell B, Cell C, Cell D, Cell E, Cell F, and Cell G will be the neighboring cells of the online BS.

For multisector cellular networks, the sector bearing parameter, indicating the direction in which the sector is pointing, will be needed to locate its neighboring sectors. By providing the cell center (lat_1 and lon_1), sector distance (d), and sector bearing (θ) in the Haversine formula (2) below, the sector edge, as shown by the dark shaded dots in Figure 16.14, can be computed. The sector distance should be chosen in a way that makes the sector edge close to the edge of the sector.

$$\text{lat}_2 = \text{asin}[\sin(\text{lat}_1) \times \cos(d/R) + \cos(\text{lat}_1) \times \sin(d/R) \times \cos(\theta)]$$

$$\text{lon}_2 = \text{lon}_1 + \text{atan2}[\sin(\theta) \times \sin(d/R) \times \cos(\text{lat}_1), \cos(d/R) - \sin(\text{lat}_1) \times \sin(\text{lat}_2)] \tag{2}$$

It is assumed that the online BS is sector edge A_1. The SON server will compute the sector edge for all sectors in the neighboring cells of the online BS, based on Haversine formula (2). Then it will compute the distance between A_1 and all other sector edges. The SON server then can identify its neighboring sectors that have the shortest distance to A_1.

Table 16.2 shows an example of a SON server database.

16.4.2 Self-Optimization

Self-optimization is the process of utilizing measurement data to optimize network performance, coverage, and capacity, and includes the following functions:

- Load balancing
- Handover performance (handover historical information and ping-pong be-havior prevention)

Table 16.2. Example of SON Server Database

BSID	Cell ID	Cell center locations	Sector bearing	Sector edge locations	Neighboring BS list	Status	Channel bandwidth, FFT size, cyclic prefix, ...
BSID_A1	Cell_A	A	15°	A_1	BSID C2, C3, D3, D2, E1, E3, A2, A3, B2, B1	Online	3.5 MHz, 512, 1/8, ...
BSID_A2	Cell_A	A	135°	B_2	BSID E1, E3, F1, F3, G1, G2, A1, A3, B2	Online	3.5 MHz, 512, 1/4, ...
BSID_A3	Cell_A	A	255°	A_3	BSID F1, F3, G2, G1, B2, B1, C3, C2, A1, A3	Online	7 MHz, 1024, 1/16, ...
BSID_B1	Cell_B	B	15°	B_1	BSID C3, C2, A1, A3, B2	Online	14 MHz, 1024, 1/16, ...
BSID_B2	Cell_B	B	135°	B_2	BSID C3, C2, A1, A3, G1, G3, B1	Online	8.75 MHz, 1024, 1/32, ...
BSID_B3	Cell_B	B	255°	B_3	BSID B1, B2, G1, G3	Offline	8.75 MHz, 1024, 1/32, ...
BSID_C1	Cell_C	C	15°	C_1	BSID D1, D3, C2, C3	Offline	10 Mhz, 2048, 1/16, ...
BSID_C2	Cell_C	C	135°	C_2	BSID B1, A1, A3, B2, B1, C3	Online	10 Mhz, 2048, 1/16, ...
BSID_C3	Cell_C	C	255°	C_3	BSID C2, A1, A3, B2, B1	Online	7 MHz, 1024, 1/16, ...

- Intercell interference mitigation
- Radio parameter optimization
- Transport parameter optimization

The self-optimization typically consists of the following steps:

1. MS, SS, BS, or WFAP collect and report performance measurements to the SON server
2. SON server performs analysis on the performance data
3. SON server sends command to fine-tune attributes in MS, SS, BS, or WFAP in order to optimize the network performance
4. Go to step 1 to continue the self-optimization process

16.5 SUMMARY

The explosive growth of smart phones and mobile electronic gadgets are taking a toll on operators' networks in many ways. Operators need to pump in more investments not only to upgrade radio access networks for better coverage and throughput, but to overhal their entire backhaul network in order to avoid data traffic jams. However, operating expenses typically represent a much bigger share of investment to mobile operators than capital expenditures. Keeping operating expenses low is, therefore, the highest priority in today's competitive and cost-sensitive mobile Internet business environment. As a result, SON and the automation of OSS/BSS are viewed as the keys to reducing operating expenses, creating and delivering new services, and achieving higher return of investment.

16.6 REFERENCES

[1] ITU-T M.3060, Principles for the Management of Next Generation Networks, 03/2006.

[2] ITU-T M.3050.0, Enhanced Telecom Operations Map (eTOM)—Introduction, 07/2004.

[3] ITU-T M-3400, Telecommunications Management Network TMN Management Functions, 02/2000.

[4] TMF GB921, Enhanced Telecom Operations Map (eTOM)—The Business Process Framework, (2005).

[5] ITU-T M.3050.1, Enhanced Telecom Operations Map (eTOM)—The Business Process Framework, 06/2004.

[6] ITU-T M-3010, Telecommunications Management Network—Principles for a Telecommunications Management Network, 02/2000.

[7] WiMAX Forum, WiMAX Network Management, NMS to EMS Interface, Release 2.0 Draft Specification, September 2008.

[8] IEEE 802.16-2009, Part 16: Air Interface for Broadband Wireless Access Systems, May 2009.

CHAPTER 17

ETHERNET SERVICES IN WiMAX NETWORKS

MAXIMILIAN RIEGEL

17.1 INTRODUCTION

Ethernet services (often called carrier ethernet services) have become common telecommunication services for establishing and connecting private networks of corporations, public authorities, or service providers offering telecommunication services on top of the infrastructure of another network operator.

Transparent Layer-2 connectivity is provided based on Ethernet technologies due to its widespread availability, high scalability in terms of bandwidth as well as network size, excellent support of various network layer protocols, and its leading-edge cost position.

The basic Ethernet technology is standardized by IEEE 802; however the Metro Ethernet Forum has established a set of specifications describing the services model and characteristics, an architectural framework, network interfaces, and the operation and management of Ethernet services. Also, the ITU-T has adopted Ethernet as a transport technology and is extending the specification framework for usage in global public telecommunications networks.

Mobile WiMAX deploys the IEEE 802.16e radio interface to serve fixed, nomadic, and mobile broadband applications over the same access network. Within its scope as universal cellular access network technology, the support of Ethernet services over a Mobile WiMAX access network is a useful extension. It covers not only the traditional mobile telecommunication services market but also allows participation in the growing carrier Ethernet market in areas or cases when an appropriate wired infrastructure is not available. In particular, the support of Ethernet services may be deployed by DSL operators to extend their access networks based on Ethernet aggregation over a wireless infrastructure to customers without a phone line. And even though it is not an established market yet, providing mobility for Ethernet Services may open new business opportunities for operators as well as deployment opportunities for Mobile WiMAX.

This chapter describes the extensions to the Mobile WiMAX specification to enable the support of Ethernet services as part of the WiMAX Forum Network Release 1.5.

17.2   ETHERNET SERVICES

The Metro Ethernet Forum established a full set of specifications on Ethernet services and on the definition of the particular service attributes to facilitate commonly understood service-level agreements between service providers and customers. Based on the concept of an Ethernet virtual connection (EVC) the MEF [1] distinguishes two kinds of Ethernet services: the Ethernet-Line (E-Line) service and the Ethernet-LAN (E-LAN) service (see Figure 17.1).

17.2.1   E-Line Service

The E-Line Service is a point-to-point connection carrying Ethernet frames between two customer interfaces of the network, denoted as user network interfaces (UNIs). It is frequently used for substituting legacy time-division multiplex (TDM) private lines by less expensive Ethernet private lines.

E-Line Service type

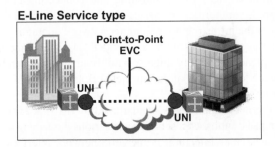

E-LAN Service type

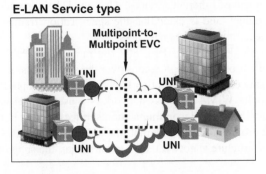

Figure 17.1.   Ethernet service types.

When multiple point-to-point connections, each carrying its distinct service are multiplexed onto a single Ethernet interface at the provider edge, the Ethernet private line service becomes an Ethernet virtual private line service. For example, it can be used by an Internet service provider (ISP) to provide Internet service to multiple customers over a single Ethernet interface, a mode that is well supported by the bandwidth hierarchy of Ethernet interfaces.

17.2.2   E-LAN Service

The E-LAN Service provides multipoint-by-multipoint connectivity for Ethernet frames across a number of customer interfaces, essentially behaving like an extension to the customer's own LAN. It is mostly deployed for creation of Transparent LAN Service (TLS), which enables full transparency for Ethernet control protocols and allows customers to establish new virtual LANs (VLANs) across their private networks without involvement of the Ethernet service provider.

Due to its inherent replication function for multicast and broadcast frames, the E-LAN service may be also used for efficient delivery of multicast services from a service provider's head-end to multiple customers.

Depending on whether the Transparent LAN Service is used by private network operators to extend corporate LANs or is used by ISPs to provide public access to Internet services to many customers, two E-LAN operational modes are distinguished:

1. Public access mode. When deploying Ethernet aggregation for public access services, the LAN service usually restricts direct communication between customers' endpoints and enforces that all Ethernet frames are passed through the ISP head-end to enable accounting and enforcement of security policies by the ISP. From the customer's perspective the service looks more like an E-Line service toward the ISP but, from the ISP side, all customers reside on the same link.

2. Enterprise LAN mode. When all connected end stations belong to the same enterprise and reside in the same security domain, direct connectivity across the TLS from any end station to any other end station is feasible. In this mode, the E-LAN service really behaves like a LAN with all stations being reachable from any interface on the same link.

Whereas the enterprise LAN mode represents standard Ethernet behavior among all participants of the LAN service, the public access mode introduces special forwarding behavior in the bridges to enforce that all customer traffic passes the ISP head-end, regardless of whether the communication peers reside on the same link or on different links.

17.3   BASIC ETHERNET SERVICES STANDARDS

The fundamental standards for Ethernet services were created in the Higher Layer LAN Protocols Working Group P802.1 of the IEEE 802 LAN/MAN Standards Committee. In extension to the IEEE 802.1D standard on media access control

(MAC) bridges enabling end stations to transparently communicate with each other across multiple LANs and defining the basic forwarding and filtering behavior in today's switched Ethernet, IEEE 802.1Q introduced the concept of segregating LAN traffic into virtual LANs, which allow the creation of multiple isolated LANs on the same bridged LAN infrastructure.

VLANs are commonly used inside larger organizations to establish separated LANs for particular applications and organizational units across widespread campuses and sites. Amendments to IEEE 802.1Q are expanding the applicability of VLANs into large operator networks and even across multiple operator networks:

1. IEEE 802.1 Q-2005 (Virtual Bridged Local Area Networks) [2] adopts from IEEE 802.1D the generic bridge architecture, the internal sublayer service, the major features of the filtering and forwarding process, the Rapid Spanning Tree Protocol, and the Generic Attribute Registration Protocol (GARP), and adds the definition of virtual LAN services, the required extensions to the filtering database and forwarding process, the definition of an extended frame format able to carry VLAN identifiers as well as priority information, the GARP VLAN Registration Protocol, and the Multiple Spanning Tree Protocol. The standard supports up to 4094 VLANs on the same bridged LAN infrastructure, which is usually sufficient for corporate networks but limits its applicability for larger provider networks.

2. IEEE 802.1ad-2005 (Amendment 4: Provider Bridges) [3] amends the IEEE 802.1Q-2005 standard to enable service providers to offer the capabilities of IEEE 802.1Q virtual bridged LANs to a number of customers with no need for alignment across the customers and only minimal interaction between the operation, administration, and maintenance (OAM) of the service provider and the OAM of the customers.

   The standard introduces another layer of traffic segregation by appending the customer VLAN identifiers (C-VID) by a service provider VLAN identifier (S-VID), allowing the service provider to encapsulate the particularities of the configuration of the customer VLANs by a VLAN ID assigned and managed by the service provider. It also addresses the extension to the frame format for inclusion of a customer VLAN tag as well as the service provider VLAN tag and reserves distinct MAC address spaces for customer L2 control protocols and service provider L2 control protocols, respectively.

   VLAN stacking (commonly called Q-in-Q) is the most widely used method in today's carrier Ethernet service deployments inside provider networks, but lacks the scalability for crossing multiple operator domains and backbone applications due to the 12 bit limit ($\sim$ 4000) of the S-VID.

3. IEEE 802.1ah (provider backbone bridges) is a new amendment to IEEE 802.1Q-2005 and IEEE 802.1ad-2005 allowing provider networks to scale up to 2^{24} ($\sim$ 16 million) Service VLANs by introduction of provider backbone bridges, which are compatible and interoperable with the provider bridges defined by IEEE 802.1ad. At the edge of a provider bridged network, the

provider backbone bridge encapsulates customer MAC frames containing the C-VID and a S-VID by assigning a 24-bit service instance identifier and putting them inside of backbone MAC frames containing the MAC addresses of the source and destination backbone bridges. At the other end of the backbone, the provider backbone bridge extracts the customer MAC frame out of the backbone MAC frame and forwards it according to the service instance identifier to the addressed provider bridged network.

Due to the encapsulation of customer MAC frames within backbone MAC frames, the protocol is commonly called MAC-in-MAC.

IEEE 802.1ah is not the only viable solution for extending Ethernet Services over large-scale networks. Other protocols like Virtual Private Line Service specified by RFC4761 and RFC4762 may be used as well for interconnection of customer and provider networks.

The frame structures for Ethernet, IEEE 802.16.1Q, 802.1ad, and 802.1ah are shown in Figure 17.2. FCS is the frame check sequence.

17.4   ETHERNET-BASED ACCESS AGGREGATION IN DSL NETWORKS

DSL networks are deploy Ethernet services based on Q-in-Q technologies in the access network to aggregate customer traffic, segregate different services, and provide even transparent LAN connectivity with support for customer-assigned VLAN-IDs.

The Broadband Forum specification TR-101 [4] describes the Ethernet-based access network architecture supporting multiple configurations to connect customer equipment and customer networks to the services of a DSL operator. The specifica-

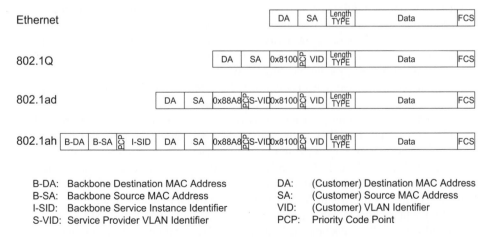

Figure 17.2.   Ethernet frame structure.

tion describes the processing and forwarding of Ethernet frames between the T-reference point toward the customer and the V-reference point towards the broadband network gateway (BNG). To facilitate different modes within the same access network infrastructure, C-VIDs and S-VIDs are used in a DSL-specific manner:

- For business TLS users, a unique S-VID is assigned at the T-interface and preserved throughout the network. For this kind of service, the DSL access network is strictly following the recommendations in IEEE 802.1ad for usage and assignment of the VLAN tags.

- For business or residential users with E-Line access to the DSL services, unique S-VIDs or a unique C-VID/S-VID combinations are assigned and preserved throughout the access network, depending on scalability requirements or usage of the C-VID by the customer.

- For residential users with shared access to the DSL services (E-LAN, public access mode), unique S-VIDs are assigned to groups of users or to all users subscribing to a particular DSL service. C-VIDs are not used in this case.

When Mobile WiMAX is used to provide access to DSL services over a wireless infrastructure, the WiMAX network emulates the behavior of the DSL access network and provides Ethernet connectivity between the customer interface and the V-interface the same way as the Ethernet based DSL aggregation network.

17.5   MOBILE WiMAX NETWORK ARCHITECTURE

The Mobile WiMAX architecture (Figure 17.3) adopts the decomposition of the Internet access network operation into several distinct operator roles. The network access provider (NAP) establishes and operates the radio access network and offers its services to one or multiple network service providers (NSP), which are in charge of customer related functions like authentication, service provisioning, and billing, and provide the backbone connectivity to services networks like the Internet or application service providers (ASP) running their particular applications.

Aligned to this role model, the Mobile WiMAX network reference architecture is built upon two logical networks entities: the access service network (ASN) providing link-layer connectivity and local mobility over the IEEE 802.16e radio interface, and the connectivity service network (CSN), comprising all subscriber-related functions for authentication, authorization, and accounting as well as the home agent or the data-path anchor for access to the services. There is a direct relation between ASN and NAP, and CSN and NSP. The ASN represents a logical access network with all the Mobile WiMAX specific functions, which belong to a NAP, and the CSN contains the Mobile WiMAX specific network functions of a NSP.

To facilitate interoperable implementations of equipment for WiMAX access networks, the ASN is decomposed into a set of base stations (BS) connected to a central control and a data forwarding instance called the ASN-Gateway (ASN-

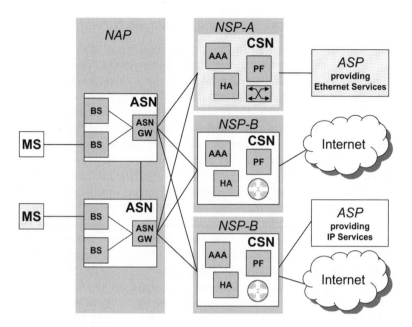

Figure 17.3. Mobile WiMAX network reference architecture.

GW). The reuse of well-known and widely deployed IP protocols and ISP principles based on a functional specification of the CSN allows NSPs to implement the required functions in their core networks based on a functional specification of the CSN.

The design of the Mobile WiMAX network architecture supports an ASN to connect to multiple CSNs for load balancing as well as network sharing purposes. To become a provider of information and communication services over Mobile WiMAX, it is sufficient to operate a NSP consisting of a single CSN with connections to one or more ASNs for leveraging the mobile access services offered by WiMAX NAPs.

Even when it is relatively easy to become a WiMAX service provider by implementing just a CSN, the overhead to establish all the contractual framework to become able to use access connectivity of WiMAX NAPs may be too complex, especially for enterprises and smaller operators just looking for some point-to-point connectivity in areas where they do not have a wired infrastructure. Ethernet Services over Mobile WiMAX offer transparent Ethernet connectivity over a wireless access network exactly in the same way as Metro Ethernet Network providers are offering today's wired Ethernet services.

Metro Ethernet Network providers may extend their businesses by implementing the CSN functions for Ethernet services in their networks and may become wireless Ethernet service providers without the need for huge investments to establish their own wireless access infrastructure.

17.5.1   Ethernet Services Support in the Access Network

Primarily designed for access to the Internet and for delivery of IP-based services, the Mobile WiMAX reference architecture is able to handle Ethernet services as well (Figure 17.4). The model operator roles with the distinction between NAP and NSP, and the definition of the ASN and the CSN are applicable also for Ethernet services. The main difference to the IP services network architecture is the modification of the data path between the mobile station (MS) and the CSN to support the transport of Ethernet frames. Fortunately, the protocols chosen for the data path for IP services are able to support Ethernet as well.

The transport of Ethernet frames over the IEEE 802.16e radio interface between the MS and the BS requires the application of the Ethernet specific part of the packet convergence sublayer (Ethernet-CS) instead of the IP-CS. The Ethernet-CS supports the transport of Ethernet frames within IEEE 802.16e MAC frames and extends the classification capabilities of the convergence sublayer to Ethernet-specific header fields like source and destination MAC addresses, Ethernet priority, and VLAN-ID, in addition to the IP header fields supported by the IP-CS.

The GRE tunneling protocol is deployed between the BS and the ASN-GW and supports the encapsulation of Ethernet frames in the same way as the encapsulation of IP packets. GRE may also be used for the connection between ASN and CSN for the MIP data path.

When Mobile IP is applied for dynamic tunnel management between the ASN-GW and the CSN, the GRE-based tunnel enables the transport of Ethernet frames in the payload. To enable this new deployment of Proxy Mobile IPv4 (PMIPv4), WiMAX makes use of an optional extension of the PMIPv4 protocol to signal the mobile host by its 48-bit MS-ID instead of an IPv4 address.

Without the demand for dynamic tunnel establishment and tunnel relocation between the ASN-GW and the CSN, any Ethernet transport protocol can be used between ASN and CSN, which is able to forward Ethernet frames based on the MS-ID instead of the destination MAC address and which preserves the customer VLAN ID assignment. In particular, when the MS is not the end station but contains a

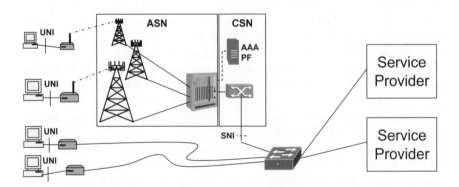

Figure 17.4.   Wireless access to Ethernet services networks.

bridge forwarding to end stations behind the MS, the Ethernet frames contain destination MAC addresses other than the MS-ID.

17.5.2 Ethernet Services Anchor in the CSN

A bridging function for forwarding Ethernet frames between MSs and the Ethernet service provider is contained in the CSN. A dedicated (virtual) bridge port is provided for each of the connected MSs to ensure compliance with standard bridging behavior. In alignment to the generic Mobile WiMAX network architecture, all payload Ethernet frames going over the radio interface are forwarded across the ASN to the bridge port in the CSN to enable the NSP to take full control of the traffic of the subscribers.

The configuration of the bridge in the CSN determines which kind of Ethernet service is provided to a particular MS. E-Line service is realized by forwarding the traffic of a MS to a single other end station, either on the wireless side or the wired side of the network. E-LAN service involves multiple other end stations, and the applied forwarding behavior in the bridge determines whether the enterprise LAN mode or the public-access mode is realized. Due to its location behind the mobility anchor in the CSN, the bridge does not have to cope with mobility issues in the access network. From the perspective of the bridge, the MS always appears on the same bridge port like in today's fixed Ethernet networks, not imposing any new requirement to the bridging functionality for support of Ethernet services over a cellular infrastructure.

The mobility in the access network in a cellular infrastructure requires management functions by a control plane, which allow the instantaneous provisioning of any kind of Ethernet services to any place in the access network. Such functions are not known in current fixed-access networks for Ethernet services.

The approach taken in Mobile WiMAX is the reuse of the existing control plane for IP services. In addition to the adaptation of the data path, there are only minor modifications necessary in the control protocols of the Mobile WiMAX access network to support Ethernet services. The following section explains the reuse of the existing control plane for Ethernet services and presents the few places where enhancements are necessary.

17.5.3 Reuse of the Mobile WiMAX Control Plane

The Mobile WiMAX control plane comprises all the functions to establish, maintain, and tear down mobile connectivity in a cellular network based on the IEEE 802.16e radio standard. The specification for IP Services [5] defines the following control plane functions. Most of the functions are applicable to Ethernet Services without any modifications, but some of them have to be amended to support Ethernet Services.

- *Network Entry Discovery and Selection/Reselection* describes the procedures of an MS to scan and detect the available WiMAX access networks, as well as to choose the preferred NAP for connecting to the home NSP. This section is applicable to Ethernet services without modifications.

- *WiMAX Key Hierarchy and Distribution* specifies the generation, derivation, and distribution of all the keying material needed in the Mobile WiMAX access network. This section is applicable to Ethernet services without modifications.

- *Authentication, Authorization, and Accounting* describes the procedures for the network access authentication and authorization according to the user profile of the subscription, and the accounting procedures for measuring and signaling the usage of the access network by a particular subscriber. Although the authentication procedures are applicable to Ethernet services without modifications, authorization was extended to enable Ethernet-specific user profiles. Not applicable to Ethernet services is the accounting model of IP services, which is based on the status of the IP session. Accounting events for Ethernet services are based on the status of the Ethernet connectivity, which fully depends on the establishment of the link-layer session and the service flows making use of the Ethernet CS.

- *Network Entry and Exit* describes the procedures for establishing initial connectivity to a Mobile WiMAX network after network entry detection and selection, and the procedures for terminating the network connectivity in an orderly fashion. This section is applicable to Ethernet services without modification.

- *QoS and SFID Management* defines the procedures for creation, modification, and deletion of the initial and further preprovisioned service flows, as well as the management of service-flow identifiers (SFIDs) and the static quality-of-service (QoS) policy provisioning. This section got a major amendment to support dynamic QoS for Ethernet services because it is not acceptable to tear down a connection completely just to install or modify the service parameters of a service flow for example, for adding another VLAN to a bundle of established VLANs on a link. Ethernet connections may carry a number of VLANs in parallel, and it is required to add, modify, or remove service flows assigned to particular VLANs while other service flows are carrying VLANs with mission-critical applications over the same link. Furthermore, Ethernet-specific attributes were added for describing the classification and the QoS parameters of service flows carrying Ethernet frames.

- *ASN Anchored Mobility Management* describes the handover of radio links across base stations inside a single ASN or across ASNs without change of the anchor ASN. This section is applicable to Ethernet services without modifications.

- *CSN Anchored Mobility Management* describes mobility management based on Mobile IP across ASNs with the mobility anchor located in the CSN. A new part was added to the Proxy MIPv4 specification to make this section applicable to Ethernet services for the transport of Ethernet frames over PMIPv4-managed tunnels. An additional attribute is used to allow signaling of the MSID in the Mobile IP protocol.

- *Radio Resource Management* is a function to increase the radio resource usage efficiency within an ASN. This section is applicable to Ethernet services without modifications.

- *Paging and Idle-Mode MS Operation* describes the control procedures for location update, paging, and entering and leaving the idle mode according to the specifications in IEEE 802.16e. Even when idle mode may be a rarely used function for Ethernet services, this section can be applied to Ethernet services without modification.

Two other functions of the Mobile WiMAX control plane are specific to IP Services and are not used for Ethernet Services:

- *IPv4 Addressing* describes the assignment of IPv4 addresses to MSs via DHCP, PMIPv4, and CMIPv4. This section is not applicable to Ethernet services.

- The section on *IPv6* defines the operation of IPv6 in the Mobile WiMAX access network. This section is not applicable to Ethernet services.

As explained in the list above, most of the existing Mobile WiMAX control plane is applied to Ethernet services without any modifications. A few functions, namely, accounting, QoS management, and CSN-anchored mobility management, required Ethernet-specific amendments, which can coexist with the IP specific functions inside the same access network. Coexistence is highly important to allow a Mobile WiMAX access network to provide IP services as well as Ethernet services concurrently over the same infrastructure to support participation in all the telecommunication markets with a single cellular infrastructure.

Support of Ethernet does not require a new network architecture or any new functional entities in the network. The support of Ethernet services is smoothly integrated into the Mobile WiMAX network architecture and can be operated parallel to the IP services in the same network.

17.6 INTERWORKING WITH DSL NETWORKS

An important usage model of Ethernet services over Mobile WiMAX is interworking with DSL networks to offer DSL services over a wireless infrastructure (Figure 17.5). The WiMAX network provides a DSL conformant T-reference point to the subscriber and connects the user interface with Ethernet to the V-aggregation point of the DSL network.

DSL-specific network functions in the Mobile WiMAX network with support for Ethernet services realize the exact network behavior for Ethernet-based aggregation according to Broadband Forum TR-101.

DSL user authentication is performed by the DSL network across the Ethernet connectivity of the WiMAX network within the PPPoETH protocol and the particular access line is signalled by DHCP options when IPoETH is deployed.

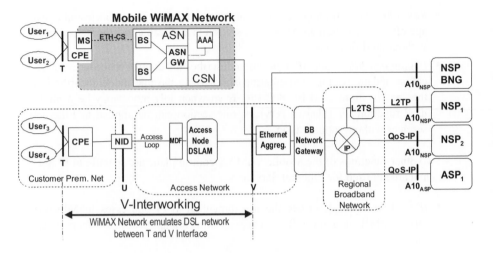

Figure 17.5.   Wireless access to Ethernet services networks.

The WiMAX security framework with PKMv2 is used to establish secure Ethernet connections over the IEEE 802.16e radio interface based on device or user identities with the credentials stored in the MS. The identities are used to manage the provisioning and the accounting of the Ethernet connectivity across the WiMAX network according to the profiles stored in the AAA server in the CSN. Essentially, there is no difference between device authentication and user authentication in the DSL interworking case. The identity is always pointing to a particular MS, and the difference between device authentication and user authentication is only whether the identity is hard-coded or soft-coded in the MS.

17.6.1   Related Standardization Activities

The specifications for Ethernet services over Mobile WiMAX as described by this chapter are part of the WiMAX Network Release 1.5 of the Network Working Group in the WiMAX Forum. Accompanying standardization topics were addressed in the 16ng working group of the IETF as well as within the REV2 project of IEEE P802.16.

The IETF has a specification on "Transmission of IP over Ethernet over IEEE 802.16 Networks" [6], which provides the generic architectural framework for Ethernet networks over IEEE 802.16 and specifies a number of enhancements to the bridging function on the network side to increase the efficiency of the radio resource usage and to reduce the power consumption of terminals when running IP over Ethernet over a cellular network.

The IEEE P802.16REV2 task group created by IEEE 802.16-2009 provided a revision of the standard, which incorporates the IEEE 802.16-2004, the IEEE 802.16e, the IEEE 802.16f, and the IEEE 802.16g standards into a single document.

Within this effort, the specification of the Ethernet-specific part of the packet-convergence sublayer was refined to make its application less cumbersome. In particular, it became possible to use the same Ethernet CS for Ethernet frames with or without IPv4 or IPv6 payload optionally tagged by VLAN IDs and priority.

17.7 SUMMARY

This chapter presents the current state of the support of Ethernet services over Mobile WiMAX. The approach of the WiMAX community is to align the Ethernet service provisioning model as closely as possible with the model for IP Services. It reduces not only the necessary standardization efforts for developing the specifications but also enables the development of WiMAX network equipment, which serves IP as well as Ethernet out of the same box. Such equipment can be deployed by any kind of WiMAX network operator, is well prepared for the evolution of services into a broader telecommunication market, and provides high commonality, which, finally, leads to better scale of economies.

With the biggest part of the investments of a network operator going into the access infrastructure, it is important to enable most flexible deployment of the access network. Participation in the growing market for Ethernet services without the prerequisites of huge investments may considerably contribute to a sustainable business case for a wireless broadband access provider. Mobile WiMAX supports an integrated approach to provide fixed, nomadic, and mobile IP services and Ethernet services out of the same network infrastructure.

17.8 REFERENCES

[1] MEF 6 - Ethernet Services Definition—Phase 1, Metro Ethernet Forum, June 2004.

[2] Virtual Bridged Local Area Networks, IEEE STD 802.1Q-2005.

[3] Amendment 4: Provider Bridges, IEEE STD 802.1ad-2005.

[4] Migration to Ethernet-Based DSL Aggregation, Broadband Forum TR-101, April 2006.

[5] WiMAX Forum Network Architecture, Stage 2 and Stage 3, Release 1.5, November 2009.

[6] RFC 5692, H. S. Jeon, M. Riegel, S. J. Jeong, Transmission of IP over Ethernet over IEEE 802.16 Networks, October 2009.

CHAPTER 18

WIMAX SYSTEM PERFORMANCE

BONG HO KIM, JUNGNAM YUN, AND YERANG HUR

18.1 INTRODUCTION

The Mobile WiMAX system is based on IEEE Std 802.16, which defines a high-throughput packet-data network radio interface capable of supporting several classes of Internet Protocol applications and services. To evaluate the performance and facilitate the design of the Mobile WiMAX system, all the aspects of performance evaluation from air link to application are required. This involves link-level, system-level, network-level, and end-to-end application-level performance simulation. This chapter describes the various levels of mobile WiMAX performance evaluation.

First, it depicts the overall structure of an end-to-end application performance simulator followed by radio performance evaluation focusing on link-level and system-level performance. This section provides a channel model and antenna-pattern model used for link-level performance evaluation, and a path-loss model, link-to-system-mapping method for system-level performance evaluation. The simulation results show the DL/UL spectral efficiency of Mobile WiMAX system for PEDB and VEHA channels with various MIMO types and HARQ considered. PHY-level abstraction for upper-layer simulation is also provided.

The following sections are dedicated to upper-layer performance evaluation. To evaluate network-level and application-level performance, we first describe the subscriber and application profile. This includes traffic mix ratio, data session attempts for applications, diurnal application traffic distribution, and application traffic model. Afterward, network performance of Mobile WiMAX is explained. Network-level performance evaluation provides the simulation results of network traffic characteristics and demand estimation based on Mobile WiMAX requirements of QoS, HO, and reattachment. Finally, in the last section, we provide simulation results of end-to-end application performance evaluation using the examples of the VoIP and TCP/IP performance-enhancement method.

18.2 DESIGN OF THE END-TO-END APPLICATION PERFORMANCE SIMULATION

To evaluate the end-to-end user perspective application performance guidelines of the following three categories must be considered as an input to the evaluation [20,25]:

1. Service provider perspective
2. System vendor perspective
3. Network provider perspective

Figure 18.1 depicts a conceptual diagram for the end-to-end system performance simulation considering these aspects. The WiMAX system simulator based on the NS-2 developed under the WiMAX Forum AWG and the wireless integrated system emulator for WiMAX (WISE-W), developed by Posdata Inc., are examples of systems adopting this simulator.

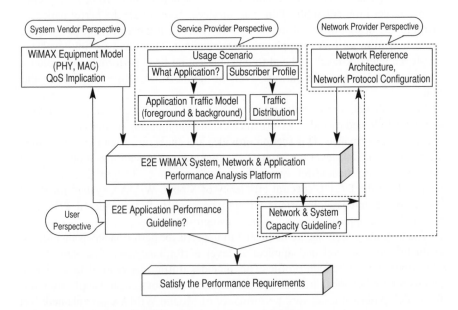

Figure 18.1. Conceptual diagram for the end-to-end application performance simulation.

One of the charters of the WiMAX AWG is to characterize the performance guidelines of certain representative WiMAX applications. The performance guidelines are intended to facilitate the development/deployment of WiMAX applications that will take full advantage of the WiMAX access technology and guide the optimal configuration/deployment of the WiMAX networks. The performance guidelines in Table 18.1 are addressed in terms of the number of parameters, including bandwidth, jitter, latency (end-to-end delay), and packet loss.

The recommended user perspective end-to-end application performance guideline that the Mobile WiMAX system and network should provide as specified in Table 18.1 must be applied during the development or deployment stage.

Table 18.1. Application performance guideline

Class #	WiMAX AWG class	Real time	Application types in each class	Bandwidth	Key performance guideline			
					Application response time (sec)	End-to-end one-way delay (msec)	Packet delay variation within a flow (msec)	Information loss (PER)
1	Interactive gaming	Yes	Interactive gaming	50–85 kbps	N/A	< 150 msec, preferred	< 100 msec	< 3%
2	VoIP, Video conference	Yes	VoIP	4–64 kbps	N/A	< 150 msec, preferred < 200 msec, limit	< 20 msec	< 1%
			Video phone	32–384 kbps	N/A	< 150 msec, preferred < 200 msec, limit Lip-synch: < 100 msec	< 20 msec	< 1%
3	Steaming media	Yes	Music/ speech	5–128 kbps	< 5 sec (initial buffering time)	TBD	< 2 sec	< 1%
			Video clips	20–384 kbps	< 5 sec (initial buffering time)	TBD	< 2 sec	< 2%
			Movie streaming	> 2 kbps	< 5 sec (initial buffering time)	TBD	< 2 sec	< 0.5%
4	Information technology	No	IM	< 250 bytes/ message	< 30 sec	< 65 msec	N/A	Zero
			Web browsing	> 500 kbps	< 4 sec/page	< 65 msec	N/A	Zero
			E-mail	> 500 kbps	—	< 65 msec	N/A	Zero
5	Media content download (store and forward)	No	Bulk data, movie download	> Mbps	—	< 65 msec	N/A	Zero
			P2P	> 500 kbps	—	< 65 msec	N/A	Zero

18.3   RADIO PERFORMANCE

The increasing demand for bandwidth and throughput has driven improvements in modulation and coding schemes, advanced antenna technologies (MIMO and beamforming), and optimized link-quality control algorithms. All these new technologies are highly coupled with accurate radio channel modeling. This section describes link-level and system-level performance evaluation methodologies, including radio channel modeling.

18.3.1   Link-Level Performance

A link-level modeling of a radio interface needs one base station and one mobile station. Even though practical wireless cellular networks are comprised of many base stations and many mobile stations, fundamental designs of the PHY layer mostly rely on the performance analysis of the link-level model.

18.3.1.1   Channel Model

A radio channel can be described as roughly a mixture of three different characteristics: path loss attenuation, shadowing, and multipath fading. These three characteristics are nearly independent and have their own cause. Each of these three phenomenon should be considered in modeling a radio channel to ensure the performance evaluation is trustworthy.

Generally, there exist two different channel models: deterministic and stochastic. Deterministic channel models are used for cell planning and stochastic channel models are widely used for simulation-based radio performance evaluation. Conventionally, the tapped delay line (TDL) model has been popular for link-level channel modeling and 6-tap ITU models were widely used for SISO systems performance evaluation [1]. Later, spatial channel models were included to capture the channels between multiple transmit and receive antennas in MIMO systems [2–4].

Large-scale fading is not considered in link-level channel modeling because link-level performance evaluation usually focuses on various SINR points of interest.

18.3.1.2   Antenna Pattern

The antenna pattern used for each BS sector is specified as

$$A(\theta) = -min\left[12\left(\frac{\theta}{\theta_{3\,dB}}\right)^2, A_m \right]$$

where $A(\theta)$ is the antenna gain in dBi in the direction θ, $-180° \le \theta \le 180°$, min [ ] denotes the minimum function, θ_{3dB} is the 3 dB beamwidth (corresponding to $\theta_{3dB} = 70°$), and $A_m = 20$ dB is the maximum attenuation.

18.3.2 System-Level Performance

The major difference between link-level simulation and system-level simulation is that system-level simulation considers point-to-multipoint communications in the multicell scenario and, hence, it includes cochannel interference and large-scale fading such as path loss and shadowing.

The COST-231 modified Hata model is widely used for the path-loss attenuation model in radio performance evaluation. The path-loss model should be different for different propagation environments, such as, microcell, macrocell, and so on.

We use the mandatory path-loss model defined in IEEE 802.16m evaluation methodology [7]. This model is applicable for test scenarios in urban and suburban areas outside the high-rise core, where the buildings are of nearly uniform height.

$$PL(dB) = 40(1 - 4 \times 10^{-3}h_{BS})\log_{10}(R) - 18 \log_{10}(h_{BS}) + 21 \log_{10}(f) + 80$$

wherein kilometers is the distance from the transmitter to the receiver, f is the carrier frequency in MHz and h_{BS} is the base station antenna height above the rooftop.

For system-wise performance evaluation, usually hexagonal 19 cells are used to introduce cochannel interference from second-tier cells or sectors and the center cell among 19 cells are evaluated. Otherwise, the 19-cell hexagonal structure is extended with seven copies of the original hexagonal network. From copied cells, virtual interferences are generated so that all 19 cells are evaluated. This is called a wrap-around model [7].

18.3.2.1 Link to System Mapping

System-level simulation should take account of instantaneous link performance in order to evaluate several cross-layer functionalities. Hence, predicting the instantaneous link PHY performance is very important and should be considered in the system-level simulation.

PHY abstraction for multicarrier systems like OFDM used in Mobile WiMAX has been a popular research area recently [8–17]. Since each subcarrier of OFDM systems faces different channel response, frequency selectivity is considered in calculating the postprocessing SINR to predict the effective SINR. Based on the calculated effective SINR, resulting BLER values can be obtained from the link abstract. Mapping the frequency-selective channel to the effective SINR of the AWGN channel is termed effective SINR mapping (ESM) [7]. There are several effective SINR mapping approaches to predict the instantaneous link performance; mean instantaneous capacity [8–10], exponential-effective SINR mapping (EESM) [11], [13–15], and mutual information effective SINR mapping (MIESM) [16,17]. Figure 18.2 shows the system-level simulation procedure with effective SINR mapping.

18.3.2.2 Simulation Parameters and Conditions

Table 18.2 shows system-level simulation conditions. We considered chase-combining HARQ for both downlink and uplink. Also, we considered SISO, MISO, and

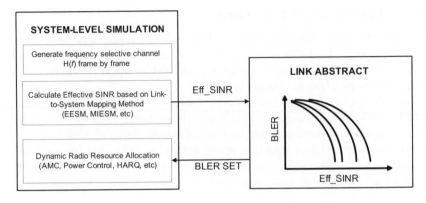

Figure 18.2. System-level simulation procedure and effective SINR mapping.

MIMO systems for downlink, and CSM (collaborated spatial multiplexing) MIMO for uplink.

18.3.2.3 Simulation Results

Figure 18.3 shows downlink spectral efficiency for PEDB 3km/h and VEHA 60km/h channels. Various MIMO types (SISO, SIMO, MIMO Matrix A and Matrix B) are presented for HARQ mode and non-HARQ mode.

Table 18.2. System-level simulation conditions

Parameter	Value	Parameter	Value
Cell layout	Two-tier, 19 cell, 3 sectors/cell (57 sectors)	BS max PA power (per antenna)	10 W (40 dBm)
BS-to-BS distance	2000 m	MS max PA power (per antenna)	200 mW (23 dBm)
Occupied bandwidth	8.75 MHz	BS Tx/Rx number of antennas	(1 or 2) /2
BS noise figure	5.0 dB	MS Tx/Rx number of antennas	1/(1 or 2)
MS noise figure	7.0 dB	Traffic model	Full buffer
Thermal noise density	−174 dBm/Hz	Number of symbols for DL/UL per frame	47 total OFDM symbols (32:15)
Carrier frequency	2.35 GHz	Frame overhead	7 OFDM symbols (4 DL, 3 UL)
BS antenna gain with cable loss	15.0 dB (cable loss: 2 dB)	DL/UL subchannelization	PUSC/UL-PUSC
Fading generation	Jakes Model	DL/UL H-ARQ	Chase combining
Channel models (multipath)	ITU-VEHA, 60 km/hr ITU-PEDB, 3 km/hr	Scheduler	Proportional fair algorithm

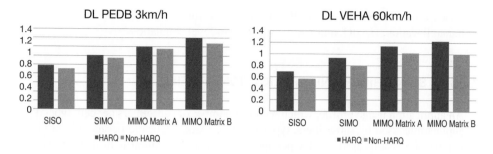

Figure 18.3. Spectral efficiency of downlink.

Figure 18.4 shows downlink spectral efficiency for PEDB 3 km/h and VEHA 60 km/h channels. Various MIMO types (SISO, SIMO, MIMO Matrix A and Matrix B) are presented for HARQ mode and non-HARQ mode.

18.3.3 PHY Abstraction for Upper-Layer Simulation

PHY abstraction for upper-layer simulation is mostly mapping the instantaneous channel to its effective SINR so that system-level simulator can perform various types of cross-layer operations, such as adaptive modulation and coding, power control, hybrid ARQ, and dynamic resource allocations. Here, all the interferences from 19 cells are considered in calculating the effective SINR.

There are two ways of generating the PHY abstraction for network-level simulation: deterministic and stochastic. As mentioned in channel modeling, deterministic abstraction is the ready-made CINR trace from 19-cell system-level simulation. In order to perform general network simulation, a great number of CINR traces should be generated in advance. Stochastic abstraction is using statistical information to generate the radio channel for network simulations. The Markov model can be a good example to realize the statistical channel model.

For the end-to-end service/application performance simulation, we use the deterministic channel model for PHY abstraction in order to reduce the simulation time.

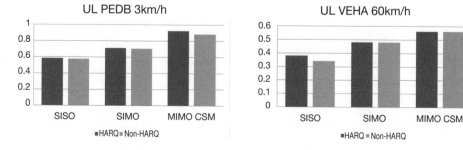

Figure 18.4. Spectral efficiency of UL in PEDB 3 km/h channel.

Based on the CINR trace, the RRM module performs link adaptation, dynamic burst allocation, HARQ, and ARQ, as in system-level simulation.

18.4   SUBSCRIBER AND APPLICATION PROFILE

An application usage profile, which includes number of subscribers per cell, type of applications used by subscribers, types of subscribers, application usage mix ratio, number of sessions used per day per application, and diurnal traffic distribution per application, has been studied with the purpose of obtaining consistent input for system evaluations such as WiMAX network dimensioning, system capacitating, and end-to-end application performance simulation with additional information, and application traffic models [23], across the WiMAX Forum working groups, especially AWG, SPWG, and NWG [21].

18.4.1   Number of Subscribers and Machines

South Korea predicts that 9.5 million users will sign up for WiBro services by 2012 and the estimated number of total WiBro subscribers has been submitted to the South Korea government on the assumption that the service is rolled out in 84 cities, focusing on WiMAX users only [1]. The following is based on the expected total number of subscribers submitted by KISDI from 2007 to 2011 [1].

Table 18.3 shows the number of estimated subscribers and machines in a cell, and the main parameters affecting the system/network load. The number of total subscriber can be estimated with series of assumptions such as total population, expected subscriber percentage, expected cell size, and size of the geographic region. The machine is also a type of subscriber that uses machine-to-machine telemetry services. Telemetry is defined as machine-to-machine messaging—low-bandwidth machine-to-machine-initiated communications monitoring or tracking of stationary objects. Vending machines, taxis, printers, parking meters, and home appliances are examples of telemetry machines. The number of expected machines is estimated based on the assumption that the total number of machines in year 2011 will be 5% of the expected total population. The assumptions may not hold for other regions or operators. Table 18.3 was prepared as an example since it is essential for the end-to-end system performance study.

Table 18.3.   Expected number of subscribers and machines

Year	2007	2008	2009	2010	2011
Number of active subscribers/cell for a major operator	160	230	260	310	320
Number of active machines/cell for a major operator	160	350	550	640	670

18.4.2   Traffic Mix Ratio and Data Session Attempt for Applications

Table 18.4 shows type of subscribers, application session attempts per application per day, and application mix ratio among 18 applications categorized in five application classes as defined by the WiMAX Forum Application Working Group (AWG). The application mix ratio represents the distribution of applications that users will most likely use [21]. Table 18.4 shows three types of subscribers: consumers, enterprises, and machines. It assumes that 64% of the total human subscribers are consumers and the rest of the human subscribers are enterprises, and the table considers two different categories of mobile terminals, which will influence traffic and symmetry. Professionals will tend to use terminals with high-resolution screens (high-complexity terminals) such as laptops and high-end PDAs for multimedia, for which the accuracy and detail of the information is crucial. On the other hand, consumer subscribers will have more interest in small, lightweight terminals, for which a high-resolution screen is not relevant. However, some exceptions are likely, when, for example, a consumer (with a small device) sends a photo to someone in the fixed network (e.g., as an e-mail attachment). Here, the recipient will have a high-resolution screen and printer and so will want the picture at high resolution. The size and resolution of the screen significantly affects the data volume of the picture or video media intended for it.

18.4.3   Diurnal Application Traffic Distribution

The diurnal application traffic distribution is important because daily traffic from certain applications may be concentrated during commuting hours, working hours (8:00 am to 6:00 pm), or evening, or may occur equally all day [19]. The number of sessions per active user during a day with the diurnal application traffic distribution can be converted into the number of application sessions during a specific time of day (Figure 18.5).

18.4.4   Application Traffic Model

With the advent of the true wireless broadband access for the fixed and mobile workforce, new applications are emerging with their own challenges and requirements. WiMAX enables last-mile point-to-point and point-to-multipoint elastic pipes that server emerging IP-based end devices. The WiMAX Forum AWG has characterized the applications in the five classes, and provided a consistent traffic model and a modeling framework across the WiMAX community so that each working group (i.e., SPWG and NWG) can use it as a reference when doing network dimensioning, system configuration or optimization, scheduling algorithm development, network sizing, end-to-end application performance simulation, and so on.

The detailed application traffic models can be found in [23] and they are categorized into two types of traffic model: foreground and background. The foreground traffic model represents a specific user behavior or interaction; in other words, application traffic is generated mainly through a user interaction with a device. One of the

Table 18.4. Application mix ratio and subscriber distribution

Class #	WiMAX AWG Class	Packet data applications	% of subscribers actively using applications			Consumer (% of total subscribers) 64%		Enterprise (% of total subscribers) 36%		% of total machines 100%
						Laptop OC 30%	Smaller device 70%	Laptop OC 70%	Smaller device 30%	
			Consumer	Enterprise	Machine	No sessions/ day/ subscriber	No sessions/ day/ subscriber	No sessions/ day/ subscriber	No sessions/ day/ subscriber	No sessions/ day/ subscriber
1	Internet gaming	Quake II	25.0%	5.0%	0.0%	0.15	0.15	0.05	0.05	0
		World of Warcraft	25.0%	5.0%	0.0%	0.15	0.15	0.05	0.05	0
		Xbox TimeSplitter2	25.0%	5.0%	0.0%	0.15	0.15	0.05	0.05	0
		ToonTown	25.0%	5.0%	0.0%	0.15	0.15	0.05	0.05	0
2	VoIP/video conference	VoIP	100.0%	100.0%	0.0%	5.71	5.71	4.44	4.44	0
		Video Conference	50.0%	100.0%	0.0%	0.30	0.30	0.27	0.27	0
		PTT	20.0%	20.0%	0.0%	0.10	5.00	5.00	5.00	0
3	Streaming Media	Music/Speech	100.0%	100.0%	0.0%	0.08	0.08	0.12	0.12	0
		Video Clip	50.0%	100.0%	0.0%	1.10	1.10	1.50	1.50	0
		Movie Streaming	100.0%	100.0%	0.0%	0.20	0.20	0.12	0.12	0
		MBS	100.0%	0.0%	0.0%	1.00	1.00	0.10	0.10	0
4	Information Technology	IM	100.0%	100.0%	0.0%	7.26	7.26	7.26	7.26	0
		Web Browsing	100.0%	100.0%	0.0%	5.00	2.00	5.00	2.00	0
		E-mail (POP3)	50.0%	50.0%	0.0%	0.65	0.65	1.50	1.50	0
		E-mail (IMAP)	50.0%	50.0%	0.0%	0.65	0.65	1.50	1.50	0
		Telemetry	0.0%	0.0%	100.0%	0.00	0.00	0.00	0.00	24
5	Media content download/backup	FTP	50.0%	100.0%	0.0%	2.00	0.10	2.00	1.10	0
		P2P	30.0%	0.0%	0.0%	0.30	0.30	0.00	0.00	0

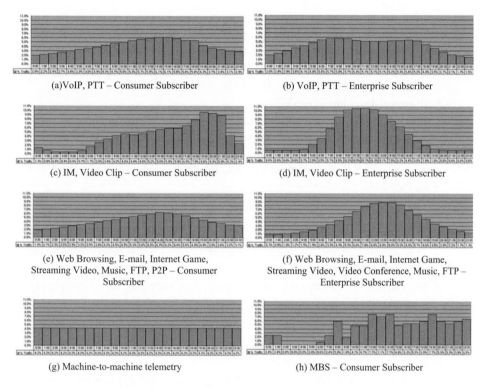

(a)VoIP, PTT – Consumer Subscriber

(b) VoIP, PTT – Enterprise Subscriber

(c) IM, Video Clip – Consumer Subscriber

(d) IM, Video Clip – Enterprise Subscriber

(e) Web Browsing, E-mail, Internet Game, Streaming Video, Music, FTP, P2P – Consumer Subscriber

(f) Web Browsing, E-mail, Internet Game, Streaming Video, Video Conference, Music, FTP – Enterprise Subscriber

(g) Machine-to-machine telemetry

(h) MBS – Consumer Subscriber

Figure 18.5. Diurnal application traffic distribution.

goals of the simulator in implementing the foreground traffic model is to evaluate a user-perspective application performance in detail. The background traffic, however, is not directly related to user interaction. Certain applications or services generate network traffic as soon as a session starts, regardless of user interaction with a device, and the amount of network traffic is significantly larger than foreground traffic for the application or service. IM application and VPN service are examples.

Each of the application traffic models can be represented in either user-level or IP-packet-level perspectives. The usages of these two levels of model are the following:

- User-level traffic model:

    User behavior or interaction with an application is modeled, and this model can be used with a simulation that includes detailed application layer, transport layer, and IP layer models on top of the Layer 1(PHY) and Layer 2 (MAC) models.

    Application performance metrics specific to the application can be evaluated (Web page download time, e-mail download time, FTP download time, etc.), and a scheduling mechanism for the application QoS in the MAC layer can be evaluated or optimized. In case the session layer protocol is proprietary or not known, detailed analysis on the protocol behavior is critical for the application performance evaluation since the interaction with lower layers, especially with the MAC, may affect the results [22].

Since the IP-level traffic will be generated according to the upper-layer protocols, the generated UL and DL traffic will be correlated.

- IP-packet-level traffic model:

This model is generally obtained from network traffic measurement and represented as statistical packet distributions, such as packet size distribution and packet interarrival time distribution at the IP layer. This model can be used with a simulation that does not include detailed protocol layers above the MAC.

The IP-packet-level traffic model can be directly applied to the PHY and MAC models, and evaluating both application performance and scheduling mechanisms in the MAC layer are not easy tasks with this type of model. In general, this model is used in system-level simulation (SLS) for the air-link resource allocation.

UL and DL traffic may not be correlated since they may be independently modeled.

18.5   NETWORK PERFORMANCE

The WiMAX Forum Service Provider Working Group (SPWG) specifies performance and capability recommendations and requirements from the perspective of network operators intending to deploy WiMAX networks. It describes business and usage scenarios, deployment models, functional requirements, and performance guidelines for the end-to-end system. This is the first of a three-stage, end-to-end network system architecture specification for broadband wireless networks based on WiMAX Forum Certified™ products, and network architecture details are specified by NWG in stage-2 and stage-3 specifications based on requirements outlined in [25]. The following two subsections describe performance requirements.

18.5.1   Requirement for Application QoS

The WiMAX network must support QoS mechanisms and protocols so that it is possible to control the packet error rates to be less than a value that is equivalent to a 10^{-6} bit error rate when measured between the BS and the SS/MS over a statistically significant measurement interval. It must support QoS mechanisms and protocols so that it is possible to control the packet delay to be less than 10 ms and the packet delay variation to be less than or equal to 2 ms, when measured between the BS and the SS/MS. It also must support QoS mechanisms and protocols that can satisfy application QoS requirements listed in Table 18.5 for communication between two user devices that are attached to two different BSs belonging to the same WiMAX access network for any usage scenario under the following conditions:

Table 18.5. Application QoS requirements

Service class #	Service class	Application layer throughput	End-to-end transport layer one-way delay	End-to-end transport layer one-way delay variation	Transport layer information loss rate
1	Real-time Games	50–85 Kbps	< 60 ms preferred	< 30 ms preferred	< 3%
2	Conversational (e.g., VoIP, Video Phone)	4–384 Kbps	< 60 ms preferred < 200 ms limit	< 20 msec	< 1%
3	Real-time Streaming (e.g., IPTV, Video Clips, Live Music)	> 384 Kbps	< 60 ms preferred	< 20 ms preferred	< 0.5%
4	Interactive Applications (e.g., Web Browsing, Email Server Aceess, IM)	> 384 Kbps	< 90 ms preferred	N/A	Zero
5	Non-Real-time Download (e.g., Bulkdata, Movie Download, P2P)	> 384 Kbps	< 90 ms preferred	N/A	Zero

- The two user devices are stand-alone and constitute the end-points for the application of interest.
- Route optimization is employed.
- One-way delay between the BS and the network-facing boundary of the WiMAX access network is under 5 ms.
- The user devices are not used during handover.

The WiMAX network must support voice-over-IP (VoIP) services. Call setup latency is particularly important for regular VoIP and for PTT scenarios. WiMAX VoIP services must support the call setup and latency requirements listed in Table 18.6 at 95% of the normal distribution.

Table 18.6. Call setup and latency requirements for VoIP services

VoIP service attribute	Limits to which WiMAX network must conform
VoIP call signaling setup from invite to alert	< 1.5 sec
Latency from answer to start of voice media stream	< 200 ms

18.5.2   Requirements for Handover and Reattachment

The WiMAX network must support inter-BS and inter-BS cluster handovers in which data loss does not exceed 1% of all MAC-layer packets transacted between an SS/MS and the WiMAX network in the interval between initiation and completion of the handover procedure. The mobility and reattachment performance requirements listed in Table 18.7 must be treated as a single requirement.

18.5.3   Network Traffic Characteristics and Demand Estimation

The IP layer application traffic characteristics and demand, in terms of bandwidth (bps and pps) in the R6 interface (between BS and ASN-GW) are is estimated with the number of subscribers per cell for year 2011 and application usage profile as described in Section 18.4, except MBS, IM, PTT, and P2P, using the end-to-end WiMAX performance simulator (WISE-W) [24]. For the sake of simplici-

Table 18.7. Mobility and reattachment performance requirements

Usage model handover support	Fixed/nomadic	Portable	Simple mobility	Full mobility
SS/MS mobility	None for Fixed. None during a session for nomadic.	5 km/h	Up to 60 km/h with no performance degradation, 60 to 120 km/h with graceful performance degradation	Up to 120 km/h with no performance degradation
Total handover latency	N/A	Best effort	<1 sec between IP subnets <150 ms within IP subnet	<FFS between IP subnets, <50 ms within IP subnet
Handover data transmission disruption time	N/A	Best effort	<150 ms1	< larger of (5 ms, one frame time)
MAC reattachment disruption time	1 sec	1 sec	Superseded by handover requirement	Superseded by handover requirement
User reattachment disruption time	Best effort, <2 sec if within the same broadcast domain	Best effort handover; best effort reattachment, <2 sec if within the same broadcast domain	Superseded by handover requirement	Superseded by handover requirement

ty, packet loss is not considered in the network and 70 msec of RTT is statically configured.

Figure 18.6 shows that the average DL and UL bandwidths are 6.5 Mbps and 3.0 Mbps, and the number of packets are 821 pps and 718 pps, respectively. The average DL-to-UL bandwith ratio is 3.7 but the median is 2.2. The higher average is caused by the larger DL-to-UL ratio in the morning and evening. The higher ratio during these times of day is less significant for allocating DL and UL radio resources known as the DL and UL symbol ratio, since the total amount of traffic during the time is low compared to the system capacity. The default symbol ratio for the system profile 1.A is 27:15 [26] and this is equivalant to the approximately 2:1 bandwidth ratio that matches the bandwidth ratio in Figure 18.6. The symbol ratio in the system, however, should be adjusted if the application usage profile for a service provider is significantly different from the one defined in the AWG.

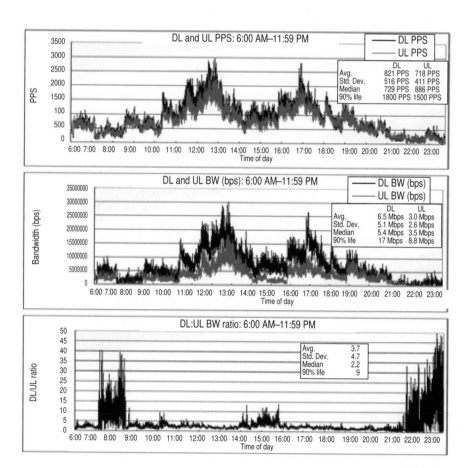

Figure 18.6. Network traffic demand expectation.

18.6   END-TO-END APPLICATION PERFORMANCE

One of the major objectives of developing the next-generation wireless communication technology is to provide a higher link capacity and increase the user-perspective application and service performance. The WiMAX system based on IEEE 802.16e has definitely higher system capacity and more sophisticated mechanisms to provide a better quality of service than the previous wireless systems such as CDMA or UMTS, and the enhanced system currently under development in the IEEE 802.16m Task Group will have even higher capacity (twice that of the IEEE 802.16e system). The wireless application/service performance limitations are known to be the narrow bandwidth of the airlink and fragile air-link quality. The tremendous amount of effort, however, is gradually overcoming the shortcomings of using wireless services. This indicates that wireless application/service performance-improvement research should be undertaken from a wider angle and over a broader area than before. The IEEE 802.16m TG is not only focusing on the PHY and MAC performance but also some level of end-to-end performance improvement, which includes the scope of the network and application to adopt the strong market request and interest. The IEEE 802.16m evaluation methodology document emphasizes the importance of the performance beyond the PHY and MAC layers, and it includes the TCP layer throughput metric as a mandatory performance measurement criterion in addition to the PHY and MAC layer throughput measurement [29].

18.6.1   VoIP Performance

We have performed analytic analyses to estimate the upper bound of the total number of the simultaneous VoIP users that is limited by the system bandwidth and performance simulation to identify the number of the simultaneous VoIP users satisfying the VoIP packet-delay requirement in Table 18.1 with same assumptions and VoIP traffic model in [23]. The assumptions include a 27:15 symbol ratio for DL/UL, PUSC for both DL and UL, four repetitions for DL_MAP, and three UL symbol reservations for the control messages.

Figure 18.7a shows that the upper bounds of the total number of VoIP users for header compression are disabled or enabled when DL and UL MCS levels are QPSK1/2 and QPSK1/2, QPSK1/2 and 16QAM3/4, and 16QAM3/4 and 16QAM3/4. The range of the number of users is from 70 to 250. This result may not be meaningful when QoS is considered in addition to the bandwidth requirement. Especially when VoIP service is provided through BE service flow other than UGS or ert-PS, the VoIP quality affect depends more on the number of VoIP users, as indicated in Figure 18.7b. The average VoIP packet delay from MS to BS when both DL and UL MCS are QPSK1/2 already exceeds 150 mesc, which is the preferred end-to-end packet delay for VOIP service, when the number of VoIP users is 60. When the DL and UL MCS are 64QAM 5/6 and 16QAM 3/4, the average packet delay starts exceeding 150 msec with 100 VoIP users. The packet delay for the DL is not a concern since the bandwidth and packet delay constraint appears in the UL first when BE service flows are used. The UL

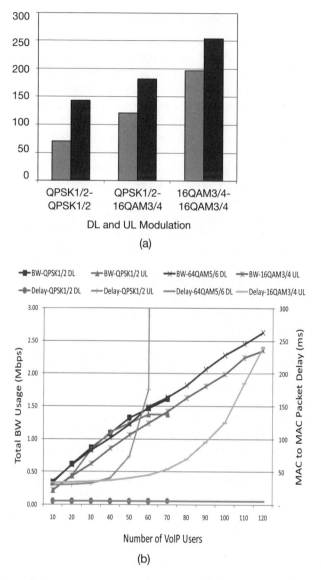

Figure 18.7. (a) Number of VoIP users: bandwidth perspective. (b) Number of VoIP users: BW usage and packet-delay perspective. DL/UL: (QPSK1/2, QPSK1/2) and DL/UL: (64QAM5/6, 16QAM3/4).

packet delay is primarily caused by the BWR procedure and it is more serious when BWR collision occurs when many VoIP users transmit BWR at the same time. The secondary cause is insufficient UL bandwidth (BW). When silence suppression is used, the packets are statistically multiplexed at the link level, and may exceed the total link bandwidth for certain instances, which causes additional delay.

18.6.2   TCP/IP Performance Enhancement Mechanism

This section describes a TCP performance enhancement mechanism—ACK Manager—for the Mobile WiMAX network that can be implemented in the Mobile WiMAX MAC or MAC/IP cross layer and provides significant improvement in data application performance. It reduces the probability of losing ACK packet information at the air link without modification to the widely used TCP protocol at the transport layer and, hence, greatly improves TCP performance. The ACK Manager does not have the significant drawbacks that were often observed with the other mechanisms intended to solve the TCP problems over wireless networks [27].

When TCP segments are transmitted by the connections in one directions that share the same physical path with the acknowledgements of connections in the opposite direction, the ACK packets, may share a common buffer in the network elements. This sharing has been shown to result in an effect called ACK compression, where ACKs of a connection arrive at the bottleneck link, behind data packets. The effect of ACK compression and the resulting dynamics of transport protocols under two-way traffic have been studied previously by many researchers, including Kalampoukas, and coworkers [30]. The degradation in throughput due to bidirectional traffic can be dropped to 66.67% of that under one-way traffic and the separation of the flow of ACKs and data for the bidirectional TCP connection [30].

18.6.2.1   TCP ACK Manager: ACK Unifier and ACK Extractor

The TCP ACK Manager consists of the ACK Unifier and the ACK Extractor and is used to reduce the drop rate or delay of the TCP ACK packet between the MS and BS. The ACK Unifier and ACK Extractor are independent features but for higher performance improvement both features can be implemented in the MS and the BS.

For a bidirectional traffic application, an MS can send and receive large packets through a single TCP connection simultaneously, and the TCP ACK packets can limit the uplink/downlink transmission throughput. For simplicity, we monitor download traffic only. If both end devices use a MTU size of 1500 bytes, a TCP ACK packet can be delivered as a separate packet or can be embedded in a 1500-byte packet from the mobile to the fixed terminal.

When a TCP ACK packet is delivered separately, it could be delayed by the large packets waiting in a queue in the MS MAC by the uplink traffic belonging to the same service flow, and can degrade the download performance. The ACK Unifier in Figure 18.8 searches for a TCP/IP packet with payload in the corresponding queue in the MS MAC; dedicated packet queues would be used per WiMAX service flow. The TCP acknowledgement information, such as ACK sequence number and ACK validation bit in the new ACK packet, are copied and the TCP checksum must be recalculated in the TCP header in the selected packet. After unifying the ACK information, the TCP ACK packet is discarded. Because the TCP-ACK information is embedded in a large MAC SDU, the probability of losing the TCP ACK information will be higher when FEC BLER increases. A 1500-byte SDU consists of 42 36-byte FEC blocks. The probability of losing a 1500-byte SDU between the

MS and BS with 5% BLER and a maximum of two link-layer retransmissions is 10% over mobile WiMAX, and this is higher than losing a separate the TCP-ACK packet (0.5% packet loss probability). To eliminate the drawback, the SDU can be delivered into two MAC PDUs; one PDU includes TCP/IP header and the other PDU includes the rest of the fragment. The early recovering mechanism is described in the ACK Extractor description.

When TCP ACK information is embedded in a large SDU by either the originating TCP or ACK Unifier, the ACK information should be recovered as soon as the TCP/IP header part is delivered without error, even before the entire SDU is delivered. The MAC PDU CRC is used for this purpose since the CRC in the IP and TCP header cannot be used. The ACK Extractor in the BS in Figure 18.8 monitors the first fragment of a MAC SDU to determine if it has the TCP/IP header and if the valid ACK field of the TCP header was marked. If it satisfies this condition, then the TCP ACK Extractor duplicates the TCP/IP header from the MAC PDU #1 and updates the length and header checksum fields in the IP and TCP header checksum fields in the extracted packet so that the IP and TCP headers carry the correct information. Also, it must invalidate the valid ACK field of the TCP header in the original fragment received to prevent the TCP protocol from treating it as a duplicate TCP ACK. It immediately transmits the recovered ACK packet to the network and also puts the payload of the MAC PDU #1 into the SDU reassembly buffer. The SDU is reassembled after receiving corresponding MAC PDUs. The TCP ACK Extractor reduces the delay of the acknowledgement packet, since it does not have to wait for the entire packet to be successfully delivered from either the MS or the BS, whichever delivers the packet.

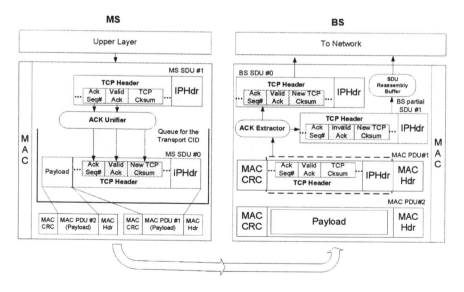

Figure 18.8. ACK Manager: ACK Unifier and ACK Extractor.

18.6.2.2 Simulation Results

A bidirectional FTP application exchanging 5 Mbyte files has been used on top of the WiMAX MAC ARQ scheme with and without TCP ACK Manager, which consists of ACK Unifier and ACK Extractor, in order to analyze the performance improvement by the ACK Manager. In the simulation, 16QAM 1/2 and 64QAM 5/6 modulation for the downlink (DL) and 16QAM 1/2 modulation for the uplink (UL) are assumed, and, for the sake of simplicity, only downlink air-link distortion was introduced. Figure 18.9 shows a downloading TCP goodput improvement ratio versus the FEC block error rate, with and without the ACK Manager. The MAC PDU error rate corresponds to the FEC block error rate. Although the FEC block error rate increases, the TCP performance degradation ratio for both 64QAM DL and 16QAM DL scenarios is very similar without the ACK Manager, and the performance dropped to about 20% when the FEC block error rate reached to 1%. This indicates that the FTP download could not take advantage of the higher DL bandwidth because the TCP performance was limited by the TCP ACK packet delay from the MS to the BS incurred by the large packets in the FTP uploading. In addition, in Figure 18.10 the offered DL bandwidth was utilized significantly by adopting the ACK Manager. When the FEC block error rate is near zero, the ACK Manager increases the performance up to 40% with 16QAM 1/2 DL and more than 90% in a 64QAM 5/6 DL scenario. The performance improvement is reduced while the DL FEC block error rate increases because the impact of the delayed ACK packet decreased because of the degraded downlink quality. The TCP goodput performance improvement by the ACK Manager diminished when the FEC block error rate reached 1%, but the ACK Manager still provided some of the air-link band-

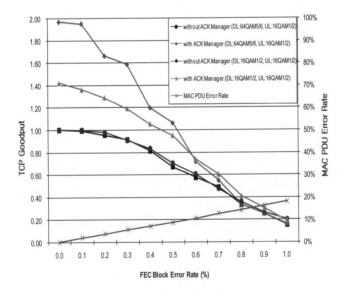

Figure 18.9. TCP goodput versus block error rate.

width saving by unifying a single TCP ACK packet with other TCP packet belonging to the same service flow.

18.8 SUMMARY

This chapter provides an overview of mobile WiMAX, subscriber and application profiles, the simulation results for the network traffic characteristics, and the demand estimation and number of VoIP users. It also introduced the TCP/IP performance enhancement method called TCP ACK manager.

One of the major objectives of developing next-generation wireless communication technology is to provide a higher link capacity, thereby increasing the application/service performance from the user perspective. Because the wireless application/service (from the user perspective) performance improvement efforts are emphasized more and more, we must consider more precise and dynamic environmental condition and system characteristics, including the type of application, application traffic model, diurnal usage scenario, and behavior of various protocol layers. This enables an operator to achieve successful cell and network capacity planning while satisfying service performance from the user perspective. As indicated in the network traffic demand estimation results, the amount of traffic into the system at different times of the day easily could be larger than a five-fold difference. The estimation of the number of VoIP users could be very different with or without considering application performance guidelines. When 16QAM3/4 MCS is used for both DL and UL, the number of VoIP users is about 100 with a 150-msec packet delay QoS guideline; whereas the number is about 200 users without considering the packet delay QoS guideline.

This chapter also introduces the ACK manager, which reduces the probability of losing ACK packet information at the airlink without modification to the widely used TCP protocol at the transport layer, and hence, greatly improves TCP performance. The ACK manager does not have the significant drawbacks that were often observed in the other mechanisms that are intended to solve the TCP problems over a wireless network.

18.8 REFERENCES

[1] Recommendation ITU-R M.1225, Guidelines for Evaluation of Radio Transmission Technologies for IMT-2000, 1997.

[2] 3GPP-3GPP2 Spatial Channel Ad-hoc Group, Spatial Channel Model Text Description, V7.0, August 19, 2003.

[3] Daniel S. Baum et al., An Interim Channel Model for Beyond-3G Systems—Extending the 3GPP Spatial Channel Model (SCM), in *Proceeding of IEEE VTC'05,* Stockholm, Sweden, May 2005.

[4] IST-WINNER II Deliverable D1.1.1 v1.0, WINNER II Interim Channel Models, December 2006.

[5] 3GPP2, CDMA2000 Evaluation Methodology, Revision 0, December 2004.

[6] ITU-R M.1225, Guidelines for the Evaluation of Radio Transmission Technologies for IMT-2000, January 1999.

[7] IEEE 802.16m-08/004r2, IEEE 802.16m Evaluation Methodology Document (EMD), July 2008.

[8] Sony, Intel, TGn Sync TGn Proposal MAC Simulation Methodology, IEEE 802.11-04/895r2, November 2004.

[9] ST Micro-Electronics, Time Correlated Packet Errors in MAC Simulations, IEEE Contribution, 802.11-04-0064-00-000n, January 2004.

[10] Atheros, Mitsubishi, ST Micro-Electronics, and Marvell Semiconductors, Unified Black Box PHY Abstraction Methodology, IEEE Contribution 802.11-04/0218r1, March 2004.

[11] 3GPP TR 25.892 V2.0.0, Feasibility Study for OFDM for UTRAN Enhancement, June 2004.

[12] WG5 Evaluation Ad-hoc Group, 1x EV-DV Evaluation Methodology—Addendum (V6), July 25, 2001.

[13] Ericsson, System Level Evaluation of OFDM—Further Considerations, TSG-RAN WG1 #35, R1-03-1303, November, 2003.

[14] Nortel, Effective SIR Computation for OFDM System-Level Simulations, TSG-RAN WG1 #35, R03-1370, November 2003.

[15] Nortel, OFDM Exponential Effective SIR Mapping Validation, EESM Simulation Results for System-Level Performance Evaluations, 3GPP TSG-RAN1 Ad Hoc, R1-04-0089, January, 2004.

[16] K. Brueninghaus et al., Link performance models for system level simulations of broadband radio access, IEEE PIMRC, 2005.

[17] L. Wan, et al., A Fading Insensitive Performance Metric for a Unified Link Quality Model, WCNC, 2006.

[18] WiMAX Forum MWG, WiBro Briefing Document, WiMAX Forum Positioning and Q&A, June 24, 2005.

[19] UMTS Forum, 3G Offered Traffic—All Services, June 2003.

[20] WiMAX Forum AWG, Application Performance Metrics and Guidelines for the Five Classes, August 2005, http://www.wimaxforum.org/apps/org/workgroup/abtg/download.php/10626/AWG_application_performance_guideline_6-3.pdf.

[21] WiMAX Forum AWG, WiMAX Application Usage Profile, April 28, 2006.

[22] Bong Ho Kim, Insup Lee, and Kelvin Chu, End-to-end Application Performance Impact on Scheduler in CDMA-1XRTT Wireless System, in *IEEE Vehicular Technology Conference,* Spring 2005.

[23] WiMAX Forum AWG, WiMAX System Evaluation Methodology v2.1, July 7 2008.

[24] Bong Ho Kim, Yerang Hur, et al., Mobile WiMAX Application Performance and Traffic Model, WiMAX GlobalCom Forum, October 11 2006.

[25] WiMAX Forum SPWG, Recommendations and Requirements for Networks Based on WiMAX Forum Certified Products Release 1.5, April 27 2008.

[26] WiMAX Forum TWG, System Profile, Release 1.0, Revision 1.2.2, November 2006.

[27] Bong Ho Kim, et al., Capacity estimation and TCP performance enhancement over Mobile WiMAX Networks, *IEEE Communication Magazine,* vol. 47 no. 6, June 2009, pp. 132–141.

[28] IEEE 802.16m, System Requirements (SRD), IEEE 802.16m-07/002r4, October 2007.

[29] IEEE 802.16m, Evaluation Methodology Document (EMD), IEEE 802.16m-08/004, March 2008.

[30] Lampros Kalampoukas, et al., Two-Way TCP Traffic over Rate Controlled Channels: Effects and Analysis, *IEEE/ACM Transactions on Networking,* vol. 6 no 6, Dec. 1998, pp. 729–743.

CHAPTER 19

FEMTOCELLS AND MULTIHOP RELAYS IN MOBILE WiMAX DEPLOYMENTS

JERRY SYDIR, SHILPA TALWAR, RAKESH TAORI, AND SHU-PING YÉH

19.1 INTRODUCTION

Wireless operators are in continual quest for higher coverage and capacity while struggling to reduce the cost of cell-site development, maintenance, and backhaul.

Providing uniform coverage is difficult due to irregularities in the topography of a coverage area, reduced signal strength or penetration loss. The common solution to the problem of providing adequate outdoor and indoor coverage is to deploy more base stations to decrease the size of the outage areas and increase link budget to improve indoor penetration.

Capacity is another key concern for operators. As the number of WiMAX users increases and as these mobile users continue to discover new bandwidth-intensive applications for their mobile terminals, the capacity needs of WiMAX operators will continue to increase. Although various techniques such as MIMO are used to increase spectral efficiency within a cell, it is envisaged that the required capacity needs will go well beyond what can be achieved by optimizing the air link.

Aggressive spectral reuse (the deployment of small cells) drives capacity across the network. While shrinking cell sizes within a single-tier topology increases spectral reuse and improves coverage and capacity, the gains come at the expense of significant increases in cell-site development and backhaul costs. Wireless backhaul solutions can reduce costs, but are not necessarily cost-effective in all situations.

Femtocells and relays are two key architectural advances that help reduce deployment cost while improving capacity and coverage.

WiMAX Technology and Network Evolution. Edited by Kamran Etemad and Ming-Yee Lai
Copyright © 2010 the Institute of Electrical and Electronics Engineers, Inc.

Femtocells are small cells, on the order of 100 m radius, deployed to provide in-door coverage within a residential or commercial building. Femtocells solve the in-building coverage problems by placing the BS, also commonly referred to as the femto access point (femto-AP or FAP), within the building. Operator costs are low-er as in-building broadband connections are used for the backhaul. Femto-APs are generally self-deployed, so the deployment costs are passed to the user.

Relay is an extension of the cellular topology in which communications between the base station (BS) and mobile station (MS) can pass through zero or more relay stations (RSs). Backhaul from the RSs to the BS is provided in-band, utilizing the operator's spectrum. RSs can be deployed to provide enhanced coverage and in-creased capacity in a variety of deployment scenarios, allowing the operator to make trade-offs between the cost of wired backhaul and the availability of spectrum used for in-band backhaul.

In this chapter, we will first introduce the concept of a multitier cellular architec-ture and its relationship to the concepts of femtocells and relays. In the first part of this chapter, we focus on the case where a femto-AP serves as a simplified WiMAX base station with a backhaul that is available through a residential broadband con-nection. We begin by pointing out the advantages of femtocell networks for opera-tors as well as end users, the associated usage and deployment scenarios, and re-quirements of network architecture. Then, we describe the capacity and coverage improvements offered by a multitier, WiMAX femtocell network. Next, we discuss the technical challenges associated with WiMAX femtocell deployments, and con-clude the femto part of the chapter with a summary of current industrial activities on femtocells.

The second part of this chapter describes a relay station and the place it is likely to carve for itself in the traditional cellular architecture. We highlight the important economic and performance benefits that relay stations could potentially offer and outline a few scenarios in which relays are likely to be deployed. The multihop re-lay standard developed by the IEEE 802.16 (Wireless Metropolitan Area Networks) working group is then used as a basis to provide an overview of the relay-enhanced cellular architecture and the key choices that can be made in developing relay sup-port within a cellular system, a precursor to what can be expected in the later releas-es of the Mobile WiMAX system. Finally, we discuss some future directions in the development of the relay systems.

19.2   MULTITIER CELLULAR ARCHITECTURE

In the traditional cellular architecture, a given coverage area is divided into smaller areas called cells. A BS, often located at the center of the cell, provides coverage to the MSs within the cell. The BS is connected to the core network via a backhaul connection, typically provided by a wired link or a point-to-point microwave link. The area within a cell can further be subdivided into sectors. Within each sector, the BS communicates with MSs that are associated with it, using what is referred to as a point-to-multipoint (PMP) link. The PMP link here refers to a specific type of multipoint link wherein a central device (BS) is connected to multiple peripheral devices (MSs). Any transmission of data that originates from the central device is received by one or more of the peripheral devices, whereas any transmission of data that originates from any of the peripheral devices is only received by the central device. Each BS manages the allocation of resources to support communications between itself and the MSs that it serves. The MSs are informed about the resource allocation by the BS. Coverage within a sector is commonly enhanced by the deployment of analog repeaters. Repeaters are simple devices that receive the signal transmitted by the BS (as well as the signals from neighboring cells, if any) and, with very little delay, indiscriminately amplify and forward this signal.

Cellular networks are most commonly deployed using a single-tier topology in which the coverage area is divided into cells that overlap only to the extent required to allow seamless handover as an MS moves between cells. It is also possible to deploy cellular networks using a multitier topology in which coverage is provided over a given area by more than one set of cells (called tiers), wherein the coverage area of one tier overlaps the coverage area of the next tier. An example of such a multitier topology is a two-tier topology in which a first tier of large cells is deployed to provide coverage for mobile users, while a second tier of small cells is deployed to provide improved coverage and capacity to less mobile users. In practice, the larger cells can be served by macro-BSs while the smaller cells can be served by BSs, RSs, or femto-APs and are typically expected to be deployed in locations where additional capacity or enhanced coverage are needed.

In a multitier topology, the different tiers may operate on the same sets of frequencies or on different sets of frequencies. The types of stations used within the tiers may also be different. Finally, the backhaul solution can be different in different tiers.

19.3 FEMTOCELLS

A femtocell is a small footprint served by a low-power base station with an IP-based backhaul that is available through a broadband connection such as cable, DSL, or fixed wireless. The femto-AP is expected to be a low-cost consumer device with a typical price target expected to be under US$200 [1] that can be installed by the subscriber in a plug-and-play fashion. The deployment of large numbers of femto-APs will provide significant advantages for cellular network operators as well as end users. Wireless operators will save on backhaul costs since femto-AP traffic is carried over wired residential broadband connections, and on deployment costs since femto-APs will support self-install and autonomous operation. Furthermore, operators will benefit from significant gains in throughput per unit area due to spatial reuse of spectrum. The consumer can expect improved data speeds, service quality, and longer battery life as it no longer be necessary to connect to outdoor macro/micro base stations. In addition, femto-APs enable the convergence of landline and mobile services since the same handheld device can be used to access the broadband wireless connection indoors and outdoors. In the future, it may also be possible to provide new services such as indoor location finding with femto-APs.

19.3.1 Femtocell Usage and Deployment

Femto-APs are typically deployed in a two-tier topology. It is expected that femto-APs will support low-mobility users requiring high data-rate connections, while an overlay macro-BS will support high-mobility users. This deployment scenario and others in which femto-AP's are expected to be provide a cost-effective solution to the coverage and capacity needs of an operator are shown in Figure 19.1.

Femto-APs can help improve coverage of indoor users who suffer from poor signal reception due to wall penetration loss. There can be a single femto-AP covering a house, or several femto-APs covering a building or an enterprise environment. Besides indoor environments, outdoor hot spots (e.g., stadiums or shopping malls) can also be served by femto-APs. Femto-APs can also help offload the macro-BS by freeing up macro resources that were originally used to serve indoor users. Finally, femto-APs can improve areal capacity by aggressively reusing the same spectrum as macro-BS.

There are two potential use cases for femto-APs. The first is for public usage, whereby all customers of a cellular service provider can connect to publicly accessible femto-APs. The typical scenario for the public case is in a coffee shop or airport. The other case is for private usage, whereby only authorized users are allowed to connect to the privately accessible femto-AP. The second is more suitable for

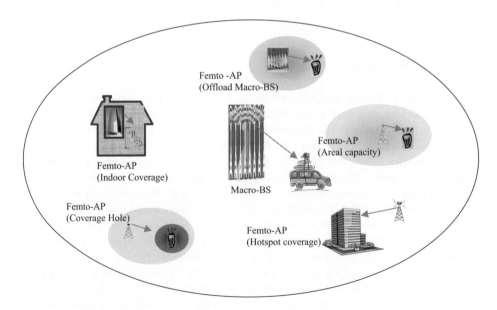

Figure 19.1. Example femtocell deployment scenarios.

home or enterprise environments. The set of users that are authorized to use a femto-AP by the femto-AP subscriber and/or service provider is typically referred to as a closed subscriber group (CSG)

19.3.2 Femtocell Network Architecture

In this section, we consider the issue of femto-AP management. In traditional cellular systems, macro/micro BSs are directly connected to the operators' network via dedicated landlines. Hundreds of BSs may be connected to a radio network controller (RNC), which is responsible for functions like radio resource management (RRM) and handoff between base stations. Once femto-APs are added to the existing network, the number of devices that an RNC needs to control is expected to increase by one or two orders of magnitude to thousands or tens of thousands. Current network control entities might not scale to handle so many devices. To integrate femto-AP deployments with the existing cellular networks, the main challenge is to come up with a scalable network architecture that integrates femto-APs, which are connected to the IP-based network from local internet service provider (ISP), into the existing cellular infrastructure.

Several possible network architecture options have been proposed by vendors and various femto-focused working groups. An example network structure for a WiMAX system with femto-APs is illustrated in Figure 19.2. The network consists of two standard components: the access service network (ASN) managed by the mobile operator, and the connectivity service network (CSN) managed by network service provider. In addition, a new component is introduced to manage the femtocell network, namely the femto-NSP, whose role is to authenticate, operate, and manage the femto-APs. The ASN includes macro/micro BSs, femto-APs, ASN gateways, and femto servers/gateways. The ASN gateways (GW) serve as interface between the ASN and CSN, enabling exchange of necessary information with servers within the CSN. The CSN is composed of servers such as the authentication, authorization, and accounting (AAA); mobile IP (MIP); home agent (HA); and policy server.

Femto servers/gateways are defined as separate entities in order to enable scaling to support a large number of femto-APs. Multiple femto servers can be employed within the ASN to support the potentially large number of femto-APs. Femto-APs connect to the femto servers through the public Internet, most likely through DSL or cable connections. In order to ensure security, the connections between the femto-APs and femto servers are established through IPSEC tunnels.

This WiMAX network architecture is flat compared to typical cellular architectures (2G/3G) since RNC functions are integrated into macro/micro BSs and Femto-APs. Thus, macro/micro BSs and Femto-APs in WiMAX networks are more autonomous and the system is less costly and easier to manage.

19.3.3 Femtocell Performance Analysis

We study the potential capacity and coverage gain from WiMAX femtocells using system-level simulations. In our simulations, we focus on downlink transmission

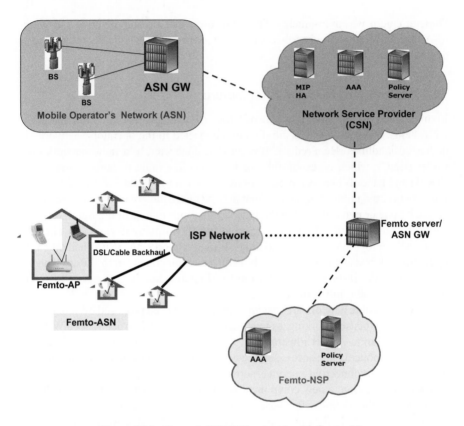

Figure 19.2. Example WiMAX network with femto-AP.

and compute long-term capacity based on path loss, slow fading, and interference. We assume the WiMAX system operates at 2.3 GHz carrier frequency with 10 MHz bandwidth. The same carrier frequency is used by all macro/micro BSs and femto-APs. The worst-case interference is modeled (frequency reuse one); that is, every macro/micro BS and femto-AP uses the whole 10 MHz for transmission.

Figure 19.3 illustrates the deployment model used in the simulation. Overlaying Femto-APs on top of the macrocell network forms a hierarchical cell structure. Macro-BSs are deployed in a hexagonal manner. There are a total of 19 macro-BSs, each with three sectors. Femto-APs (FS) are deployed within a hexagon with radius of four times the macrocell radius. The locations of femto-APs are randomly selected from a grid with 20 meters separation. Therefore, a minimum distance of 20 meters between neighboring femto-APs is guaranteed. We assume each femto-AP is located at the center of a house, where a house is approximated by a disk area with 10 meters radius. Equal numbers of indoor and outdoor users are deployed within a hexagon with radius equals to two times the macrocell radius. Indoor (outdoor) users are uniformly distributed inside (outside) the houses. Statistics are collected

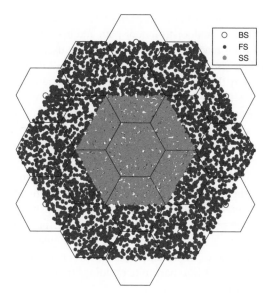

Figure 19.3. Hierarchical deployment model used for simulations.

only from users covered by base station or femto-APs inside the center cell. Macro-BSs and femto-APs outside of the center cell are deployed only to provide more accurate estimate of interference.

For each user in the center cell, we compute SINR and corresponding spectral efficiency (SE). The SE is estimated assuming a 3 dB gap to theoretical capacity. The following equation shows how we estimate the spectral efficiency:

$$SE \le \frac{1}{2}\log_2\left(1+\frac{SINR}{2}\right), \quad SE = 0.5k, \quad k = 0,1,...,12$$

When *SE* equals zero, the corresponding user is in outage. The percentage of users who are not in outage is called the coverage of the system. The other important metric, system capacity, is computed as follows. There are a total of thirteen different modulation and coding schemes (MCS) available in our simulation. For each user, the MCS mode with maximum achievable SE is selected. To compute the average user rate, we multiply MCS-based SE by total BW available. The mean data rate that a macro-BS/femto-AP can support is the mean of average rates of all users associated with a particular macro-BS/femto-AP. Then, the system capacity is given by the sum of the mean data rate of the macro BS and all femto-APs located within the center cell. We also compute the areal capacity gain, which is equal to dividing the system capacity with femto-APs deployed by the system capacity without femto-APs.

The ITU channel models [4] are used to model path loss and slow fading. We use the indoor office scenario to model the channel from femto-APs to users in the same building. The channel from macro base station to outdoor users is modeled by a vehicular scenario. The outdoor-to-indoor path and pedestrian scenario is used to model the channel from macro base station to indoor users and the channel from femto-APs to outdoor users. The channel from femto-APs to users in different buildings is modeled by the outdoor-to-indoor and pedestrian scenario plus an additional penetration loss from walls.

Two different macrocell settings are used in our simulation. The first one is the large cell scenario, in which the macrocell radius is 1500 meters and the macro-BS transmission power is 46 dBm. The other one is the small cell scenario, where the macrocell radius is 500 meters and the macro-BS transmission power is 36 dBm. In both cases, the macro-BS antenna has a 15 dB gain with antenna height 30 meters. The main difference in the above two scenarios is that when there are no femto-APs deployed, indoor coverage for the large cell can be further improved by increasing macro-BS transmission power, whereas the small cell performance improvement is limited due to interference amongst macro-BSs. Figure 19.4 shows coverage holes for indoor users in the large cell scenario (part a), whereas small cell users receive enough power from macro-BS to penetrate buildings (part b). Shadow fading is not simulated here for clarity.

We also consider different femto-AP deployment densities to evaluate the effect of cochannel interference. There are about 10 femto-APs per sector for the sparse deployment and about 100 femto-APs per sector for the dense deployment. Lastly, both public and private femto-AP usages are tested. When publicly accessible femto-APs are deployed, the user selects the macro-BS or Femto-AP that provides maximum received signal power as its home cell. For the private usage case, a user can access a femto-AP only when they are in the same house. Thus, indoor users can choose between macro-BSs and the private femto-AP in their house, whereas outdoor users only select from macro-BSs.

Simulation results are summarized in Table 19.1 and Table 19.2. Table 19.1 shows that nearly 100% coverage is obtained for indoor users in the large cell scenarios, but this coverage improvement reduces as the deployment gets dense and the cell size gets smaller as more femto-APs are deployed. Femto-to-femto interference becomes the biggest stumbling block for indoor coverage when private femto-APs are densely deployed. In fact, in the case of dense femto-AP deployments, increasing cochannel interference reduces outdoor coverage as well. This effect is more serious for privately accessible femto-APs since outdoor users close to houses cannot be covered by private femto-APs. In large cell, sparse deployments, the effect of cochannel interference from femto-APs is not noticeable since femto-APs are distant from each other. Hence, indoor coverage is greatly improved for both public and private scenarios. The coverage performance also suffers when the femto-AP transmission power is too high. In order to achieve better coverage, further interference mitigation techniques, such as advanced power control, fractional time or frequency reuse, and multiantenna techniques, should be considered.

Table 19.2 shows that high areal capacity gain can be achieved with femtocells (up to 300 times in dense deployments). The areal capacity gain comes from two aspects. The first one is that femto-APs reuse the same bandwidth as the macro-BSs,

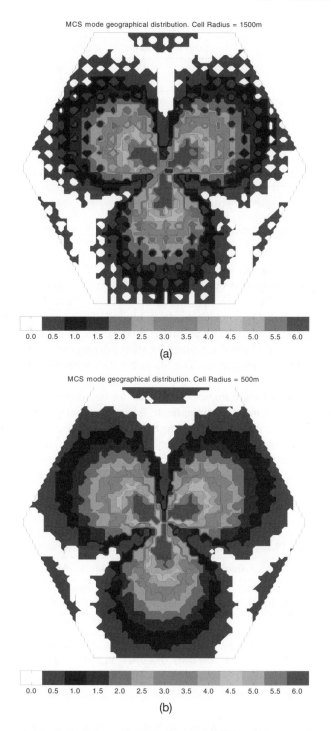

Figure 19.4. Geographical distribution of MCS modes in a cell.

Table 19.1. Indoor/outdoor coverage in different simulation scenarios

| | | | Large cell scenario | | | | Small cell scenario | | | |
| | | | Sparse deployment | | Dense deployment | | Sparse deployment | | Dense deployment | |
Coverage (%)			Public	Private	Public	Private	Public	Private	Public	Private
FS Tx Power	0 dBm	Indoor	99.7992	99.7992	98.3728	93.787	97.9310	93.7931	89.8050	70.0355
		Outdoor	75.4035	75.2498	72.9183	68.5603	71.3465	69.1297	67.8608	55.4464
	10 dBm	Indoor	100	100	98.8116	94.0828	99.0038	94.3295	91.3121	70.3901
		Outdoor	75.3267	73.1745	71.4397	57.4319	71.0181	64.1215	64.9468	33.4971
	20 dBm	Indoor	100	100	99.0385	94.1568	99.387	94.4061	91.1348	69.9468
		Outdoor	74.6349	68.6395	67.2374	37.7432	69.4581	51.6420	65.7658	13.5954
Coverage (%)		Indoor	70.48		70.86		80.15		79.34	
w/o femto-AP		Outdoor	76.10		74.32		72.33		73.55	

so the available bandwidth per unit area increases. The other comes from better spectral efficiency supported by femto-APs, as shown in Figure 19.5. The reason for higher spectral efficiency is that most users associated with femto-APs experience little signal attenuation, which results in high SINR and correspondingly high SE for these users. The spatial reuse gain is proportional to the number of femto-APs per sector, which is about ten times for sparse deployments and about a hundred times for dense deployments. The SE gain varies with the density of femto-APs.

Figure 19.5 shows two examples of SE CDF for two different types of deployment scenarios. Deployment of public femto-APs only slightly degrades the SE for macrocell users, while it greatly boosts the SE of femtocell users. When the density of femto-APs increases, the cochannel interference from neighboring femto-APs gets stronger and reduces the supportable SE for each user. The SE gain is highest in the large cell sparse deployment, and is the lowest in small cell dense deployment.

Table 19.2. Areal capacity gain in different simulation scenarios

| | | Large cell scenario | | | | Small cell scenario | | | |
| | | Sparse deployment | | Dense deployment | | Sparse deployment | | Dense deployment | |
Areal capacity gain*		Public	Private	Public	Private	Public	Private	Public	Private
FS Tx Power	0 dBm	54.479	54.672	356.8325	364.1799	34.1512	34.3671	152.0866	151.5067
	10 dBm	57.764	59.1746	350.6903	384.0944	36.1507	37.5628	143.5709	153.5189
	20 dBm	57.4989	60.7799	319.2529	392.5468	34.9872	39.1827	134.7428	152.4803
System capacity without femto-AP		25.921 Mbps		27.377 Mbps		27.621 Mbps		27.551 Mbps	

*Areal capacity gain = (system capacity with femto-APs deployed)/(system capacity without femto-APs).

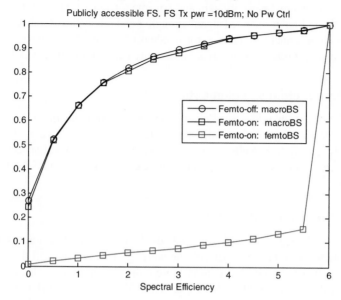

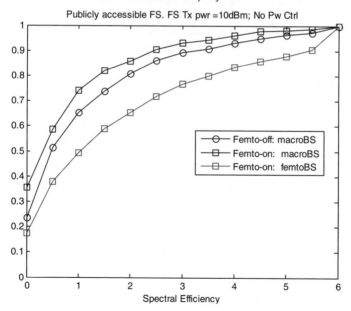

Figure 19.5. CDF of SE for macrocell and femtocell users.

Finally, we note that when private femto-AP deployments get denser, there is a trade-off between areal capacity gain and coverage. Higher femto-AP transmission power results in higher received signal strength and higher SE for indoor users, but causes more cochannel interference for neighboring femto-APs. Hence, increasing femto-AP transmit power increases areal capacity gain but lowers coverage.

19.3.4   Key Technology Elements

The simulation results in the previous section serve as an indication that femtocells have the potential to provide significant capacity and coverage gains. In order to successfully deploy femtocells in the real world, several technical aspects need to be addressed. In this section, we discuss the elements for practical femtocell deployments.

19.3.4.1   Network Architecture

As discussed in Section 19.3.2, the flat all-IP network architecture adopted in WiMAX networks serves as a scalable network option for femtocell deployments. In such a distributed-control structure, more radio-resource management needs to be implemented at femto-APs. Therefore, WiMAX femto-APs need to be more autonomous and powerful. Self-organization requirements for femto-APs will be detailed later in this section. In addition, the large neighbor-cells list that needs to be kept at a BS for timely handover can become difficult to manage. The network architecture also needs to consider infrastructure support for seamless mobility during handover. Management protocols used in DSL systems, like the TR-069 CPE (customer premises equipment) WAN Management Protocol, can be adopted for efficient management of large-scale femtocell networks.

19.3.4.2   Interference Management

In a hierarchical overlay network, where WiMAX femto-APs operate on the same frequency band as macro-BSs, cochannel interference becomes an important factor that limits overall network performance. When femto-APs are installed indoors, however, walls help to alleviate the interference between macro-BSs and femto-APs. As the number of femto-APs increase, the accumulated interference needs to be dealt with. At a minimum, power control is required in femto-APs to avoid performance degradation in mobile terminals served by macro/micro base stations. To guarantee close to 100% coverage, further interference mitigation strategies such as fractional frequency reuse (FFR) could be applied. In order to apply these more advanced interference mitigation strategies, good synchronization is essential, as well as efficient means for exchanging messages between macro-BSs and femto-APs.

19.3.4.3   Synchronization

Synchronization between the mobile station and base stations is required in order to have successful handovers. Furthermore, synchronization is essential for interference management in outdoor systems using time-division duplex (TDD), such as the 2.5 GHz systems planned for the initial release of mobile WiMAX. There are multiple synchronization options, including GPS, IEEE 1588, or over-the-air synchronization using a preamble transmitted by a macro-BS. GPS is the most accurate and provides both time and location data. Localization may be a mandatory feature if operators want to avoid customers moving femto-APs outside of their houses. However, GPS is more expensive and relies on the GPS system availability. For indoor femtocells, GPS is not suitable since it requires line of sight from the satellite, which is difficult to achieve indoors. IEEE 1588, Precision Timing Protocol (PTP), may be a more suitable solution for femto-APs. It uses a master–slave structure. There is a master clock in the network providing a timing reference to the slave clocks at femto-APs. The timing signal is transmitted over IP/Ethernet backhaul. IEEE 1588 is a low-cost, stand-alone solution that achieves submicrosecond accuracy. Field trial results in [3] show that, under certain conditions, IEEE 1588 can achieve satisfactory frequency accuracy over the networks. Application of IEEE 1588 to TDD WiMAX, which requires both time and frequency synchronization, is yet to be demonstrated. Finally, over-the-air network synchronization may be feasible with special preamble designs that can provide sufficient estimation accuracy indoors. However, some femtocell network operators may only deploy femtocells without a macronetwork overlay; hence, more than one synchronization option needs to be available in such a system.

19.3.4.4   Security and Performance

In traditional cellular systems, base stations are connected directly to the operator's network. With the registration and authentication process, the cellular operator can, thus, easily prevent unauthorized users from accessing its own network to ensure security. However, femto-APs utilize the local ISP networks, which may be different from the operators' network and are much more difficult to protect. Public IP networks can be accessed by almost everyone, including hackers who attempt to eavesdrop on conversations or control the femto-AP. Therefore, in addition to a more sophisticated registration and authentication process, encryption of IP packets is necessary. Another issue with the IP network is that a cellular operator has no control over the channel and cannot prioritize voice packets from femto-APs. To ensure system and user performance, collaboration and service-level agreement between cellular and landline operators will be required. For example, with higher priority given to voice packets from femto-APs, end-to-end QoS can be guaranteed.

19.3.4.5   Self-Organization and Autonomous Operation

WiMAX networks require higher levels of self-organization at both macro/micro BSs and femto-APs because of the flat network architecture. Cooperation is required among BSs and femto-APs for successful handover and RRM information exchange. The communication between BSs/femto-APs can be over a landline or even over the air. Latency can be an issue when sending control signals over the IP network. Besides the self-organization functions shared with macro/micro BSs, a femto-AP requires even higher autonomy since it should be a plug-and-play device that can integrate itself into the mobile network without user intervention. A configuration function in the device should be capable of adjusting parameters under various environments since the locations of femto-APs cannot be planned in advanced, as for macro/micro BSs. Also, the large number of femto-APs deployed within the network makes manual maintenance virtually impossible. The possibility of updating firmware and software of the femto-APs could also be an important requirement.

19.3.5   Industrial Activities

As a relatively new technology, femtocells have drawn lots of attention from the wireless industry. The Femto Forum [5], was launched in July 2007 to promote femtocell deployments worldwide. More than 60 companies have joined this not-for-profit organization, including mobile operators, telecom hardware and software vendors, content providers, and startups. The Femto Forum has formed a partnership with the NGMN (next-generation mobile networks) Alliance. Their goal is to ensure that NGMN can incorporate femtocells from the very beginning of their deployments. Unlike 2G/3G, wherein femtocells are overlaid on top of the existing cellular networks, NGMN can evaluate the cost/performance ratios and carefully optimize the network by simultaneously deploying femto-APs and macro/micro-BSs. The Femto Forum also partners with the Broadband Forum. Knowledge from the Broadband Forum helps develop efficient network management schemes for femtocells.

Within the WiMAX Forum, the Service Providers Working Group (SPWG) has identified deployment scenarios and associated system requirements to support WiMAX femtocell systems [6]. The specification includes requirements for mobile devices, femto-AP's and network elements, femtocell management systems, and the end-to-end network architecture. The WiMAX Femtocell system will support legacy mobile devices as well as enhanced mobile devices that are optimized for femtocell operation. The road map includes two phases of femtocell development. The first phase, having a deployment timeline of 2010, is based on WiMAX Release 1.0 or 1.5 system profile and network specifications of Release 1.6. The second phase, being planned for 2012/2013 deployment, is part of WiMAX Release 2.0 based on the IEEE 802.16m advanced air interface. The first phase will rely mostly on network-level solutions to enable basic femtonetwork operation, whereas the second phase will benefit from femtocell-specific optimizations on the air interface (IEEE 802.16m) and assistance from femto-enhanced mobile devices.

In parallel, there are several standard activities in this area, including 3GPP [7] and IEEE 802.16m, which is the advanced interface task group for WiMAX. 3GPP has recently published a femtocell standard in collaboration with the Femto Forum and Broadband Forum to support interoperability of femto-APs with femto gateways from different vendors. In IEEE 802.16m, the System Description Document that defines air-interface functions includes support for femtocell operation, including the network entry procedure, handover support, synchronization, interference mitigation, and self-organization [8]. The detailed definition of the air interface is expected to be issued this year.

Major WiMAX service providers such as Sprint-Nextel and KT are preparing for the introduction of femto-APs into their networks. This action has accelerated the development efforts of global manufacturers. Also, femto-APs will provide new business opportunities for wireline operators to enter the mobile virtual network operator (MVNO)-based wireless markets by fully utilizing their own broadband wireline access infrastructures.

19.4 RELAY

19.4.1 Relay Architecture

The traditional cellular architecture can be enhanced using devices called relay stations (RSs), which intelligently relay data between the BS and MSs, wirelessly. The BS communicates with a multitude of RSs within its coverage using the PMP mode mentioned above and the RSs in turn also communicate with MSs associated with them using the PMP link. The BS maintains overall control over the RSs and MSs associated with it, although the implementation of individual control functions (e.g., scheduling) may be centralized at the BS or distributed between the BS and RSs. The RSs operate using a store-and-forward paradigm. The RSs receive the data selectively in specific time/frequency allocations indicated by the BS, decode and process the data, and subsequently, transmit (relay) this data in different allocations that occur later in time.

Figure 19.6 illustrates an example of a sector enhanced by using RSs. The area immediately surrounding the BS is the BS's coverage area. The MSs in this area associate with the BS and are served directly by the BS. Each RS has a coverage area within which it serves the MSs. The coverage areas of the BS and RSs can overlap to varying degrees.

As Figure 19.6 shows, the path between the BS and an MS can consist of one or more hops. An MS may receive data directly from the BS or from one of the RSs. The RSs within a sector can be logically thought of as being arranged in a hierarchical structure with the BS at the top of the hierarchy. RSs at the first level of the hierarchy are attached directly to the BS. RSs at the second level are attached to the first-level RSs, and so on. In its simplest form, the topology of the sector enhanced by using RSs, is a tree with a single path between the BS and each MS. In a more complex topology, for example, a mesh topology, more than one path can exist between the BS and one or more MSs, RSs at the same level may communicate with one another, and a given RS may appear in different levels of the hierarchy on different paths.

From a control perspective, there exists a hierarchical relationship between the BS, RSs, and MSs. The BSs maintain overall control over the RSs and the MSs. The MSs, in turn, are controlled by a combination of the BS and RSs. Individual control functions can be implemented in a centralized manner, that is, within the BS, or can be distributed between the BS and RSs.

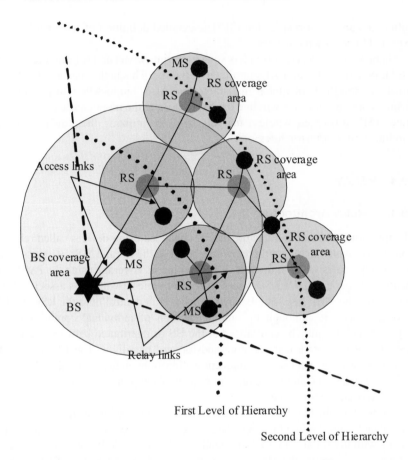

Figure 19.6. A sector of base station (BS) coverage enhanced using relay stations (RSs).

RSs can operate according to one of two major relay modes. The first of these modes is referred to as transparent relay. We use the term transparent relay because in this mode of operation the presence of the RSs is essentially transparent to the MSs. In transparent mode, only the BS transmits downlink (DL) synchronization and control channels to the MSs directly, whereas the RSs may relay selected unicast data transmissions to the MSs as determined by the BS. RSs do not transmit DL synchronization in transparent mode.

The second relaying mode is referred to as the nontransparent relay. The name nontransparent is used because in this mode the MSs are aware of the existence of the RSs. In this mode, both the BS and RSs transmit DL synchronization and control channels and MSs within a sector associate with either the BS or one of the RSs. In this mode, the RSs operate more like base stations, except that overall control of the sector is maintained by the BS and RSs relay data between the MSs and BS.

In the literature, the terms relay, tree, and mesh are used quite loosely. Here, we will use the term relay to express the act of relaying; the term relay station (RS) shall refer to a device that relays, whereas the terms mesh and tree shall refer to specific topologies as explained above. We refer to the BS and RSs as infrastructure stations and to the link between the BS and an RS, or between two RSs, as a relay link, whereas the link between the BS and an MS, or between an RS and an MS, is referred to as an access link. The relay and access links within a relay-enhanced sector can either share the resources of a single RF carrier, or operate on different RF carriers. Although relaying can be performed by the mobile stations as well (sometimes referred to as client relay), in this chapter we will exclude the case of client relay and confine the discussions to infrastructure relay, in which the relaying is performed by the RSs.

19.4.2   Advantages

The deployment of RSs offers performance/cost benefits over the use of BSs in a traditional cellular network. Performance benefits include improvements in coverage and/or increases in capacity. The deployment of RSs within a macrocell enhances the coverage and capacity in areas where the capacity of the direct link between the BS and MSs is low. Such areas can exist at the cell edge (e.g., MS1 in Figure 19.7), in the shadows of large objects such as tall buildings (e.g., MS8), within the buildings themselves, or underground. RS deployment enhances coverage in areas where the capacity of the direct link between the BS and MS is zero (e.g., MS2 in a coverage hole or MS7 beyond the edge of the cell). RS deployment enhances capacity throughout the cell due to increased frequency reuse. See Section 19.4.5 for an analysis of coverage and capacity improvements.

It is important to understand that performance improvements from relay are due to the deployment of additional infrastructure stations, in this case RSs. In principle, the performance gains provided by the deployment of RSs can also be achieved by the deployment of additional BSs (decreasing the size of the cells or creating a multitier topology in which the second tier is served by BSs). The deployment of additional BSs may provide even better access link capacity than RSs because the air link resources need not be used to support the relay links. However, as we discuss next, the deployment of BSs with dedicated wired backhauls (or dedicated point-to-point microwave backhauls) is not necessarily a cost-effective solution.

The cost benefits of relay are realized due to a reduction in the cost of the equipment, site development and maintenance, and the cost of provisioning and operating the backhaul connections. Compared to a traditional BS, the equipment cost associated with an RS is likely to be lower due to the (expected) lower complexity and lower cost of the chassis and the power amplifier. It is likely that the RS antennas will be deployed on top of the buildings or on lamp posts and, therefore, are likely to be less expensive to develop and maintain when compared to traditional cell sites with tall towers. These differences in cost are expected to decrease over time as the coverage area of BSs becomes smaller, however.

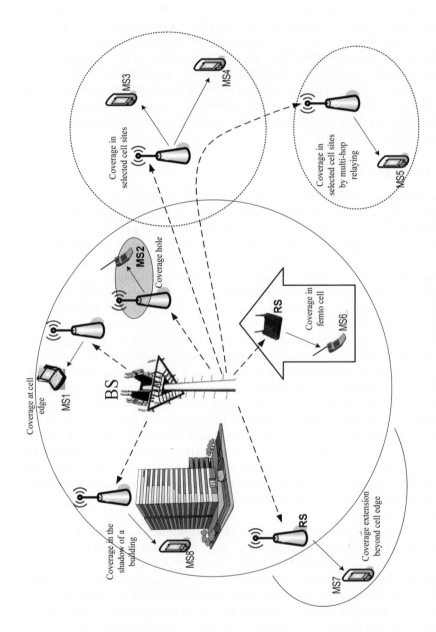

Figure 19.7. Example RS deployment scenarios.

The primary advantage of deploying RSs in terms of the cost is expected to come from the differences in the cost of the backhaul. When an RS is deployed instead of a BS, there are no direct backhaul costs. There is no cost for provisioning the wired connection and there are no monthly charges for the backhaul. There is no need to purchase, set up, and maintain microwave link equipment, and no need to purchase the rights to additional spectrum in which this equipment operates. RSs are also expected to be less costly to deploy than BSs with wireless backhaul, because they do not require strict line-of-sight channel conditions on the relay link, allowing greater flexibility in site selection than a BS with wireless backhaul. On the other hand, it should be understood that when RSs are deployed using the in-band model of deployment, relay link communications and access link communications share the same frequency allocation. Readers looking for additional information on the economic benefits of the deployment of RSs are referred to the IEEE 802 tutorial [10] which contains results of various studies carried out by the study group (within the IEEE 802.16 WG) dealing with the economic impact of RS deployment.

When deploying a wireless broadband network within each cell, an operator can deploy a BS with wired backhaul, a BS with wireless backhaul, or an RS. Though the per-site costs of using a BS with wired/wireless backhaul are expected to be higher than the costs of using an RS, it should be clearly understood that the operator must dedicate a part of the spectrum that could have been used for access link communications for the relay link communications. By deploying RSs, the operator is essentially trading off the cost of BS sites against the cost of access link capacity to reduce the overall cost of deploying and operating the network.

19.4.3 Deployment Models

In order to discuss the deployment of relay stations, we will outline a few example deployment scenarios that are expected to be used during different phases of the evolution of the network. Figure 19.7 illustrates some of these scenarios. Additional, relevant information on RS deployment scenarios can be found in the IEEE 802.16j Usage Models document [11].

The first example deployment scenario is relevant during the initial build-out of the network. As a network is first deployed, RSs can be used to reduce costs when the subscriber base is being established and the system capacity is not an issue. In this scenario, the operator selects some cell sites for the deployment of BSs, while deploying RSs within other cell cites. As the network adds subscribers and requires more capacity, selected RSs are replaced by BSs to increase the capacity of the network.

Another deployment scenario involves the usage of RSs to enhance coverage within the network. RSs can be deployed in a regular pattern to improve coverage across the sector, or RSs can be deployed in specific locations to provide coverage in the shadow of large buildings or other physical obstructions.

Yet another deployment scenario involves the deployment of RSs to provide cost-effective wireless backhaul in a dense cell deployment. RSs can be used to reduce the cost of such a deployment because they allow the operator to trade off the cost of backhaul with the usage of some of their air-link resources for supporting re-lay links.

Another potential deployment scenario is for providing in-building coverage within a home or small office (e.g., MS6 in Figure 19.7), often referred to as a fem-tocell. Coverage within a femtocell can be provided by a small BS or by an RS. The RSs can be used in the case where a wired backhaul is not available.

There are a number of other, more novel deployment scenarios in which RSs can be used. For example, in-vehicle coverage can be provided by a mobile RS that is deployed within a vehicle such as a train, bus, or ferry to MSs in the moving vehi-cle. RSs can also be used to provide temporary coverage during a concert or in the aftermath of a disaster when the physical infrastructure may have been disabled.

19.4.4 IEEE 802.16j: The First Relay Standard

The use of relay has been specified for IEEE 802.16 systems by the IEEE 802.16 WG. At the time of the writing of this chapter, IEEE 802.16j amendment [12] had been approved for publication by IEEE SA Standards Board. In this chapter, we will refer to [12] as the 16j draft.

19.4.4.1 Legacy Support

The 16j draft is an amendment to the upcoming IEEE 802.16-2009 standard. The 16j draft specifies relay support for 802.16 networks without modifying the MS specifications of IEEE 802.16-2009. This means that, as far as the MS is concerned, the access link through a relay station (the RS-MS link) is identical to the BS-MS link. This restriction had a large influence on the scope and architecture of 16j. Packet headers and access link protocols could not be modified. Only strictly back-ward-compatible changes could be made to the PHY layer and the MAC layer con-structs that are visible to the MS, though the relay link protocols could be different. The 16j frame structure had to be designed in such a way that it allowed access and relay link communications to be multiplexed within an RF channel in a way that is transparent to the MSs.

19.4.4.2 Protocol Architecture

For the relay link, 16j defines a complementary set of tightly coupled protocols that extend the 16e protocols across multiple hops. A high-level representation of the protocol architecture is illustrated in Figure 19.8. This representation is not a part of the 16j draft, but it is included here for illustrative purposes. The protocols between the RS and MS are defined in 16e, whereas the relay link protocols (BS-RS link in the figure) are defined in 16j. In some cases, the 16j protocol simply comprises the relaying the 16e messages between the BS and MS, whereas in other cases, new messages have been defined in the 16j protocol to allow additional communications between the RSs and the BS. An important assumption in 16j is that the BS has/ob-tains all the information that is relevant for operations, such as network entry (of MSs and RSs), RS selection, path selection, and handover, and, thus, maintains

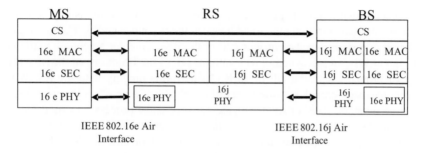

Figure 19.8. General structure of the 16j protocols.

overall control of the entire BS cell. Necessary signaling, not the corresponding algorithms, have been defined (PHY allocations and MAC messages) on the relay link to facilitate the aforementioned level of BS control.

19.4.4.3 Relaying Modes

The 802.16j draft supports both transparent and nontransparent relay modes and specifies both centralized and distributed scheduling and security models. It supports topologies that are greater than two hops, but restricts the topology to a tree structure. The hop count here is measured in terms how many hops the MS is away from the BS. The 16j draft standard supports both fixed and mobile RSs. In the remainder of this section, we will describe the basic operation of nontransparent relay and provide a brief overview of transparent relay.

In case of 802.16j, the MSs cannot distinguish between BSs and nontransparent RSs (in that the MSs perceive the nontransparent RS as just another BS). The addition of nontransparent RSs can improve coverage, extend the range, and/or increase capacity. The 16j draft defines both centralized and distributed scheduling for nontransparent relay. Here, we restrict the description to nontransparent relay with distributed scheduling.

The major types of transmissions that occur in nontransparent relay mode are illustrated in Figure 19.9. Both the BS and RS support a PMP link for communicating with the MSs (access link) and a PMP link for communicating with the other RSs (relay link). The access links of the BS and RSs within a sector are distinct from the perspective of the MSs. The MSs within the network associate with either the BS or one of the RSs. MSs that associate with the BS receive all downlink transmissions directly from the BS and transmit their uplink transmissions directly to the BS. MSs that associate with one of the RSs receive all of their downlink transmissions from this RS and transmit their uplink transmissions to this RS. The RS relays MS data to/from the BS through zero or more intermediate RSs. The length of the path between the BS and MS can be two hops or more. Routing decisions are generally less frequent than in transparent relay and are decoupled from the scheduling decisions. The BS and RSs within a sector can operate on the same RF carrier or on different RF carriers.

In a sector with nontransparent RSs, the BS and the RSs establish the topology of the sector by determining the route between the BS and each RS. The topology is modified in response to the entry or exit of the RSs. To enter a network with nontransparent RSs, an MS synchronizes with the DL of the BS or one of the RSs. If an MS synchronizes with the DL of an RS, it performs ranging with the RS. When ranging has been completed, the RS notifies the BS of a new MS that is trying to enter the network. The remainder of the network entry process is carried out between the BS and MS, with the RS relaying messages between them. As part of this process, security associations and connections are established between the BS and MS. The route for these connections is determined according to the topology that has been established within the sector. The RSs along the path of each connection store the connection identifier associated with that connection.

Figure 19.9 also illustrates the frame structure used in nontransparent mode for the case in which the topology is limited to two hops. We see that the BS and RS transmit DL synchronization and control channels at the same time-frequency location in the frame. The transmissions within the DL access zone are performed according to the 16e protocol. From the perspective of the 16e MS, the DL access zone appears identical to a 16e DL zone in 16e radio frame. In the DL relay zone, the BS transmits data to the RSs attached to it. From the perspective of the 16e MSs, this zone appears to be a 16e DL zone with no allocations (appears to be empty). The data that the RS transmits to its MSs in the DL access zone was received from the BS in the DL relay zone of a previous frame. Similarly, the UL access zone is used for communications on the access link (MS-RS or MS-BS). Transmissions within this zone are performed in accordance with the 16e specification and to a 16e MS this zone appears like a 16e UL zone in a 16e radio frame. The UL relay zone is used for RS-to-BS transmissions. To the MS, this zone appears to be a 16e UL zone with no allocations.

The BS and each RS independently perform MCS selection and schedule the transmissions on the PMP link between themselves and their downstream neighbors (MSs and RSs). MAC protocol data units (MPDUs) are transmitted between the BS and MS along the path that was selected when the connection was established. Each RS along this path receives the MPDUs and determines the next hop. The RS may decrypt and manipulate the MPDUs in order to aggregate individual MPDUs into larger MPDUs or to split large MPDUs into multiple smaller MPDUs for the sake of achieving greater efficiency.

As an MS moves around the sector, it may move from the coverage area of one of the infrastructure stations (BS or one of the RSs) to the coverage area of another infrastructure station that is in the same sector or in a different sector. This triggers a handover procedure that, from the perspective of the MS, is identical to the handover procedure specified in 16e. The handover procedure is always controlled by the BS, even if the MS is handing over to/from an RS.

The 16j draft also defines the transparent relay mode. The hop count in the case of a transparent relay operation is limited to two. To enter the system, MSs synchronize with the DL of the BS and receive the DL control information. Security associations and connections are established between the BS and MS. The RSs are not

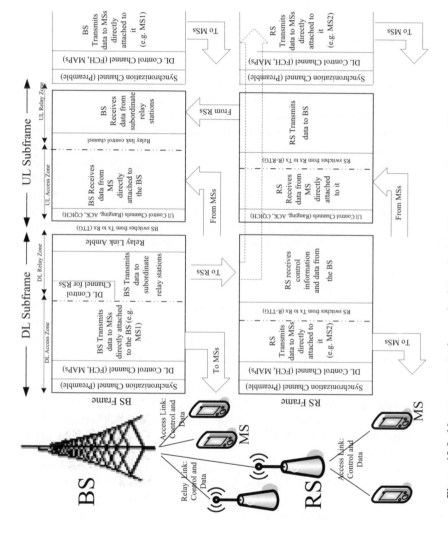

Figure 19.9. Nontransparent relay mode. Operations and frame structure for the two-hop case.

471

aware of the security associations and connections and do not have access to the keys or connection identifiers. In transparent relay mode, the BS schedules the usage of all air-link resources within the sector. When scheduling data transmissions, the BS determines whether the data is to be transmitted directly to the MS or whether one of the RSs should be used to relay the data. Handover is not performed between RSs within a sector because the BS dynamically determines which RS should be used to relay data for each MS.

19.4.4.4 Advanced Concepts

In the previous section, we have described the major operations of transparent and nontransparent relay in 16j. The 16j draft also specifies the operation of all basic 16e protocols for relay, such as HARQ, ARQ, bandwidth request, connection management, sleep, and idle mode. In addition to specifying basic relay operations in transparent and nontransparent relay mode, the 16j draft also specifies a number of advanced relaying techniques. We mention them here briefly, to provide a more complete picture of the scope of the draft. Cooperative transmission between the BS and RSs is supported. The BS and/or RSs can use virtual MIMO techniques to transmit data to an MS. Mobile RSs (MRSs) are also supported in the 16j draft. Handover procedures are specified for the RSs and for the MSs associated with them. Finally, the concept of RS groups is specified in the draft, wherein multiple RSs collaborate in transmitting data to an MS appearing to the MS as a single virtual BS.

19.4.5 Coverage and Capacity Analysis

The theoretical underpinnings for the performance improvements that can be achieved due to relay are based on the increase in effective capacity of a multihop wireless link as the number of hops or distance between the communicating nodes varies. Oyman showed that for any distance that separates two nodes communicating using wireless protocols, there exists an optimal number of hops that maximizes the effective capacity of the link between the two nodes [9]. The effective capacity of the link takes into account the capacity of the individual hops and the resources that are used to transmit the same data multiple times, once for each hop, as well as the additional frequency reuse that results when multiple RSs transmit at the same time on the same frequency.

RS deployment enhances coverage by providing a usable link to MSs in areas of the sector where the capacity of the direct link between the BS and MS is zero (e.g., MS2 in a coverage hole in Figure 19.7). RS deployment enhances capacity in areas where the MSs could communicate directly with the BS, but where the effective capacity of the path through the relay (BS-RS-MS) is significantly higher than the capacity of the direct (BS-MS) link.

We present here some results from a simulation study in order to highlight the manner in which relays can improve capacity and coverage in the deployment model used in the simulation. It should be noted that the simulation results are highly dependent on the specific deployment model and simulation parameters.

Nineteen BSs, each with three sectors, are deployed in a hexagonal grid. The site-to-site distance between BSs is 1500 meters. BS transmission power is 46 dBm, with a 17 dB gain antenna at a height of 32 meters. Two RSs are deployed within each sector as illustrated in

Figure 19.10, where distance r is equal to 3/8 of the site-to-site distance and the angle φ is 26°. The distance and angle were selected to optimize the coverage areas of the two RSs and to minimize interference on the relay link. RSs are assumed to have two antennas. An omnidirectional antenna with 0 dB gain is used for access link transmissions. A 20° beam-width antenna with 20 dB gain is aimed at the BS and is used for relay link transmissions. Both RS antennas are mounted at a height of 32 meters and transmit power is 36 dBm on both access and relay links. MSs have omnidirectional 0 dB gain antennas, are positioned at a height of 1.5 meters, and transmit with a power of 23 dBm. The simulation was done at 2.5 GHz using a 10 MHz channel and a 1/8 cyclic prefix. ITU channel models [13] are used to model path loss and slow fading on the access link. The COST–Walfish–Ikegami models [14] are used to model the relay link.

We simulate three different modes of operation. In the first mode, RSs are not used. Only the BSs are deployed in this mode and all MSs associate with a BS. We refer to this as the without-relay mode and use the results from this mode as a reference for comparison. The second mode is the relay-division mode. In this mode, two RSs are deployed and the two RSs and BS share the airlink resource in a time/frequency division manner. Only one station (BS or RS) is allowed to transmit or receive at a time. Finally, the third mode is referred to as relay-simultaneous mode. In this mode, two RSs are deployed and the BS and both RSs transmit to (and receive from) their respective MSs at the same time and frequency.

We first examine coverage. In Figure 19.11 we plot the cumulative distribution function of the access link SINR for three modes. We see from the CDF curves that coverage is enhanced in the case of relay because a larger percentage of users

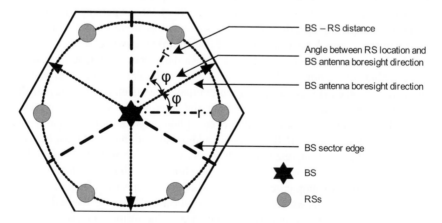

Figure 19.10. Position of RSs within the sector for simulation purposes.

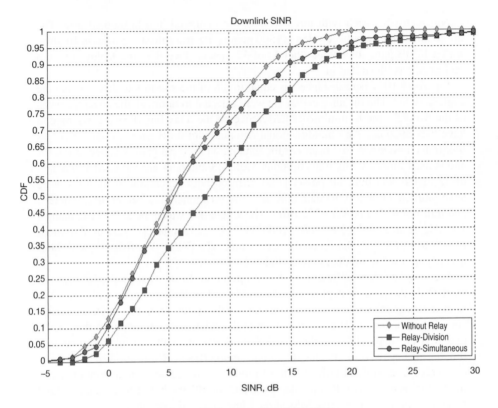

Figure 19.11. Downlink SINR CDF.

achieve a minimum SINR required to support a useable link. For example, we see
that in without-relay mode around 12% of the MSs experience an SINR of less then
0 dB, whereas in Relay-division mode around 5.5% of MSs experience an SINR of
less than 0 dB. Although the difference in this example is not dramatic, it illustrates
the mechanism by which the deployment of RSs improves coverage.

Next, we examine capacity. Table 19.3 shows the spectral efficiency measured
in the downlink and uplink for the three cases that we simulated. Spectral efficiency
is measured in bits/s/Hz/sector. When we compare the spectral efficiency of the
without-relay mode with relay-division mode, the latter provides a modest increase
in spectral efficiency. This indicates that even without any additional spatial reuse,
the increase in access link SINR offsets the additional resources that must be used
to transmit the data over the relay link. The effective capacity of the two-hop link is
larger than the effective capacity of the one-hop link even though the data is transmitted twice in the two-hop case. When we examine the spectral efficiency of the
relay-simultaneous mode, we see that there is a substantial increase above the without-relay and relay-division modes. This increase is due to the increase in spatial
reuse. Although we see in Figure 19.11 that MSs in this case receive lower SINR

Table 19.3. Spectral efficiency for relay modes

	DL SE (bits/s/Hz/sector)	UL SE (bits/s/Hz/sector)
Without-relay mode	0.87	0.61
Relay-division mode	1.28	0.99
Relay-simultaneous mode	2.64	1.96

than in the relay-division mode, we have three times the reuse, resulting in higher spectral efficiency.

The relay-division and relay-simultaneous modes represent the extreme cases of scheduling strategies within a relay sector. A more sophisticated strategy is to determine which MSs will benefit more from relay-division mode, which will benefit from relay-simultaneous mode, and which will benefit from without-relay mode, using a fractional frequency reuse scheme between the stations within a sector. Such a scheduling strategy can be tuned to make the desired trade-off between coverage and capacity, favoring relay-division or without-relay modes to provide better coverage, or relay-simultaneous mode to favor capacity.

19.4.6 Relay Evolution in Next-Generation Mobile WiMAX

The 16j draft is the first attempt at specifying relay support in a broadband wireless system (the Mobile WiMAX system). Relay support in 16j was designed to be transparent to a 16e MS. Future enhancements related to relay support (which are targeted in the IEEE 802.16m task group) are likely to focus on relaxing some of the restrictions placed on 16j and the need for simplicity. In future relay systems, MSs can be made aware of the existence of the relay stations, which is likely to help the MSs make more informed network entry or handover decisions, for instance.

Looking a step further, perhaps beyond the timelines of the IEEE 802.16m task group, MSs can be made to operate as relay stations, relaying data for other MSs. Other potential directions might be to move from a tree topology toward a more general mesh or to define more distributed control functions to allow the network to operate in a more efficient manner as cell sizes decrease and the number of RSs increases. In this situation, self-organization capabilities will make the network easier to operate and more distributed handover and network entry decisions will make the network more efficient. Other improvements might include improvements to relay link efficiency and the ability for an RS to receive and transmit at the same time.

19.5 SUMMARY

Femtocells and relays are two recent advances in cellular network architecture that promise significant gains in coverage and capacity. Relays and femto-APs typically provide targeted coverage in areas where signal from the macro BS is weak or where there is a concentrated need for throughput. Femtocells and relays utilize different solutions to the problem of backhaul. Femtocells utilize wired backhaul available within the building in which they are deployed, whereas relays utilize the operator's spectrum. As wireless operators continue to build out their broadband wireless networks and wireless customers begin to expect broadband speeds from their mobile wireless devices, femtocells and relays will provide a cost-effective means for operators to deliver the required service. We also expect that the concept of multitier cellular topologies will evolve over time to generalize and combine the specific concepts of femtocells and relays into a more general multitier architecture.

19.6 REFERENCES

[1] picoChip, The Case for Home Base Stations, April 2007.

[2] WiMAX Forum™ Technical Working Group, WiMAX Forum™ Mobile System Profile Release 1.0 Approved Specification, (revision 1.4.0), May 2007.

[3] J. Edison, *Measurement Control and Communication Using IEEE 1588,* Springer London, 2006.

[4] Guidelines for Evaluation of Radio Transmission Technologies for IMT-2000, Rec. ITU-R M.1225, January 1997.

[5] Femto Forum, http://www.femtoforum.org.

[6] Requirements for WiMAX Femtocell Systems v1.0.0, WiMAX Forum, Service Providers Working Group, 2009-04-08.

[7] H. Claussen, Performance of Macro- and Co-channel Femtocells in a Hierarchical Cell Structure, IEEE PIMRC 2007.

[8] IEEE 802.16m-08/003r8, TGm System Description Document (SDD), draft version 2009-04-10.

[9] O. Oyman, J. N. Laneman, and S. Sandhu, Multihop Relaying for Broadband Wireless Mesh Networks: From Theory to Practice, *IEEE Communications Magazine,* June 2007.

[10] IEEE 802.16 WG Study Group on Mobile Multihop Relay, IEEE 802 Tutorial: 802.16 Mobile Multihop Relay, March 2006. http://www.ieee802.org/16/sg/mmr/docs/80216mmr-06_006.zip.

[11] Usage Model Ad Hoc Group J. Sydir (editor), Harmonized Contribution on 802.16j (Mobile Multihop Relay) Usage Models, September 2006. http://www.ieee802.org/16/relay/docs/ 80216j-06_015.pdf.

[12] IEEE P802.16j-2009: Air Interface for Fixed Broadband Wireless Access Systems: Multihop Relay Specifications, June 2009.

[13] ITU-R M.1225, Guidelines for the Evaluation of Radio Transmission Technologies for IMT-2000.

[14] IST-2003-507581 WINNER, D5.4 v. 1.4, Final Report on Link Level and System Level Channel Models, November 2005, https://www.ist-winner.org/.

CHAPTER 20

WiMAX SPECTRUM REQUIREMENTS AND REGULATORY LANDSCAPE

REZA AREFI AND JAYNE STANCAVAGE

20.1 INTRODUCTION

Access to spectrum is crucial to any wireless broadband deployment; without spectrum, a network could not be deployed. Spectrum policy, at the national and/or international level, can have a profound impact on the viability of any deployment. Spectrum policy can affect many different areas of a WiMAX deployment, including the operator's business model, the cost of devices to consumers, the quality and types of services delivered to customers, and even whether the service can be offered at all.

20.2 WiMAX SPECTRUM REQUIREMENTS

20.2.1 Deployment Scenarios

WiMAX systems can be deployed in various scenarios, including both as a fixed and mobile service. WiMAX is typically deployed in a wide-area network with cellular topology to deliver fixed, nomadic, and mobile applications to end users. A desktop communication box connected to a computer is an example of a fixed application. A nomadic application, as defined by Recommendation ITU-R F.1399, is a "wireless access application in which the location of the end-user termination may be in different places but it must be stationary while in use." A laptop computer could be used as a nomadic device delivering nomadic applications.

WiMAX Technology and Network Evolution. Edited by Kamran Etemad and Ming-Yee Lai
Copyright © 2010 the Institute of Electrical and Electronics Engineers, Inc.

Handsets are examples of devices delivering mobile applications. Regulators typically allow nomadic applications to be deployed in both fixed and mobile allocations. The main distinction between fixed services and mobile services is in the ability of devices to perform handover functions in the network as they move from one point to another.

With the proliferation of mobile devices with more advanced capabilities and processing power, many regulators are currently enabling mobile services to be deployed in what were previously fixed-services bands. For example, the European Commission recently harmonized the 3400–3800 MHz band [1]. In this decision, the Commission permits fixed, nomadic, and mobile services to be deployed in the band.

20.2.2 Minimum Required Spectrum

The amount of spectrum required per operator can vary substantially. A minimum of 30 MHz of usable spectrum per operator is required according to a 2007 WiMAX Forum report [2]. However, even more spectrum is often required in order to enable true broadband services and provide for future growth. The total amount of spectrum available impacts the economical viability of a business model and scalability of the network, including all the major parameters of a business case, the quality of service and range of services offered, and the average revenue per user (cost).

Furthermore, the amount of spectrum necessary can vary based upon a number of other factors. For example, more spectrum is necessary to deliver triple play media (voice, video, and data). Delivery of high-definition television broadcasts would require even more spectrum. Finally, the amount of spectrum necessary for an operator will also grow over time. Bandwidth requirements and user expectations are dramatically higher than they were even a few short years ago. Given that many licenses are awarded for periods of ten to twenty years with renewal rights, operators typically require certainty regarding the amount of spectrum available.

20.3 REGIONAL AND INTERNATIONAL REGULATIONS AND REGULATORY BODIES

20.3.1 International Telecommunications Union

The International Telecommunications Union (ITU) is an international organization of the United Nations for information and communication technology issues. There are three sectors of the ITU: the ITU-Radio communication Sector (ITU-R), the ITU-Telecom Standardization Sector (ITU-T), and the ITU-Development Sector (ITU-D). ITU-D focuses on equitable, sustainable, and affordable access to communication technologies for developing nations. Figure 20.1 depicts these three sectors of the ITU and their relationship in the overall ITU structure.

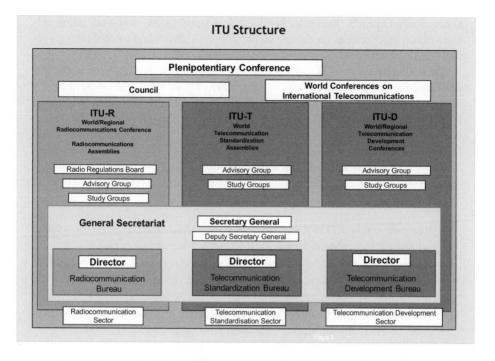

Figure 20.1. ITU structure.

Members of the ITU include member states (national administrations) and sector members, such as companies and international organizations. The ITU is a consensus-driven organization in which member states work together to gain consensus on various topics within the ITU to develop recommendations that can be applied globally.

The ITU-R has three major responsibilities. The first is to develop ITU-R recommendations on the technical characteristics of, and operational procedures for, radio communication services and systems. Standards are specified in the recommendations. The second responsibility is to draft the technical bases for radiocommunication conferences. The third is to compile handbooks on spectrum management and emerging radiocommunication services and systems (for example, the land mobile handbook provides information about land mobile services).

There are several ITU groups that are most relevant to those interested in WiMAX technology. ITU-D Study Group 2 focuses on the development and management of telecommunication services and networks and ICT applications. ITU-R Study Group 5 focuses on Terrestrial Services. Within ITU-R Study Group 5, there are two relevant working parties. Working Party 5A focuses on terrestrial services, including all land mobile services except IMT. ITU-R Working Party 5D focuses on international mobile telecommunications (IMT) systems (both IMT-2000 and IMT-Advanced). ITU-R Joint Task Group 5-6 (JTG 5-6) is a joint task group between Study Group 5 (terrestrial services) and Study Group 6 (broadcasting services) to conduct studies on the use of the 790–862 MHz band by mobile applications and by other services.

There are two types of ITU-R documents. The first type is comprised of the Radio Regulations, which are international treaties. These regulations governing the allocation and use of spectrum are negotiated every three to four years during the World Radio Conference. The second category of documents includes ITU-R recommendations and reports. These are the output of various working parties within the ITU-R , which are approved at the study group level. ITU-R recommendations could be adopted as national rules by administrations if they so choose. IEEE 802.16 standards are incorporated into several ITU-R recommendations including the following:

- Recommendation ITU-R F.1763, "Radio interface standards for broadband wireless access systems in the fixed service operating below 66 GHz"
- Recommendation ITU-R M.1801, "Radio interface standards for broadband wireless access systems, including mobile and nomadic applications, in the mobile service operating below 6 GHz"
- Recommendation ITU-R M.1457, "Detailed specifications of the radio interfaces of International Mobile Telecommunications-2000 (IMT-2000)"

The ITU-R is also currently developing recommendations on IMT-Advanced. According to the ITU-R, "IMT-Advanced systems support low to high mobility applications and a wide range of data rates in accordance with user and service demands in multiple user environments. IMT Advanced also has capabilities for high quality multimedia applications within a wide range of services and platforms, providing a significant improvement in performance and quality of service." [6] The IEEE, WiMAX Forum, and sector member companies are participating in the development of the next generation of WiMAX systems that meet the requirements of IMT-Advanced.

Some of the most important ITU-R reports and recommendations for IMT-Advanced include:

- Recommendation ITU-R M.1645, "Framework and overall objectives of the future development of IMT-2000 and systems beyond IMT-2000"
- Report ITU-R M.2133, "Requirements, evaluation criteria and submission templates for the development of IMT-Advanced"
- Report ITU-R M.2134, "Requirements related to technical performance for IMT-Advanced radio interface(s)"
- Report ITU-R M.2135, "Guidelines for evaluation of radio interface technologies for IMT-Advanced"

The proposed timeline for IMT-Advanced has already been established. According to the current schedule, the submission of candidate technologies were to have been completed as of October 2009. Registered evaluation groups must deliver their evaluation reports to the ITU-R by June 2010. By October 2010, Working Party 5D should decide on the framework and key characteristics of IMT-Advanced. The development of the recommendation of initial radio interface specifications should be completed by February 2011.

20.3.1 Regional Regulatory Bodies

Regional regulatory bodies are typically composed of government representatives, operators/service providers, communication equipment manufacturers, and other relevant organizations. These groups address regional issues and often harmonize views and develop common proposals for World Radio Conferences. Some examples of regional groups include the African Telecommunications Union (ATU), the Arab Spectrum Management Group (ASMG), the Asia-Pacific Telecommunity (APT), European Conference of Postal and Telecommunications Administrations (CEPT), Inter-American Telecommunication Commission (CITEL), Regional Commonwealth in the Field of Communications (RCC), and Southern African Development Community (SADC).

20.4 WiMAX SPECTRUM BANDS

20.4.1 Common Broadband Wireless Bands

The spectrum most commonly allocated for broadband wireless access (BWA) technologies are the 2.3–2.4 GHz, 2.5–2.69 GHz, and 3.3–3.8 GHz bands. Technology-neutral regulators prefer to allocate spectrum to broadband wireless access rather than identify the spectrum for specific technologies. However, in some countries, the regulators may limit access to certain bands to certain technologies such as international mobile telecommunications.

Although allocations vary from country to country, spectrum that can be utilized for WiMAX deployments is available globally. The 2.5–2.69 GHz band is being utilized for WiMAX deployments in numerous countries in Africa and the Middle East, as well as North and South America. Additionally, many European countries are expected to allocate spectrum in this band during 2009–2010. In Asia, numerous countries have either already allocated or are in the process of allocating spectrum that can be used for Mobile WiMAX in the 2.3–2.4 GHz band. The 3.4–3.6 GHz band has been utilized for WiMAX deployments in many countries through the world, primarily for fixed and nomadic services. However, over time, allocations in this band are likely to evolve to mobile allocations.

Currently, WiMAX Forum Certified™ devices based upon the IEEE 802.16d-2004 standard are available for the 3400-3600 MHz band. WiMAX Forum Certified™ devices based upon the IEEE 802.16e-2005 standard are available for the 2300–2400 MHz, 2496–2690 MHz, and 3400–3600 MHz bands. More information is available at the WiMAX Forum website (http://www.wimaxforum.org).

20.4.2   IMT Bands

Beginning in 1992, the ITU has identified spectrum for IMT in the Radio Regulations (RR), which are binding international treaties. This identification does not preclude the use of these bands by any application of the services to which they are allocated and does not establish priority in the Radio Regulations. However, in practice, many governments either explicitly or implicitly reserve the spectrum for IMT technologies.

Members of the ITU voted at the Radiocommunications Assembly of 2007 to expand the IMT-2000 family to include a new radio interface based upon Mobile WiMAX, Release 1. This decision provided Mobile WiMAX with access to spectrum in those countries which identify the spectrum for IMT.

Frequency arrangements for IMT bands, as identified by World Radio Conferences of the years 2000, 2003, and 2007, and their associated band plans, are summarized in Recommendation ITU-R M.1036, frequency arrangements for implementation of the terrestrial component of International Mobile Telecommunications (IMT) in the bands identified for IMT in the Radio Regulations (RRs). These bands, together with their associated footnotes in the Radio Regulations, are listed in Table 20.1. Figures 20.2 and 20.3 depict the frequency arrangements for the bands around 2 GHz and the 2.6 GHz band. Frequency arrangements for other IMT bands are currently under consideration and will be included in the next revision to Recommendation ITU-R M.1036.

20.4.3   Digital Dividend

Many regulators worldwide are determining how to best address the "digital dividend" spectrum. As television broadcasters move from analog to digital channels, some spectrum may become unused. This spectrum, which is in the sub-1-GHz range, has excellent propagation characteristics that may be extremely useful in providing access to rural areas. A decision at the World Radio Conference 2007 identifies 689–862 MHz for IMT in Region 2, which includes all of the Americas, and in nine countries of Region 3 (Bangladesh, China, Korea, India, Japan, New

Table 20.1. IMT bands and their associated footnotes in the Radio Regulations

Band (MHz)	Footnotes identifying the band for IMT
450–470	5.286AA
698–960	5.313A; 5.317A
1710–2025	5.384A, 5.388, 5.388A, 5.388B
2110–2200	5.388
2300–2400	5.384A
2500–2690	5.384A
3400–3600	5.430A, 5.432A, 5.432B, 5.433A

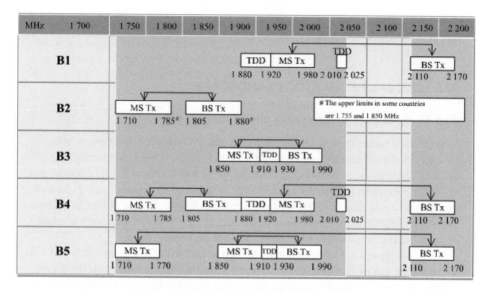

Figure 20.2. Frequency arrangement options for IMT bands around 2 GHz.

Zealand, Papua New Guinea, Philippines, and Singapore). In Region 1 (which includes Europe, Africa, the Middle East west of the Persian Gulf and including Iraq, the former Soviet Union, and Mongolia), the spectrum identified for IMT was 790–862 MHz. These bands, also called 700 MHz or 800 MHz, are of interest for future Mobile WiMAX networks.

20.4.4 Future Bands

Most current WiMAX products operate in the 2.3-2.4 GHz, 2.5-2.69 GHz, and 3.3-3.8 GHz spectrum bands. As a member of the IMT-2000 set of standards, WiMAX can also be deployed in all other existing IMT bands such as the ones below 1 GHz.

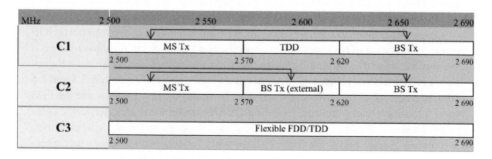

Figure 20.3. Frequency arrangement options for the 2.6 GHz band.

Any future allocations for BWA and/or IMT are likely to be at higher frequencies due to existing services in the lower frequencies.

20.5 GLOBAL REGULATORY LANDSCAPE

20.5.1 Spectrum Engineering and Management

Utilization of radio frequencies by various broadcast (such as television or radio) or two-way communication systems (such as walkie-talkies) grew significantly larger as more systems were introduced. Two factors contributed to the increase in demand for spectrum: (1) increase in capacity in response to demand and (2) innovative new systems that required more spectrum in bands already in use, or in new bands. Both these factors resulted in utilization of new bands, which were generally in higher frequencies where more unused spectrum was available.

This process did not have a simple solution as migration to higher bands was not an easy task; this migration usually involved additional technology cost. On the other hand, reuse of the same spectrum by various users in the same or adjacent areas was not always practical or feasible due to interference issues.

As demand for spectrum increased and more technologies were developed, having access to spectrum became increasingly important. Governments, as owners of the airwaves, started managing this valuable resource by providing licenses to various users of the spectrum. Licensing of spectrum took different forms in different countries and changed over the past decades. However, governments still license most spectrum to users, on an exclusive basis, for a limited amount of time in exchange for a fee. The exclusivity of the licenses is put in place to prevent interference and to protect the investment of the licensees.

The licensing of the spectrum does not solve all interference concerns so there are rules in place that determine the conditions under which the license should be used. These rules provide limits on factors such as transmitter power or out-of-band emissions, with the purpose of limiting the impact of any given system on systems operating using a different license in adjacent bands or areas. Enforcement of such regulations may come in different forms. Some governments prefer to implement guard bands (chunks of unused spectrum separating one license from another), whereas others allow operator coordination to minimize interference.

License-exempt utilization of spectrum, whereby users do not apply to obtain a license for operation of their services in a specific band, emerged in response to various needs of the scientific community and industry. An example of one such band is the so-called Industrial, Scientific, and Medical (ISM) Band, generally between 2.4 and 2.5 GHz in a near-globally harmonized manner. Nowadays, many systems use the ISM band for commercial purposes, including RLANs (radio local area networks, commonly known as Wi-Fi), microwave ovens, garage door openers, and cordless phones.

License-exempt bands are to be utilized with the understanding that no protection from interference is guaranteed. Technologies used in this band often utilize techniques that detect interference and avoid it by various means such as delayed or suppressed transmission. As can be easily inferred, such techniques work efficiently as long as channel utilization is not excessive.

20.5.2 Spectrum Policy Regimes

Many regulators worldwide are moving from existing "command and control" technology-specific policies to an approach based upon technology neutrality. Technology neutrality allows a service provider to deploy the technology that best suits their business model, subject to minimal technical requirements regarding interference to adjacent services. In some cases, operators may decide to offer a new technology that offers lower costs, better spectral efficiency, or higher throughput if the operator believes that the advantages of the new technology outweigh the costs. In other cases, the operator might choose to continue to deploy their existing technology. Regardless of what a specific operator decides, new technologies can spur competition in pricing and services offered.

The United States Federal Communications Commission (FCC) supports technology neutrality and providing greater benefits to consumers. In 2006, the FCC issued a ruling regarding the 2.6 GHz band [4]. The purpose of the order was "designed to provide both incumbent licensees and potential new entrants in the 2495–2690 MHz band with greatly enhanced flexibility to encourage the efficient and effective use of spectrum domestically and internationally, and the growth and rapid deployment of innovative and efficient communications technologies and services. "As part of the ruling, the FCC issued a new band plan that restructured the band into upper- and lower-band segments for low-power operations, and a midband segment for high-power operations. This band plan created opportunities for existing systems or devices to migrate to compatible bands based on marketplace forces, reduced the likelihood of interference caused by incompatible uses, and, at the same time, opened the majority of the band to innovative use. This globally harmonized band has since turned into a flagship band for Mobile WiMAX.

The European Commission also supports the principle of technology neutrality. The European Commission has mandated a technology-neutral approach in the following bands as part of their WAPECS (Wireless Access Policy for Electronic Communications Services) initiative:

- 470–862 MHz;
- 880–915 MHz/925–960 MHz
- 1710–1785 MHz/1805–1880 MHz
- 1900–1980 MHz/2010–2025 MHz/2110–2170 MHz
- 2500–2690 MHz;
- 3400–3800 MHz

In response to this initiative, the European Conference of Postal and Telecommunications Administrations (CEPT) analyzed the frequency bands and chose a block-edge mask (BEM) as the best and technically least restrictive mechanism to apply a technology neutral approach. Block-edge masks apply to the edges of a block of spectrum that is assigned to an operator, irrespective of the number of channels occupied by the chosen technology that the operator may deploy in their block. These masks might also include emissions limits within the block of spectrum (i.e., in-block power) in addition to emissions outside the block (i.e., out-of-block emission) [3].

Mobile stations out of band emissions limits, however, are usually driven by global roaming requirements and are generally more complex than BEMs for base stations. Figure 20.4 depicts the WiMAX mobile station mask (10 MHz channel) for the 2.6 GHz band.

In their decision of 13 June 2008, the European Commission harmonized the 2500–2690 MHz frequency band in Europe. This Commission decision offers regulators the ability to provide more TDD spectrum than the original 50 MHz provided by the band plan in ECC DEC 05 (05), which is based on option C1 in Figure 20.3. Between the TDD and FDD blocks, a restricted-use 5 MHz block applies. However, this separation can be modified through operator negotiation. Service providers can still utilize the restricted 5 MHz channel, but operation is subject to a restricted BEM. This gives operators more flexibility in order to potentially utilize the band. The restricted BEM approach is more advantageous than mandating an unnecessary and inflexible guard band arrangement.

20.5.3 Spectrum Assignment Regimes

Spectrum assignments are the actual awarding of spectrum to an individual licensee or service provider. The typical mechanisms for awards are auctions, beauty con-

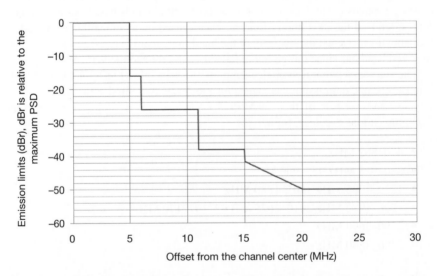

Figure 20.4. WiMAX mobile station power spectral density (PSD) emission limit (10 MHz channel) for the 2.6 GHz band.

tests, or first-come, first served. Auctions are usually held when the demand for a particular spectrum block exceeds the supply. Auctions are a market-based approach that allows the spectrum to be granted to the user that values the spectrum most. Some countries allocate spectrum via "beauty contests." In this regime, conditions for the spectrum are issued and potential licensees apply. Government authorities then determine which potential licensees best meet the requirements. Aspects frequently considered in beauty contests include technical considerations, financial viability of the business model, and build-out commitments. First-come, first-served assignments are typically reserved for cases in which there is not as much demand for a spectrum block as there is supply.

Since licenses are typically awarded for long durations, it is important that sufficient spectrum be allocated for future needs. Of course, due to continuous improvements in technology as well as increasing consumer expectations, there are likely to be cases in which the amount of spectrum originally allocated is insufficient. Therefore, more and more regulators are allowing secondary markets (i.e., spectrum trading) in order to allow spectrum to be put to the highest value use.

One important consideration in the spectrum allocation process is the creation of a band plan that allows operators access to sufficient spectrum and accommodates terminal-to-terminal interference. Although paired (e.g., frequency division duplex) spectrum has been most common in the past, there is increased interest in unpaired spectrum, which could accommodate time-division duplex (TDD) operation. Given the asymmetric nature of data communications, there are advantages to TDD operation. It is important to note that many regulators allow TDD operation in paired bands in addition to unpaired bands.

Many regulators worldwide are currently considering the allocation of spectrum in the 2.5–2.69 GHz band. As of today, perhaps the most innovative and technology-neutral method is the market-based approach proposed by the United Kingdom Office of Communications (OFCOM) [5]. Their proposal to allocate the band is based upon a two-round auction. The results of the bidding (demand) in the first round will determine the ratio of paired versus unpaired spectrum. After this ratio is determined, a band plan will be developed that will maintain the 120 MHz duplex gap utilized by European countries while minimizing the number of restricted channels between FDD and TDD operations. During the second round (assignment round), potential licensees will bid on actual spectrum blocks. Furthermore, the OFCOM proposal provides for a generous spectrum cap (80 MHz of usable spectrum) per operator in order to not negatively impact any operator's business model. Spectrum trading is also permitted.

The channel size permitted is also an important factor in spectrum assignments. Many BWA technologies currently support 5 and 10 MHz channels. Due to anticipated bandwidth considerations, many technologies are also planning to support wider channel sizes (e.g., 20 MHz up to 100 MHz). For this reason, many regulators allow various channel sizes to be deployed and allocate spectrum in multiples of 5 MHz (e.g., 20 MHz, 30 MHz, and 60 MHz)

20.6 SPECTRUM SHARING

Due to scarcity of spectrum suitable for many wireless communication services, sometimes different systems or services need to use a certain band on a shared basis. These systems could belong to the same service or different services. Sharing of the UHF band by the terrestrial and broadcast services is one example of band sharing by systems belonging to different services, whereas sharing of a band identified for IMT by various IMT technologies is an example of sharing within a service.

Systems sharing a band could do so on a coprimary or primary/secondary basis. Rules governing each case in terms of protection of other systems are different. A secondary service in a band is obligated to protect the operation of the primary service in a manner defined by the ITU-R. ITU-R Radio Regulations define the allocation of services in various bands.

20.6.1 Coexistence Scenarios

Coexistence scenarios depend on the systems involved. They are generally different when the systems involved belong to the terrestrial service (such as a WiMAX TDD system and a 3G FDD system) versus the case in which they belong to different services (such as a WiMAX system and a broadcasting satellite system).

Generally, coexistence scenarios are differentiated based on whether the base stations or the mobile stations are causing or being affected by interference. These scenarios, therefore, are determined by various factors such as duplex method of the systems involved and assumed location and mode of operation of transmitters and receivers.

Due to transmitter and receiver filtering imperfections, two systems do not have to operate in overlapping frequencies to create interference problems. Operation in an adjacent spectrum could potentially create interference as well. Any time a transmission from a source has the opportunity of entering another system's receivers, there would potentially be a coexistence scenario that could lead to interference problems.

The degree and severity of interference, its impact, and its potential remedies, depend on not just the scenarios but on the characteristics of the systems involved. These characteristics are mainly governed by transmitter and receiver performance and RF filtering.

Transmitters are characterized by their output power, linearity of their power amplifiers, and sharpness of their filters. The higher the output power, the greater the chances a transmitter could cause interference. The less linear a power amplifier, the stronger the intermodulation harmonics it will create. These harmonics, falling in adjacent bands, are essential in determining the level of out-of-band emissions created by a transmitter. Lastly, the level of out-of-band emissions are also impacted by the isolation RF filters could provide in suppressing RF splatter from amplifiers into adjacent bands.

Receivers are characterized by their sensitivity performance in the presence of additional noise caused by interference, linearity of their low-noise amplifiers, and sharpness of their filters. Aside from degree of susceptibility to additional noise due to interference, receiver amplifier linearity and filter performance highly affects the level of interference impacting the system.

In both transmitters and receivers, better performance is usually directly related to cost. Better filters, especially those whose use is not physically prohibitive, are more expensive to implement. This cost–performance trade-off is the main element of the next section dealing with interference mitigation techniques.

20.6.2 Interference Mitigation

There are generally two types of techniques to mitigate interference: (1) those that can be done unilaterally by one operator and (2) those that require coordination among all parties involved. Unilateral interference mitigation techniques include the use of better receiver or transmitter filters, shaping coverage of transmitting antennas, and the use of more linear power amplifies. Interference mitigation techniques that must be coordinated between operators include coordinated frequency planning and coordinated site/ antenna placement.

In both cases, there are cost and complexity issues involved with implementation of each technique. It should be noted that not all techniques have the same level of effectiveness or complexity. These techniques and the effectiveness associated with them are discussed in detail in Report ITU-R M.2045.

20.7 SUMMARY

Access to spectrum is crucial to the successful deployment of any wireless technology. As the demand for greater bandwidth increases, demand for access to spectrum also increases. Sound spectrum policies and regulations at the national and international levels are needed to enable widespread, affordable broadband services.

Spectrum assignment activities for the primary spectrum bands currently of interest for WiMAX deployments (2.3–2.4 GHz, 2.5–2.69 GHz, and 3.3–3.8 GHz) are occurring around the world. During 2009-2011, even more assignments are expected to occur in the 2.3–2.4 GHz band (primarily in Asia), the 2.5–2.69 GHz band (primarily in Europe, Africa, and the Americas), and the 3.3–3.8 GHz band (worldwide). These spectrum assignments, as well as increased flexibility for existing licenses, will enable even more WiMAX deployments.

20.8 REFERENCES

[1] 2008/411/EC Commission Decision of 21 May 2008, European Commission.

[2] A Review of Spectrum Requirements for Mobile WiMAX TM Equipment to Support Wireless Personal Broadband Services, WiMAX Forum, 2007.

[3] Report from CEPT to the European Commission in Response to the Mandate to Develop Least Restrictive Technical Conditions for Frequency Bands Addressed in the Con-

text of WAPECS, European Conference of Postal and Telecommunications Administrations (CEPT), December 2007.

[4] Order on Reconsideration and Fifth Memorandum Opinion and Order and Third Memorandum Opinion and Order and Second Report and Order, Federal Communications Commission, 2006.

[5] Award of Available Spectrum: 2500–2690 MHz, 2010–2025 MHz, and 2290–2300 MHz, OFCOM, April 2008.

[6] http://www.itu.int/ITU-R/index.asp?category=information&rlink=imt-advanced &lang=en, accessed May 2010.

■■■ Index